Transport and Chemical Transformation of Pollutants in the Troposphere

---

Series editors: Peter Borrell, Patricia M. Borrell, Tomislav Cvitaš, Kerry Kelly and Wolfgang Seiler

Springer

Berlin
Heidelberg
New York
Barcelona
Budapest
Hong Kong
London
Milan
Paris
Santa Clara
Singapore
Tokyo

Transport and Chemical Transformation of Pollutants in the Troposphere

Volume 2

Heterogeneous and Liquid Phase Processes

Laboratory Studies Related to Aerosols and Clouds

With 60 Figures and 39 Tables

Peter Warneck (Mainz)
Editor and Coordinator

HALIPP Steering Group

Philippe Mirabel Straßburg
G. Arthur Salmon Leeds
Christiaan Vinckier Leuven
Cornelius Zetzsch Hannover

Springer

Editor

PROF. DR. PETER WARNECK

Max-Planck-Institut für Chemie
Abteilung Biogeochemie
Saarstraße 23
D-55122 Mainz
Germany

The coverpicture is a schematic diagramm for the oxidation of sulfur (IV) catalysed by a one-electron metal oxidant (see: R. van Eldik, Chapter 2).

ISBN 3-540-60792-7 Springer-Verlag Berlin Heidelberg New York

Cataloging-in-Publication Data applied for

Die Deutsche Bibliothek – CIP-Einheitsaufnahme
Transport and chemical transformation of pollutants in the troposphere :
[an account of the work of EUROTRAC]. – Berlin ; Heidelberg ; New York ; Barcelona ; Budapest ; Hong Kong ; London ; Milan; Paris ; Santa Clara ; Singapore ; Tokyo : Springer 1996
Vol. 2. Heterogeneous and liquid phase processes : laboratory studies related to aerosols and clouds / Peter Warneck, ed. and coordinator. 1995
ISBN 3-540-60792-7 (Berlin ...)

Cover Design: Struve & Partner, Heidelberg

SPIN: 10514776 30/3136 – 5 4 3 2 1 0 – Printed on acid free paper

Transport and Chemical Transformation of Pollutants in the Troposphere

An account of the work of EUROTRAC

Series editors: Peter Borrell, Patricia M. Borrell, Tomislav Cvitaš, Kerry Kelly and Wolfgang Seiler

Volume 1 Transport and Chemical Transformation of Pollutants in the Troposphere

An Overview of the Work of EUROTRAC

Volume 2 Heterogeneous and Liquid Phase Processes

Laboratory Studies Related to Aerosols and Clouds

Volume 3 Chemical Processes in Atmospheric Oxidation

Laboratory Studies of Chemistry Related to Tropospheric Ozone

Volume 4 Biosphere-Atmosphere Exchange of Pollutants and Trace Substances

Experimental and Theoretical Studies of Biogenic Emission and of Pollutant Deposition

Volume 5 Cloud Multi-phase Processes and High Alpine Air and Snow Chemistry

Ground-based Cloud Experiments and Pollutant Deposition in the High Alps

Volume 6 Tropospheric Ozone Research

Tropospheric Ozone in The Regional and Sub-regional Context

Volume 7	Tropospheric Modelling and Emission Estimation *Chemical Transport and Emission Modelling on Regional, Global and Urban Scales*
Volume 8	Instrument Development for Atmospheric Research and Monitoring *Lidar Profiling, DOAS and Tunable Diode Laser Spectroscopy*
Volume 9	Exchange and Transport of Air Pollutants over Complex Terrain and the Sea *Field Measurements and Numerical Modelling: Aircraft, Ship, Ocean Platform and Laboratory Measurements*
Volume 10	Photo-oxidants, Acidification and Tools: Policy Applications of EUROTRAC Results *The Report of the EUROTRAC Application Project*

Foreword by the Series Editors

EUROTRAC is the European co-ordinated research project, within the EUREKA initiative, studying the transport and chemical transformation of pollutants in the troposphere. The project has achieved a remarkable scientific success since its start in 1988, contributing substantially both to the scientific progress in this field and to the improvement of the scientific basis for environmental management in Europe. EUROTRAC, which at its peak comprised some 250 research groups organised into 14 subprojects, brought together international groups of scientists to work on problems directly related to the transport and chemical transformation of trace substances in the troposphere. In doing so, it helped to harness the resources of the participating countries to gain a better understanding of the trans-boundary, interdisciplinary environmental problems which beset us in Europe.
The scientific results of EUROTRAC are summarised in this report which consists of ten volumes.

Volume 1 provides a general overview of the scientific results, prepared by the Scientific Steering Committee (SSC) and the International Scientific Secretariat (ISS) of EUROTRAC, together with brief summaries of the work of the fourteen individual subprojects prepared by the respective subproject coordinators.
Volumes 2 to 9 comprise detailed overviews of the subproject achievements, each prepared by the respective subproject coordinator and steering group, together with summaries of the work of the participating research groups prepared by the principal investigators. Each volume also includes a full list of the scientific publications from the subproject.

The final volume, 10, is the complete report of the Application Project, which was set up in 1993 to assimilate the scientific results from EUROTRAC and present them in a condensed form so that they are suitable for use by those responsible for environmental planning and management in Europe. It illustrates how a scientific project such as EUROTRAC can contribute practically to providing the scientific consensus necessary for the development of a coherent atmospheric environmental policy for Europe.

A multi-volume work such as this has many contributors and we, as general editors, would like to express our thanks to all of them: to the subproject coordinators who have borne the brunt of the scientific co-ordination and who have contributed so much to the success of the project and the quality of this report; to the principal investigators who have carried out so much high-quality scientific work; to the members of the International Executive Committee (IEC) and the SSC for their enthusiastic encouragement and support of EUROTRAC; to the participating governments in EUROTRAC, and in particular the German Government (BMBF) for funding, not only the research, but also the ISS publication activities; and finally to Mr. Christian Witschel and his colleagues at Springer Verlag for providing the opportunity to publish the results in a way which will bring them to the notice of a large audience.

Peter Borrell (Scientific Secretary, ISS)
Patricia May Borrell
Tomislav Cvitaš
Kerry Kelly
Wolfgang Seiler (Director, ISS)

EUROTRAC ISS
Fraunhofer Institute (IFU)
Garmisch-Partenkirchen

Preface by the HALIPP steering group

HALIPP was established as a cooperative research project to meet the need for laboratory studies of heterogeneous and liquid phase chemical processes in the troposphere. These processes include the interaction of trace species with cloud drops, reactions in the aqueous phase of clouds, and processes occurring on the surfaces of aerosol and ice particles. Laboratory studies of reaction mechanisms and rate processes are essential in developing a quantitative and sufficiently detailed description of atmospheric chemistry for use in computer-based simulation models. In contrast to gas-phase reactions, which had been studied for many years with the result that a substantial data base existed even prior to EUROTRAC, heterogeneous and liquid phase reactions had been largely neglected, and the associated mechanisms and reaction rates were poorly known. There was, however, a growing recognition that such processes cannot be ignored in any reasonably complete description of tropospheric chemistry. A particularly urgent problem appeared to be the acidification of precipitation due to unknown oxidation pathways of sulfur dioxide and nitrogen oxide in clouds. The aims of HALIPP were to identify important heterogeneous and liquid phase processes and to provide detailed information on reaction mechanisms and rate parameters by means of appropriate laboratory studies.

Eight years of HALIPP research have witnessed significant advances in the field, especially in the chemistry related to clouds. It should be stressed that progress has been achieved both in *knowledge* and in the *understanding* of heterogeneous and liquid phase chemical processes. About 20 research groups with a complementary scientific background have contributed to HALIPP. A close mutual collaboration and exchange of experience, and a favourable distribution of research tasks assigned to the individual groups were essential in making HALIPP a success. In many cases it was necessary to develop new experimental techniques to achieve the goals of the subproject. Joint LACTOZ/HALIPP workshops organised annually under the auspices of the European Commission (Directorate–General XII) facilitated cooperation and the exchange of ideas among the individual research groups.

We take this opportunity to express our appreciation to the HALIPP research groups for a successful research effort. We also wish to thank the EUROTRAC Scientific Secretariat, especially Dr. P. Borrell, for moral support and liaison with other EUROTRAC groups. It is a pleasure to acknowledge on behalf of the Principal Investigators financial support from the national funding agencies as well as from the European Union. Finally, we wish to thank Dr. G. Angeletti from the European Commission for supporting the HALIPP steering group.

Peter Warneck Mainz
Philippe Mirabel Strasbourg
G. Arthur Salmon Leeds
Christiaan Vinckier Leuven
Cornelius Zetzsch Hannover

Table of Contents

EUROTRAC
(A EUREKA Environmental Project)

Transport and Transformation of Pollutants in the Troposphere, Volume 2.
Heterogeneous and Liquid Phase Processes

List of Contributors

Mr. P. Amels
TH Darmstadt
Anorg. Chemie III
Petersenstraße 18
D-64287 Darmstadt
Germany

Dr. S. Ashty
University of Leuven
Laboratory of Organic Synthesis
Department of Chemistry
Celestijnenlaan 200F
B-3001 Leuven
Belgium

Dr. S. Barlow
University of Leeds
Cookridge Radiation
Research Centre
GB-Leeds LS16 6QB
United Kingdom

Dr. W. Behnke
Fraunhofer-Institut
I T A
Physikalische Chemie
Nikolai-Fuchs-Straße 1
D-30625 Hannover
Germany

Dr. H.-J. Benkelberg
Max-Planck-Institut für Chemie
Abt. Biogeochemie
Postfach 3060, D-55020 Mainz
Saarstraße 23, D-55122 Mainz
Germany

Mr. J. Berglund
University of Lund
Inorganic Chemistry I, Chemical Center
P.O. Box 124
S - 22100 Lund
Sweden

Dr. M. Bizjak
National Institute of Chemistry
Analytical Chemistry
Hajdrihova 19, P.O.B. 30
SLO-61000 Ljubljana
Slowenia

Dr. A. Bongartz
Forschungszentrum Karlsruhe GmbH
Inst.f. Meteorologie u. Klimaforschung
Bau 326: Atmosphärische Aerosolforschung
Postfach 3640
D-76021 Karlsruhe
Germany

Dr. J. Dierssen
J.W. Goethe-Universität
Zentrum für Umweltforschung
Postfach 111932-196, D-60054 Frankfurt
Georg-Voigt-Straße 14, D-60325 Frankfurt
Germany

Prof. Dr. R. van Eldik
Universität Erlangen-Nürnberg
Institut für Anorganische Chemie
Egerlandstraße 1
D-91058 Erlangen
Germany

Prof. Dr. L. I. Elding
Lund University
Inorganic Chemistry 1
P.O.Box 124
S-22100 Lund
Sweden

Dr. M. Elend
Fraunhofer-Institut
I T A
Physikalische Chemie
Nikolai-Fuchs-Straße 1
D-30625 Hannover
Germany

Prof. H. Elias
Anorganische Chemie III
TH Darmstadt
Petersenstraße 18
D-64287 Darmstadt
Germany

Dr. M. Exner
Universität Essen
Institut für Physikalische Chemie
Universitätsstraße 5-7
D-45141 Essen
Germany

Dr. F. F. Fenter
Ecole Polytechnique Fédérale
de Lausanne (E.P.F.L.)
Lab. de Pollution Atmosphérique et Sols
CH-1015 Lausanne
Switzerland

Dr. M. Fischer
Max-Planck-Institut für Chemie
Abt. Biogeochemie
Postfach 3060, D-55020 Mainz
Saarstraße 23, D-55122 Mainz
Germany

Dr. P. Foster
Université Joseph Fourier
IUT, Chemie, GRECA
1, Rue Francois Raoult
F-38000 Grenoble
France

Dr. M. Geißler
Universität Erlangen-Nürnberg
Institut für Anorganische Chemie
Egerlandstraße 1
D-91058 Erlangen
Germany

Dr. C. George
Centre de Géochimie de la Surface
C N R S
28 rue Goethe
F-67083 Strasbourg Cedex
France

Dr. I. Grgic
National Institute of Chemistry
Hajdrihova 19
SLO-61115 Ljubljana
Slowenia

Dr. C. Guillard
Ecole Centrale de Lyon
Equipe CNRS Photocatalyse
B.P. 163
F-69131 Ecully Cedex
France

Dr. L. Gutzwiller
Ecole Polytechnique Fédérale
de Lausanne (E.P.F.L.)
Lab. de Pollution Atmosphérique et Sols
CH-1015 Lausanne
Switzerland

Dr. U. Götz
Anorganische Chemie III
TH Darmstadt
Petersenstraße 18
D-64287 Darmstadt
Germany

Dr. H. Herrmann
Universität -GH- Essen, Fb 8
Institut für Physikalische Chemie
Universitätsstraße 5
D-45141 Essen
Germany

Dr. C. Hoang-Van
Ecole Centrale de Lyon
URA au CNRS Photocatalyse, Catalyse et Environnement
B.P. 163
F-69131 Ecully Cedex
France

Prof. J. Hoigné
EAWAG
Überlandstraße 133
CH-8600 Dübendorf
Switzerland

Dr. V. Hudnik
National Institute of Chemistry
Analytical Department
Hajdrihova 19
SLO-61000 Ljubljana
Slowenia

Dr. V. Jacob
Université Joseph Fourier
IUT, Chemie, GRECA
1, Rue Francois Raoult
F-38000 Grenoble
France

Mr. H.-W. Jacobi
Universität -GH- Essen, Fb 8
Institut für Physikalische Chemie
Universitätsstraße 5
D-45141 Essen
Germany

Dr. W. Jaeschke
J.W. Goethe-Universität
Zentrum für Umweltforschung
Postfach 111932-196, D-60054 Frankfurt
Georg-Voigt-Straße 14, D-60325 Frankfurt
Germany

Dr. J. Kames
Forschungszentrum Karlsruhe GmbH
Inst.f. Meteorologie u. Klimaforschung
Bau 326: Atmosphärische Aerosolforschung
Postfach 3640
D-76021 Karlsruhe

Dr. H.-U. Krüger
Fraunhofer-Institut
I T A
Physikalische Chemie
Nikolai-Fuchs-Straße 1
D-30625 Hannover
Germany

Prof. J. Lagrange
EHICS École Européene des Hautes Études des Industries Chimique
Lab. de Cinétique et Analyse
1, Rue Blaise Pascal
F-67000 Strasbourg Cedex
France

Dr. P. Lagrange
EHICS École Européene des Hautes Études des Industries Chimique
Lab. de Cinétique et Analyse
1, Rue Blaise Pascal
F-67000 Strasbourg Cedex
France

Mr. V. Lepentsiotis
Universität Erlangen-Nürnberg
Institut für Anorganische Chemie
Egerlandstraße 1
D-91058 Erlangen
Germany

Dr. L. Magi
Centre de Géochimie de la Surface
C N R S
28 rue Goethe
F-67083 Strasbourg Cedex
France

Ms. T. Malone
University of Leeds
Cookridge Radiation
Research Centre
GB-Leeds LS16 6QB
United Kingdom

Dr. F. Marme
Ecole Centrale de Lyon
URA au CNRS Photocatalyse, Catalyse et
Environnement
B.P. 163
F-69131 Ecully Cedex
France

Dr. D. Mas
Ecole Centrale de Lyon
URA au CNRS Photocatalyse, Catalyse et
Environnement
B.P. 163
F-69131 Ecully Cedex
France

Dr. S. McGowan
University of Leeds
Cookridge Radiation
Research Centre
GB-Leeds LS16 6QB
United Kingdom

Dr. P. Mirabel
Centre de Géochimie de la Surface
C N R S
28 rue Goethe
F-67083 Strasbourg Cedex
France

Ms. S. A. Murray
Cookridge Radiation Research Center
Cookridge Hospital
GB-Leeds LS16 6QB
United Kingdom

Dr. M. Novic
National Institute of Chemistry
Hajdrihova 19
SLO-61115 Ljubljana
Slowenia

Dr. C. Pallarès
EHICS École Européene des Hautes
Études des Industries Chimique
Lab. de Cinétique et Analyse
1, Rue Blaise Pascal
F-67000 Strasbourg Cedex
France

Prof. Dr. W. Pasiuk-Bronikowska
Polish Academy of Sciences
Institute of Physical Chemistry
Laboratory of Process Kinetics
Kasprzaka 44/52
PL-01-224 Warsaw
Poland

Dr. J.-C. Petit
CNRS Laboratoire de Combustion et
Systèmes Réactifs
1C, Avenue de la Recherche Scientifique
F- 45071 Orléans Cedex 2
France

Dr. P. Pichat
Ecole Centrale de Lyon
URA au CNRS Photocatalyse, Catalyse et
Environnement
B.P. 163
F-69131 Ecully Cedex
France

Mr. M. Poje
Boris Kidric Institute of Chemistry
Analytical Chemistry
Hajdrihova 19, P.O.B. 30
SLO-61115 Ljubljana
Slowenia

Mr. J. L. Ponche
Centre de Géochimie de la Surface
C N R S
28 rue Goethe
F-67083 Strasbourg Cedex
France

Mr. G. Raabe
Universität -GH- Essen, Fb 8
Institut für Physikalische Chemie
Universitätsstraße 5
D-45141 Essen
Germany

Mr. A. Reese
Universität -GH- Essen, Fb 8
Institut für Physikalische Chemie
Universitätsstraße 5
D-45141 Essen
Germany

Dr. M. J. Rossi
Ecole Polytechnique Fédérale
de Lausanne (E.P.F.L.)
Lab. de Pollution Atmosphérique et Sols
CH-1015 Lausanne
Switzerland

Dr. K. J. Rudzinski
Polish Academy of Sciences
Institute of Physical Chemistry
Laboratory of Process Kinetics
Kasprzaka 44/52
PL-01-224 Warsaw
Poland

Dr. G. A. Salmon
Cookridge Radiation Research Center
Cookridge Hospital
GB-Leeds LS16 6QB
United Kingdom

Mr. V. Scheer
Fraunhofer-Institut
I T A
Physikalische Chemie
Nikolai-Fuchs-Straße 1
D-30625 Hannover
Germany

Prof. U. Schurath
Forschungszentrum Karlsruhe GmbH
Inst.f. Meteorologie u. Klimaforschung
Bau 326: Atmosphärische Aerosolforschung
Postfach 3640
D-76021 Karlsruhe
Germany

Dr. A. Schäfer
Max-Planck-Institut für Chemie
Abt. Biogeochemie
Postfach 3060, D-55020 Mainz
Saarstraße 23, D-55122 Mainz
Germany

Dr. B. Servajean
Ecole Centrale de Lyon
URA au CNRS Photocatalyse, Catalyse et
Environnement
B.P. 163
F-69131 Ecully Cedex
France

Dr. U. Steingens
Anorganische Chemie III
TH Darmstadt
Petersenstraße 18
D-64287 Darmstadt
Germany

Dr. K. Tabor
Ecole Polytechnique Fédérale de
Lausanne (EPFL)
Lab. de PollutionAtmosphérique et Sols
CH-1015 Lausanne
Switzerland

Mr. J. Thlibi
CNRS Laboratoire de Combustion et
Systèmes Réactifs
1C, Avenue de la Recherche Scientifique
F- 45071 Orléans Cedex 2
France

Dr. N. Van Hoof
University of Leuven
Laboratory of Analytical and
Inorganic Chemistry
Celestijnenlaan 200F
B-3001 Leuven
Belgium

Prof. Dr. C. Vinckier
University of Leuven
Laboratory of Analytical and
Inorganic Chemistry
Celestijnenlaan 200F
B-3001 Leuven
Belgium

Dr. S. Volkwein
J.W. Goethe-Universität
Zentrum für Umweltforschung
Postfach 111932-196, D-60054 Frankfurt
Georg-Voigt-Straße 14, D-60325 Frankfurt
Germany

Dr. K. Wannowius
TH Darmstadt
Anorg. Chemie III
Petersenstraße 18
D-64287 Darmstadt
Germany

Prof. Dr. P. Warneck
Max-Planck-Institut für Chemie
Abt. Biogeochemie
Postfach 3060, D-55020 Mainz
Saarstraße 23, D-55122 Mainz
Germany

Dr. G. Wenger
EHICS École Européene des Hautes
Études des Industries Chimique
Lab. de Cinétique et Analyse
1, Rue Blaise Pascal
F-67000 Strasbourg Cedex
France

Dr. J. E. Williams
University of Leeds
Cookridge Radiation
Research Centre
GB-Leeds LS16 6QB
United Kingdom

Dr. J. Wohlgemuth
J.W. Goethe-Universität
Zentrum für Umweltforschung
Postfach 111932-196, D-60054 Frankfurt
Georg-Voigt-Straße 14, D-60325 Frankfurt
Germany

Dr. N. D. Wood
University of Leeds
Cookridge Radiation
Research Centre
GB-Leeds LS16 6QB
United Kingdom

Mr. C. Wunderlich
Universität Bonn
Institut für Physikalische Chemie
Wegelerstraße 12
D-53115 Bonn
Germany

Dr. R. Zellner
Universität Essen
Institut für Physikalische Chemie
Universitätsstraße 5-7
D-45141 Essen
Germany

Prof. Dr. C. Zetzsch
Fraunhofer-Institut
I T A
Physikalische Chemie
Nikolai-Fuchs-Straße 1
D-30625 Hannover
Germany

Dr. J. Ziajka
Polish Academy of Sciences
Institute of Physical Chemistry
Laboratory of Process Kinetics
Kasprzaka 44/52
PL-01-224 Warsaw
Poland

Chapter 1

1.1 Highlights of the Results

Aqueous phase oxidation of sulfur(IV) by non-radical reactions

* Stopped-flow techniques have been used to determine rate coefficients, pH dependence and the general mechanism for the non-radical oxidation of sulfur(IV) by hydrogen peroxide, peroxomonosulfate and other hydroperoxides.

* Rate coefficients for the oxidation of S(IV) by O_3, H_2O_2 and ClO^- have been determined under high ionic strength conditions applicable to deliquescent aerosol particles.

The role of transition metal ions in atmospheric waters

* Sulfur(IV) oxidation in the presence of transition metals (Fe, Mn, Co) has been demonstrated to feature autocatalysis. For iron as a catalyst it has finally been proven (90 years after its discovery) that the oxidation is a radical-driven chain reaction, and a mechanism has been established. An important step in the metal ion redox cycle is the interaction of SO_5^- with the bivalent species to regenerate the trivalent ion, which initiates the chain by formation of a sulfito-complex.

* The catalytic activity of carbon in the oxidation of sulfur(IV) has been demonstrated to arise from traces of iron in soot.

* The predominant fate of the hydroperoxy radical, HO_2, and its anion O_2^-, which do not oxidise sulfur(IV) directly, is reaction with ions of trace metals, copper being most efficient. The reactions produce H_2O_2, which can then oxidise S(IV). The reaction of O_2^- with ozone, which would provide a source of OH radicals, probably cannot compete with trace metals.

* The interaction of nitrogen oxide with sulfur(IV) in the presence of iron or manganese ions produces hydroxylaminedisulfonate, which hydrolyses to form hydroxylaminemonosulfonate under redox cycling of the metal ion. The role of such processes in atmospheric water drops remains to be further explored.

Photochemical sources of radicals in the aqueous phase

* The yields of radicals from the photodecomposition of NO_3^-, NO_2/HNO_2, H_2O_2, $FeSO_4^+$ and $FeOH^{2+}$ have been determined. The Fe(III)-hydroxo-complex was found to be the most favourable as a source of OH radicals in terms of quantum yield and abundance in cloud water.

* Photodecomposition of Fe(III)-oxalato complexes provides an efficient source of HO_2/O_2^- radicals in atmospheric water drops.

Free radical reactions in the aqueous phase

* Pulse radiolysis, steady-state radiolysis and laser flash photolysis have been employed to determine rate coefficients for the reactions involved in the chain oxidation of sulfite and hydrogen sulfite initiated by OH radicals and propagated by SO_5^- and SO_4^- radicals.

* The reactivities of SO_5^- and SO_4^- with Fe^{2+} and Mn^{2+}, which are important reactions in the oxidation of sulfur(IV) catalysed by transition metals, have been determined.

* The oxidation of formaldehyde initiated by OH radicals has been studied by steady-state and laser flash photolysis. The products are formic acid and hydrogen peroxide. In the presence of H_2O_2 a chain reaction can occur.

* The oxidation of hydroxymethanesulfonate, *i.e.* the addition compound of HSO_3^- and formaldehyde, initiated by OH, has been thoroughly investigated and the main reaction steps characterised.

* The nitrate radical, NO_3, which could enter cloud drops from the gas phase, reacts rapidly with aqueous chloride, HSO_3^- and SO_3^{2-} by charge transfer, and with organic substrates by H atom abstraction. The consequences of these NO_3 reactions have not yet been fully explored.

Gas-liquid interactions

* Four experimental techniques, involving a train of water drops, a liquid water jet, a wetted wall flow tube and a water drop floating in an uprising

air current, have been developed for the measurement of uptake coefficients, which are required for the calculation of gas-liquid uptake rates. Aqueous phase chemical analysis is used in all cases to quantify the mass transfer. Mass accommodation coefficients are derived by extrapolation to vanishing concentrations and contact times.

* Mass accommodation coefficients have been determined for important solutes: Values for SO_2, HNO_3, HCl and NH_3 are of the order of 0.1, that is about 10 % of the gas molecules that strike the droplet surface enter into solution; values for formic and acetic acid, HONO and N_2O_5 lie in the range 0.01–0.065. The uptake coefficient for N_2O_5 increases with increasing NaCl salt content from 0.013 to 0.032. The uptake of these gases meets little resistance.

* The uptake coefficient for ozone is 4.5×10^{-3}, whereas that for CO_2 is one order of magnitude smaller. The small value for ozone may seriously impede the supply of this important oxidant in cloud drops.

* Henry's law coefficients for a number of organic nitrates and peroxynitrates that are unstable in water have been determined.

Aerosol surface reactions and photocatalysis

* The activity of solids such as silicon dioxide, titanium dioxide, muscovite, haematite, ilmenite, fly ash, in catalysing the oxidation of organic compounds has been explored. TiO_2, which is known to generate OH radicals photocatalytically from adsorbed water, was found particularly effective in promoting the oxidation of alkanes, simple aromatics, naphthalene, ß-pinene. Fe_2O_3 was shown to be much less photocatalytically active in this regard.

* The possibility that photocatalytic activity arises from the formation of Lewis acid centres was explored. Only TiO_2 and Al_2O_3 showed a noticeable production of reactive sites, the response on SiO_2 and desert sand was minor. In all cases the Lewis acid centres were destroyed in the presence of water. A higher activity on fly ash was attributed to the presence of carbonaceous material, which reacts with water to produce CO_2 and H_2O_2.

* Compounds such as naphthalene, which are more or less permanently adsorbed on particles, are strongly susceptible to photocatalytic degradation in sunlight. In the presence of water, Fe_2O_3 forms complexes with organic compounds, which are photolytically decomposed, whereby the dissolution of iron in the aqueous phase is promoted.

Sea salt and nitrogen oxides

* The reaction of nitrogen pentoxide, N_2O_5, with deliquescent sodium chloride to produce $NaNO_3$ and $ClNO_2$ has been studied with submicron droplets suspended in an aerosol chamber and with a flow tube reactor the walls of which are wetted with NaCl solution. The yield of $ClNO_2$ approaches 100 % at high NaCl concentrations. At pH < 2, the yield of $ClNO_2$ decreased appreciably and NOCl was formed as a product. The results are explained by a chemical mechanism featuring NO_2^+ and NO_3 as intermediates.
* The solubility of $ClNO_2$ is much less than that of N_2O_5 so that $ClNO_2$ is released to the gas phase, where it undergoes photodecomposition to form NO_2 and Cl atoms. The consequences for tropopheric chemistry are an increase in the residence time of NO_x and an enhancement of radical chemistry due to extremely reactive Cl atoms.
* A Knudsen cell reactor with mass spectrometric detection of products was used to study the displacement of chloride and bromide from sodium halide crystals by reaction with HNO_3 and nitrogen oxides. HNO_3 produced 100 % of HCl and HBr from NaCl and NaBr, respectively; NO_2 gave HONO as the main product from NaCl (traces of water were present), no products were observed with NaBr; N_2O_5 produced 60 % $ClNO_2$, whereas the products with NaBr were Br_2 and HONO; $ClONO_2$ reacting with NaCl produced Cl_2 with high yield, with NaBr the products were Br_2, BrCl and Cl_2. Uptake coefficients ranged from about 10^{-7} for NO_2 reacting with NaCl to 3×10^{-2} for HNO_3 reacting with NaCl and NaBr.

1.2 Original aims of the EUROTRAC Sub-project HALIPP

HALIPP was established as a cooperative research project to meet the need for laboratory studies dedicated to the development of a quantitative description of heterogeneous and liquid phase chemical processes in the troposphere. In terms of the original work statement the objectives were:

(i) to study processes that generate and control oxidants and products in atmospheric water drops;
(ii) to study the transfer of free radicals and gases across the gas-liquid interface of atmospheric droplets;
(iii) to study the contribution of aerosols to the chemistry of trace gases and radicals, including the photochemical surface production of active species.

Major gaps and uncertainties existing at the beginning of HALIPP were related to the mechanisms responsible for the oxidation of sulfur(IV) species in the aqueous phase of clouds; the values of mass accommodation coefficients controlling the gas-liquid transfer of materials; and the role of truly heterogeneous reactions on surfaces of aerosol and ice particles in atmospheric chemistry. In each field of study the aim was to delineate the physico-chemical mechanisms and to quantify the individual rate processes. Although the ultimate aim is to provide data for computer models simulating atmospheric transport and chemical transformation of trace compounds, the implementation of such data in the models was not included as part of HALIPP activities.

Chapter 2

Review of the Activities and Achievements of the EUROTRAC Subproject HALIPP

P. Warneck[1], P. Mirabel[2], G. A. Salmon[3], R. van Eldik[4], C. Vinckier[5], K. J. Wannowius[6] and C. Zetzsch[7]

[1]Max-Planck-Institut für Chemie, Mainz, Germany
[2]Laboratoire de physico-chimie de l'atmosphère – Centre de Géochimie de la Surface, 28, rue Goethe, 67083 Strasbourg Cedex, France
[3]Cookridge Radiation Research Centre, The University of Leeds, Cookridge Hospital, Leeds, LS16 6QB, UK
[4]Institute for Inorganic Chemistry, University of Erlangen-Nürnberg, Egerlandstr. 1, D-91058 Erlangen, Germany
[5]Laboratory of Analytical and Physical Chemistry, Department of Chemistry, KU Leuven, Celestijnenlaan 200F, 3001 Heverlee, Belgium
[6]Institut für Anorganische Chemie, Technische Hochschule Darmstadt, Petersenstraße 18, D-64287 Darmstadt
[7]Fraunhofer-Institut für Toxikologie und Aerosolforschung, Nikolai-Fuchs-Straße 1, D-30625 Hannover, Germany

2.1 Introduction

In order to facilitate the assessment of scientific results obtained by HALIPP it will be helpful to reiterate the major problems and uncertainties in the field of heterogeneous chemistry that existed in 1987 when HALIPP was established. This background information is given below. The subsequent sections review the current status of heterogeneous and liquid phase processes. To put HALIPP results into a proper perspective, reference will be made to work performed outside HALIPP as well as to results reported by HALIPP contributors. An assessment of the progress achieved is given in 2.8.

Chemical reactions in cloud and fog drops have drawn attention ever since the subject of tropospheric chemistry was first developed in the early 1950s [1]. At that time it was primarily the oxidation of sulfur dioxide, which was thought to occur in clouds. Later, when the emphasis in atmospheric chemistry shifted toward gas-phase reactions, clouds were often neglected. At the start of HALIPP two decades of research had made it abundantly clear that gas-phase reactions alone were inadequate to destroy SO_2 at the rate indicated by field observations. Thus it was necessary to turn the attention again toward reactions in the aqueous phase of clouds. Today it is believed that clouds convert between 50 and 80 % of SO_2 to H_2SO_4 in the troposphere [2] and that this process contributes substantially to the formation of acid rain. A major problem was, and to some extent still is, to identify the most important aqueous phase oxidation reactions. Sulfite and hydrogensulfite are formed by the dissolution of SO_2 in water. Two oxidation pathways that were known in 1987 are reactions with hydrogen peroxide and ozone [3]. With regard to H_2O_2 it was not clear whether this oxidant is primarily absorbed from the gas phase, or whether it can also be formed in the aqueous phase. Another process oxidising dissolved SO_2, which was discovered at the beginning of the century [4], involves ions of certain transition metals as catalysts. Until recently, the mechanism for this process had remained obscure, however. It has also been proposed [5, 6] that similar to gas phase chemistry the OH radical might play a role in the aqueous phase by triggering a chain reaction. Graedel *et al.* [7] had called attention to the potential of iron(III)-hydroxo complexes to provide a photolytic source of OH radicals in cloud drops, and this effect needed quantification. Quantum yields for the photodecomposition of other ionic species also were required. Peroxomonosulfate, HSO_5^-, which was implicated as an intermediate product in the radical-propagated chain oxidation of sulfur(IV), needed study as an oxidant of dissolved SO_2. Sulfite and hydrogensulfite interact with aqueous formaldehyde to form hydroxymethanesulfonate; reactions of this compound with H_2O_2 or ozone were known to be negligible, but reactions with radicals had not been explored. A considerable range of problems thus existed with regard to the aqueous chemistry in clouds, and appropriate laboratory studies were needed to solve these problems.

Reactions in the aqueous phase of clouds are predicated on the uptake of gases by cloud drops. If chemical reactions are absent or slow, the concentration of a gaseous solute is determined by the equilibrium partioning established by Henry's law. Most of the Henry's law coefficients required were known in 1987 when HALIPP started. In the case of rapid chemical reactions, the rate of uptake of a speies from the gas phase becomes important because it may limit the consumption of the species [8]. Uptake rates are determined by the mass accommodation coefficient, which is the fraction of molecules entering into the aqueous phase compared with the total number of molecules striking the drop's surface. The fraction must be experimentally determined. In 1987, the problem had just been recognised and methods suitable for the laboratory measurement of mass accommodation coefficients were under development. It was also considered necessary to study uptake rates for radicals, especially HO_2, which by its reaction with dissolved ozone was thought to provide a source of OH radicals in the aqueous phase [9].

Important natural sources of particulate matter in the troposphere are sea spray, mineral dusts from arid soils, sulfates and nitrates produced in the oxidation of gaseous sulfur and nitrogen compounds, respectively, and condensable organic oxidation products derived from terpenes and other biogenic emissions [10]. Primary anthropogenic emissions contribute fly ash from coal-fired electric power plants, mineral dusts from steel works or cement production, and soot from combustion processes. Atmospheric aerosol particles thus contain a complex mixture of water-soluble salts, insoluble minerals, and organic compounds. All aerosol particles attach some liquid water as a function of relative humidity. Especially salt particles undergo deliquescence and turn into solution drops, and the larger of these particles generate cloud and fog drops when the relative humidity rises beyond 100 %. The tropospheric residence time of particles due to rainout is of the order of a few weeks [11]. The total surface of particles present in a vertical column of the troposphere corresponds approximately to that of the ground surface underneath the column.

When HALIPP was initiated, there were several indications that surface reactions on aerosol particles can add significantly to atmospheric chemistry in special cases. One such process that was recognised is the oxidation by photocatalysis of organic compounds adsorbed on minerals. Two effects may contribute to this phenomenon. One is the bathochromic shift of the ultraviolet absorption spectrum of the adsorbed compound toward longer wavelengths so that it can be photodecomposed by solar radiation [12]. Another effect is the photolytic generation of active sites on the surface of minerals that may trigger chemical reactions. In 1987 it had been demonstrated that polyaromatic hydrocarbons are photolytically degraded when they are associated with particles containing oxides of transition metals such as titanium, iron, or zinc, which liberate OH radicals from water adsorbed on the surface of the solid [13]. Such processes needed more study in order to assess their potential in causing chemical transformations during the long-range transport of particles in the troposphere.

Additional chemical reactions involving aerosol particles in the atmosphere derive from the interaction of gaseous species with the liquid water associated with aerosol particles and with dissolved electrolytes. For example, field observations and smog chamber studies had revealed the generation of HONO from nitrogen oxides reacting on the wetted surface of particles and chambers [14].

The mechanism of this reaction was, and still is, not sufficiently well established. A similar phenomenon in this category is the release of chlorine atoms from sea salt particles reacting with the gaseous components in a smog chamber [15]. In this case it was originally thought that the reaction of nitrogen oxides with seasalt produces nitric acid, that HNO_3 liberates HCl, and that chlorine atoms derive from the reaction of OH radicals with HCl. This reaction did not, however, explain the high rate of Cl formation observed in smog chambers. Molecular chlorine, a photolabile precursor of Cl atoms, was identified as a product arising from a heterogeneous reaction of ozone with HCl on glass and silica surfaces in the absence of NO_x [16]. In the presence of NO_x, nitrogen pentoxide (N_2O_5), and its precursor, nitrogen trioxide (NO_3) were thought most likely to be involved. The reaction of NO_2 with ozone is the source of NO_3 in smog chambers as well as in the atmosphere, and N_2O_5 results from association of NO_3 with NO_2. It had also been recognised, however, that the reaction of these oxides with NaCl may generate other chlorine-containing compounds that may serve as a secondary photolytic source of atomic chlorine. One such case that was known is the interaction of NO_2 with dry sea salt [17], which releases NOCl. These processes appeared to be potentially quite important to tropospheric chemistry, and their study demanded some priority.

2.1.1 References

1. C. E. Junge, *Air Chemistry and Radioactivity*, Academic Press, New York 1963.
2. J. Langner, H. Rodhe, *J. Atmos. Chem.* **13** (1991) 255–263.
3. S. A. Penkett, B. M. R. Jones, K. A. Brice, A. E. J. Eggleton, *Atmos. Environ.* **13** (1979) 323–337.
4. A. Titoff, *Z. Phys. Chem.* **45** (1903) 641–683
5. W. L. Chameides, D. D. Davies, *J. Geophys. Res.* **87** (1982) 4863–4877.
6. W. J. McElroy, *Atmos. Environ.* **20** (1986) 323–330.
7. T. E. Graedel, C. J. Weschler, M. L. Mandich, *Nature* **317** (1985) 240–242.
8. S. E. Schwartz, J. E. Freiberg, *Atmos. Environ.* **15** (1981) 1129–1144.
9. W. J. McElroy, *Atmos. Environ.* **20** (1986) 427–438.
10. P. Warneck, *Chemistry of the Natural Atmosphere*, Academic Press, San Diego 1988.
11. R. Jaenicke, *Ann. N. Y. Acad. Sci.* **338** (1980) 317–329.
12. S. Gäb, S. Nitz, H. Parlar, F. Korte, *Chemosphere* **4** (1975) 251–256.
13. H. Güsten, in: W. Jaeschke (ed) *Chemistry of Multiphase Atmospheric Systems*, NATO ASI Ser. G 6, Springer, Heidelberg 1986, pp. 567–594.
14. C. Kessler and U. Platt, in: B. Versino and G. Angeletti (eds) *The Physico–Chemical Behaviour of Atmospheric Pollutants, 3rd. European Symp.* Reidel, Dordrecht, Holland 1984, pp. 412–422.
15. C. Zetzsch, in: R. Zellner (ed) *Formation, Distribution and Chemical Transformation of Air Pollutants*, DECHEMA Monographs **104** VCH Verlagsges. Weinheim, Germany 1987, pp. 187–212.
16. C. Zetzsch, G. Pfahler, W. Behnke, *J. Aerosol Sci.* **19** (1988) 1203–1206.
17. B. J. Finlayson-Pitts, *Nature* **306** (1983) 676–677.

2.2 Sulfur(IV) Oxidation by Ozone, Hydroperoxides and Hypochlorous Acid

2.2.1 Sulfur(IV) in the aqueous phase

Whereas gas phase sulfur(IV) exists uniquely as SO_2 in the atmosphere, dissolution of SO_2 in water partly converts it to sulfurous acid and the corresponding ionic dissociation products HSO_3^- or SO_3^{2-}. Their relative concentrations depend on the pH. The chemistry of sulfur(IV) has been recently reviewed with respect to its atmospheric relevance [1]. At the level of atmospheric S(IV) concentrations the pyrosulfite anion, $S_2O_5^{2-}$, is of no significance. It is known [2, 3] that the bisulfite ion exists in two tautomeric forms (O or S bonded hydrogen atom). The rate of isomerization, which presumably is pH dependent, is fast compared to the rate of S(IV) oxidation.

2.2.2 Ozone as oxidant

The aqueous reaction of ozone with S(IV) has been extensively studied during the last twenty years [4–13]. A number of studies were carried out heterogeneously, *i.e.* with suspended water drops [11–13], but the present brief review deals exclusively with the homogeneous system. The pH range generally was 1–5 (approximately), and all the results agree that the overall reactivity rises with increasing pH. The reaction rate is proportional to the concentrations of ozone and S(IV). The system is complicated by two factors: One arises from the generation of acidity, which causes the reaction rate to decrease with time unless the pH is controlled by means of a buffer; the other complication is due to the decomposition of ozone in pure water. Decomposition is slow in acidic solutions, but its rate rises with increasing pH because ozone reacts with the hydroxyl ion. The reaction generates HO_2/O_2^- radicals and the ozonide anion as intermediates in the formation of OH radicals. Whereas the mechanism of ozone decomposition in alkaline solutions is well accepted [15, 16], that in acidic solutions appears to involve additional reactions that are not well understood [16, 17]. Since hydroxyl radicals react with sulfur(IV) species initiating a chain reaction, the decomposition of ozone may interfere with the direct reaction between O_3 and sulfur(IV). This effect can be suppressed by the addition of radical scavengers. Organic acid buffers also may act as a scavenger unless secondary organic peroxyradicals propagate the chain reaction.

Penkett *et al.* [5] and Lagrange *et al.* [9] both found rate laws depending approximately inversely on the root of $[H^+]$ concentration, and they interpreted this behaviour to arise from a chain reaction propagated by SO_3^- and SO_5^- radicals. In contrast, Erickson *et al.* [4] and Hoigné *et al.* [8] favoured a polar mechanism involving an electrophilic attack of ozone on the sulfur atom, and they interpreted the observed pH dependence to arise from the shift in dissociation equilibrium between HSO_3^- and SO_3^{2-} combined with a faster rate of the reaction with SO_3^{2-} compared with HSO_3^-. Hoigné [8] in particular worked with tertiary butanol as OH scavenger. In the presence of oxygen, the *t*–butyl radical produced leads to the formation of a *t*–butylperoxy radical, which is not expected to initiate any fast chain reactions, however. The individual rate coefficients derived by Hoigné *et al.* [8] for the reactions $O_3 + SO_3^{2-}$ and $O_3 + HSO_3^-$ at 22 °C were $(1.0 \pm 0.1) \times 10^9$ and $(3.2 \pm 0.2) \times 10^5$ $dm^3\ mol^{-1}s^{-1}$, respectively, whereas Erickson *et al.* [4] had found $(2.2 \pm 0.4) \times 10^9$ and $(3.1 \pm 0.2) \times 10^5$ $dm^3\ mol^{-1}\ s^{-1}$ at 25 °C, in good agreement. These data have been widely accepted by the atmospheric science community. On the other hand it should be noted that although tertiary butanol is a fast scavenger for OH radicals, it would not be a good scavenger for sulfuroxy radicals, if these were involved in a chain reaction initiated by ozone interacting with sulfur (IV). Botha et al. [10], who did stopped-flow experiments, noted that the rate law proposed by Penkett *et al.* [5] provides a good description of the overall trend of the data, whereas the rate law given by Hoigné *et al.* [8] gave a better fit to the detailed kinetic behaviour. Botha *et al.* [10] thus chose to combine both rate laws, *i.e.* they used the $[H^+]^{-1/2}$ dependence with individual reaction rates for HSO_3^- and SO_3^{2-}. Although the authors interpret their results by a chain reaction mechanism, they note that the empirical fitting procedure does not provide proof for or against such a mechanism.

The temperature dependence of the reaction has been repeatedly studied [5, 9, 10]. The activation energies obtained appear to vary with pH, for example $E_A = 16.9 \pm 0.9$ kJ mol^{-1} at pH 2, and $E_A = 25.6 \pm 0.2$ kJ mol^{-1} at pH 4 [10]. The source of the variation must be associated with the mechanism, for example the shift in the HSO_3^-/SO_3^{2-} dissociation equilibrium. Several studies have explored the influence of metal ions such as Fe^{2+}, Fe^{3+}, Mn^{2+}, Cu^{2+} and Cr^{3+} on the reaction [6, 10, 18]. Only Harrison *et al.* [18] reported a strong increase in reaction rate at pH 4.5 after the addition of Fe^{3+} or Mn^{2+}. The other authors found no catalytic effects.

In conclusion, it appears that up to now there is no definite proof for a chain reaction resulting from the decomposition of ozone. Thus the interpretation offered by Hoigné *et al* [8] , who considers the reaction to proceed by oxygen atom transfer, currently provides the simplest description in terms of a useful rate law.

2.2.3 Hydroperoxides as oxidants

The oxidation of S(IV) by hydroperoxides has been studied for a variety of compounds of the general composition ROOH. These include H_2O_2 [5, 19–27], CH_3OOH [27, 28], C_2H_5OOH [29], *t*–C_4H_9OOH [29], $HOCH_2OOH$ [27, 29], HC(O)OOH [27, 29], $CH_3C(O)OOH$ [28, 29], $C_2H_5C(O)OOH$ [29], $^-O_3SOOH$ [29–33], ONOOH [23, 29, 34], O_2NOOH [29].

Hydrogen peroxide is considered an important oxidant for the aqueous phase oxidation of sulfur (IV) in the atmosphere. Methylhydroperoxide, CH_3OOH [45], hydroxymethylhydroperoxide, $HOCH_2OOH$ [45], and peroxonitric acid, O_2NOOH [46, 47], have been detected in trace amounts in field campaigns. On the basis of model calculations [48] it was claimed that appreciable amounts of peroxomonosulfate, $^-O_3SOOH$, should exist in the aqueous phase in the atmosphere. The relevance of the other peroxo-compounds to the chemistry of the atmosphere is less obvious.

The chemistry of the peroxo-nitrogen compounds is not yet fully understood. A few years ago peroxonitrous acid was considered to play a role in biological systems. The state of knowledge regarding ONOOH has been reviewed recently [35]. A few studies have been published on peroxonitric acid [36–39]. The aqueous chemistry of O_2NOOH and the reactions involved are rather incompletely understood, however [29, 39]. Peroxonitrous and peroxonitric acid behave in a different way in aqueous solution: the anion of peroxonitrous acid, $ONOO^-$, is rather stable whereas peroxonitrate, O_2NOO^-, turned out to be very unstable. The opposite is true for the conjugate acids. Nevertheless, the oxidation of sulfur(IV) at moderate sulfur(IV) concentrations is faster than the decomposition rates of the peroxides [29, 34]. Recently an unforeseen synthetic route to O_2NOOH has been published which could be of significance to atmospheric chemistry [37].

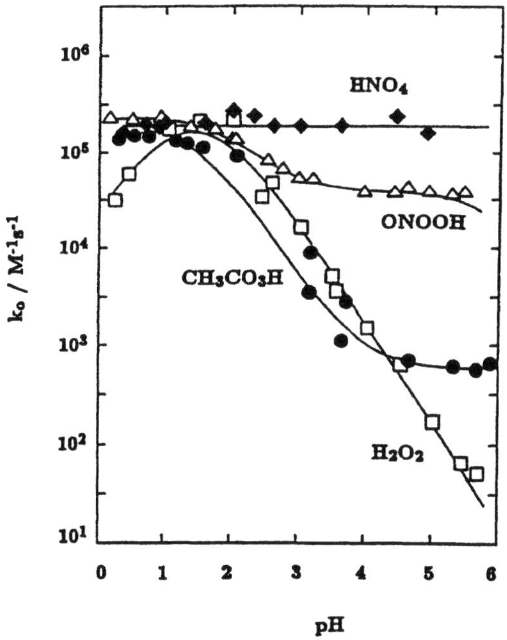

Fig. 1: pH profile for the buffer-corrected second order rate constant k_0 for various systems S(IV)/ROOH (H_2O_2: $T = 285$ K, $I = 0.5$ M $NaClO_4$; $CH_3C(O)OOH$: $T = 285$K, $I = 0.5$ M $NaClO_4$; ONOOH: $T = 293$K, $I = 1.0$ M Na_2SO_4; O_2NOOH: $T = 285$K, $I = 1.0$ M $NaClO_4$) (from [27]).

For small concentrations of the reactants the rate of S(IV) oxidation is first order with respect to both sulfur(IV) and the peroxo-compound ROOH (1).

$$-d[S(IV)]/dt = k[S(IV)][ROOH]_T \qquad (1)$$

where $[ROOH]_T$ indicates the total concentration of all forms of ROOH. The second order rate constant k exhibits a marked dependence on pH (see Fig. 1 [27]). The pK_A s of the sulfur(IV) species ($pK_{S1} = pK_A(SO_2) = 1.37$ and $pK_{S2} = pK_A(HSO_3^-) = 6.37$ (I = 1.0M)) as well as the pK_A s of the hydroperoxides control the concentrations of the various protonated/deprotonated species.
The rate of S(IV) → S(VI) conversion is given by the sum of the contributions of the individual reactant pairs (2).

$$-d[S(IV)]/dt = k_2[HSO_3^-][ROOH] + k_{SO3}[SO_3^{2-}][ROOH] + k'_{SO3}[SO_3^{2-}][ROO^-] \qquad (2)$$

Table 1: Rate constants for the oxidation of sulfur(IV) by hydroperoxides ($pK_{R1} = pK_A(ROOH_2^+)$; $pK_{R2} = pK_A(ROOH)$).

+ROOH	No.	T K	I M	pH–range [a]	pK_{R2}	pK_{R1}	$k_H \cdot 10^{-7}$ [b] $M^{-2}s^{-1}$	k_{HOH} $M^{-2}s^{-1}$	k_{SO3} $M^{-1}s^{-1}$	k'_{SO3} $M^{-1}s^{-1}$	Ref.
H_2O_2	i	298	0.5	0–13	11.3	1.4	3.9	0.16	0.19	0.001	23, 24, 29
		285	0.5	0–13			1.7	0.04	–	–	27, 29
		298	0.5				10.6 [c]	–	0.19	–	25
		298							0.2		19
		285					27				20
		298					5.6				21
CH_3OOH	h	285	1.0	0–13	11.4	1.6	0.45	0.41	0.02	–	27, 29
		296		3.5–5			1.7				28
C_2H_5OOH	k	285	1.0	2.5–6.5	–	–	0.78	–	–	–	27, 29
t-C_4H_9OOH	j	285	1.0	0–14	12.4	0.8	0.58	0.001	0.004	–	27, 29
$HOCH_2OOH$	g	285	1.0	0–10	10.5	1.2	0.90	0.50	17	–	27, 29
$^-O_3SOOH$	f	285	1.0	0–13	9.1	0.4	0.62	3.4	2300	16	29, 33
HC(O)OOH	c	285	0.5	4.5–6	7.0	–	2.6	36.0	–	–	27, 29
$CH_3C(O)OOH$	e	285	1.0	0–13.5	7.7	0.3	4.9	10.0	690	0.48	27, 29
		291		3–6			3.5	11.0	–	–	28
$C_2H_5C(O)OOH$	d	285	1.0	4.5–7	7.6	–	2.7	6.9	–	–	27, 29
ONOOH	b	293	1.0	0–13	6.8	<0	2.0	720	600	0.033	29, 34
O_2NOOH	a	285	1.0	0–3.5	5.0	<0	0.6	3500	–	–	29

[a] pH range covered by experiments. [b] Based on proton activities. [c] Calculated for I=0.5M $NaClO_4$.

The rate constants k_{SO3} [19, 29] and k'_{SO3} [29] are independent of pH ($k_{SO3} > k'_{SO3}$) whereas for the rate constant k_2 a strong pH dependence is observed. In addition, the reaction of HSO_3^- with ROOH is subject to general acid catalysis (3) [29, 33].

$$k_2 = k_H[H^+] + k_{HX}[HX] + k_{HOH}[H_2O] \tag{3}$$

Studies at constant proton concentrations instead of constant proton activities, however, could not verify a catalytic effect of acetic acid [25]. Buffers are often used in laboratory studies to maintain the pH during the kinetic runs. The buffer effect has to be eliminated, however, by extrapolation to zero buffer concentration (4) [27, 29, 33, 34].

$$k_o = k_2 - k_{HX}[HX] = k_H[H^+] + k_{HOH}[H_2O] \tag{4}$$

In principle, kinetic studies do not allow a decision between the two alternatives of whether SO_2 becomes involved as a reactant or whether the reaction proceeds with HSO_3^- by proton catalysis [33]. Generally, the second case is assumed to hold. The decrease in rate when the pH < 2 [21, 24, 33], which is observed in the system $S(IV)/H_2O_2$ and in some others (see Fig. 1), is ascribed to the formation of the protonated hydroperoxide, $ROOH_2^+$, which is a poor nucleophile and therefore unreactive towards sulfur(IV) [33].

Table 1 summarises the rate constants for some hydroperoxides used as oxidants for sulfur(IV).Fig. 2 [27] shows that the third order rate constant k_H is nearly unaffected by the nature of the hydroperoxide whereas $k_P = k_{HOH}[H_2O]$ strongly increases with increasing acid strength of the hydroperoxide (5) [29].

$$\log k_P = \alpha \, pK_a(ROOH) + \beta \tag{5}$$

As a consequence, the plateau in the pH profile (see Fig. 1) which is observed in the pH range 2–6 [27, 29] becomes more pronounced in the sequence of the hydroperoxides.

$$H_2O_2 < {}^-O_3SOOH < CH_3C(O)OOH < ONOOH < O_2NOOH \tag{6}$$

Whereas for the hydroperoxides labelled g–k in Table 1 the oxidation rate in the range pH 2–6 is quite comparable, peroxonitric and peroxonitrous acid are much more effective oxidants in the pH region where $k_H[H^+] < k_P$ [27]. Even low concentrations of the hydroperoxide may contribute significantly to the sulfur(IV) oxidation.

The general acid catalysis observed for the HSO_3^- reaction [27, 29] obeys a BrØnsted type relationship (7) for each peroxo-compound ROOH.

$$\log k_{HA} = \alpha \, pK_a(HA) + \beta' \tag{7}$$

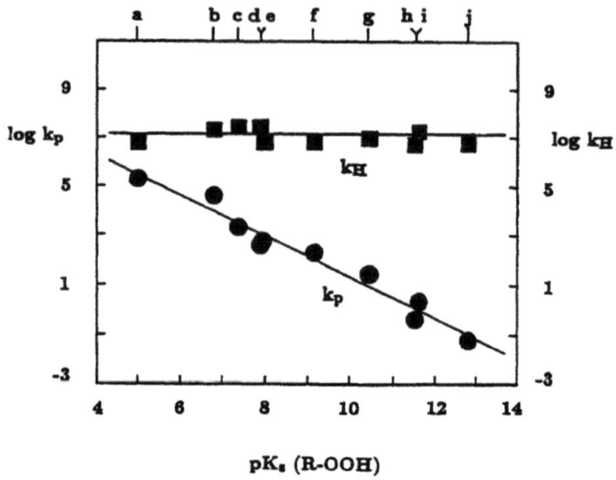

Fig. 2: Oxidation of bisulfite by various hydroperoxides: Dependence of the rate constants k_H and $k_P = k_{HOH}[H_2O]$ on the acid strength of the peroxo compound ROOH (The letters used to label the hydroperoxides are the same as shown in Table 1 in column 2)(from [29]).

The various kinetic findings can be understood best with the assumption that the type of mechanism changes when HSO_3^- is replaced by the SO_3^{2-}-anion.
The sulfite anion can be regarded as a good S-nucleophile, which attacks the hydroperoxides at the peroxo-group [40]. Displacement at the oxygen centre leads either to RO^--ions (8) [40] or OH^--ions (9) [32, 33]. In the latter case an intermediate compound, $ROSO_3^-$, should be formed. Its hydrolysis yielding the ions RO^- and SO_4^{2-} is considered to be a fast consecutive step [29, 32].

$$ROOH + SO_3^{2-} \rightarrow RO^- + SO_4^{2-} + H^+ \quad (8)$$
$$ROOH + SO_3^{2-} \rightarrow ROSO_3^- + OH^- \quad (9)$$
$$ROSO_3^- + H_2O \rightarrow RO^- + SO_4^{2-} + 2\,H^+ \quad (10)$$

In the system $ROOH/HSO_3^-$ the peroxo-compound ROOH acts as nucleophile towards the bisulfite ion. In a fast pre-equilibrium a peroxosulfurous acid species, $ROOSO_2^-$, is formed [33]. The intramolecular redox reaction, which is rate determining and subject to general acid catalysis, leads to a second intermediate, $ROSO_3^-$. In some $ROOH/HSO_3^-$ systems a consecutive step (assigned to the hydrolysis of the species $ROSO_3^-$) is observed experimentally [28, 32, 34].

$$ROOH + HSO_3^- \rightleftharpoons ROOSO_2^- + H_2O \quad (11)$$
$$ROOSO_2^- \rightarrow ROSO_3^- \quad (12)$$
$$ROSO_3^- + H_2O \rightarrow ROH + SO_4^{2-} + H^+ \quad (13)$$

With regard to the influence of the ionic strength disparent results have been published [23, 25]. Some authors report that there is no ionic strength effect [23], others find a marked influence of the ionic strength on the rate of S(IV) oxidation [25]. This problem seems to be connected to the question whether the variation of the ionic strength is done at constant pH or at constant $-\log[H^+]$. The constant A in (14) depends on the concentration of the electrolytes as well as on the type of the salts used.

$$pH = -\log[H^+] + A \quad (14)$$

Unfortunately, it is not always clear whether the kinetic parameters reported refer to hydrogen ion activity or concentration. In view of this uncertainty a factor of up to 2 may occur between data reported by different groups. The use of kinetic data for modelling purposes may thus require a critical assessment and selection of data depending on the application.

2.2.4 Hypochlorous acid as oxidant

Hypochlorous acid is considered to play a certain role at least in the chemistry of the marine atmosphere. Hypochlorous acid, HOCl, is a strong oxidant. Because of its good electrophilicity at the chlorine atom HOCl is an even stronger oxidant than hydrogen peroxide. The reactions of hypochlorous acid occur at rates near the diffusion controlled rate limit. Experimental studies are, therefore, often restricted to the range pH > pK_A(HOCl) where most of the hypochlorous acid is present in its anionic form, which is unreactive, however.

The rate of the oxidation of sulfite ions is given by the rate law [41, 42]

$$-d[Cl(I)]/dt = k[H^+][OCl^-][SO_3^{2-}] = k'[HOCl][SO_3^{2-}] \quad (15)$$

Mechanistically, the sulfur interacts with hypochlorous acid by forming the anion of chlorosulfonic acid as intermediate which is hydrolyzed to chloride and sulfate [43].

$$H^+ + OCl^- \rightleftharpoons HOCl \quad (16)$$
$$HOCl + SO_3^{2-} \rightarrow Cl{-}SO_3^- + OH^- \quad (17)$$
$$Cl{-}SO_3^- + H_2O \rightarrow SO_4^{2-} + Cl^- + 2\,H^+ \quad (18)$$

The chemistry of hypochlorous acid [44] and related non-metal compounds suggests, however, that the reaction of hypochlorous acid with nucleophiles such as the SO_3^{2-} ion may be catalysed by protons as well.

$$-d[Cl(I)]/dt = k'[HOCl][SO_3^{2-}] + k''[H^+][HOCl][SO_3^{2-}] \quad (19)$$

An extrapolation from the alkaline pH range to a pH of atmospheric relevance is difficult, however, because

(i) the second term in (19) could be dominant, or

(ii) a change in mechanism might occur as in the system S(IV)/ROOH.

The rate law of the reaction of HSO_3^- with hypochlorous acid is not known.

Table 2: General acid catalysis for the oxidation of the HSO_3^- ion by hydrogen peroxide (from [27]; HA = H^+, HX, H_2O).

HA	T / K	k_{HA}/ M^{-2} s^{-1}
H_3O^+	298	3.9×10^7 [a)]
	285	1.7×10^7 [a)]
$ClCH_2COOH$	285	9.2×10^5
HCOOH	285	9.0×10^4
CH_3COOH	285	2.4×10^4
$H_2PO_4^-$	298	4.7×10^3
HSO_3^-	298	2.0×10^3
$HTRIS^+$	298	45
NH_4^+	298	12
H_2O	298	0.16
	285	0.04

a) Based on proton activities.

2.2.5 References

1. C. Brandt, R. van Eldik, *Chem. Rev.* **95** (1995) 119–190.
2. R. M. Golding, *J. Chem. Soc.* (1960), 3711–3716.
3. D. A. Herner, R. E. Connick, *Inorg. Chem.* **25** (1986) 2414–2417.
4. R. E. Erickson, L. M. Yates, R. L. Clark, D. McEwen, *Atmos. Environ.* **11** (1977) 813–817.
5. S. A. Penkett, B. M. R. Jones, K. A. Brice, A. E. J. Eggleton, *Atmos. Environ.* **13** (1979) 123–137.
6. H. G. Maahs, *J. Geophys. Res.* **88** (1983) 10721–10732.
7. L. R. Martin, in: J. G. Calvert (ed) *SO_2, NO and NO_2 Oxidation mechanisms: Atmospheric Considerations*, Butterworth, Boston, MA 1984, pp. 62–100.
8. J. Hoigné, H. Bader, W. R. Haag, J. Staehelin, *Water Res.* **19** (1985) 993–1004.
9. J. Lagrange, C. Pallares, P. Lagrange, *J. Geophys. Res.* **99** (1994) 14595–14600.
10. C. F. Botha, J. Hahn, J.J. Pienaar, R. van Eldik, *Atmos. Environ.* **28** (1995) 3207–3212.
11. T. V. Larson, N. R. Horike, H. Harrison, *Atmos. Environ.* **12** (1978) 1597–1611.
12. R. L. Steele, W. A. Gertler, U. Katz, D. Lamb, D. F. Miller, *Atmos. Environ.* **15** (1981) 2341–2352.
13. D. F. Miller, W. A. Gertler, M. R. Whitbeck, D. Lamb, *J. Geophys. Res.* **91** (1986) 14439–14444.
14. R. E. Bühler, J. Staehelin, J. Hoigné, *J. Phys. Chem.* **88** (1984) 2560–2564.
15. J. Staehelin, R. E. Bühler, J. Hoigné, *J. Phys. Chem.* **88** (1984) 5999–6004.
16. K. Sehested, H. Corfitzen, J. Holcman, H. Fischer, E. Hart, *Environ. Sci. Technol.* **25** (1991) 1589–1596.
17. K. Sehested, H. Corfitzen, J. Holcman, E. Hart, *J. Phys. Chem.* **96** (1992) 1005–1009.
18. H. Harrison, T. V. Larson, C. S. Monkman, *Atmos. Environ.* **16** (1982) 1039–1041.
19. P. M. Mader, *J. Am. chem. Soc.* **80** (1958) 2634–2639.
20. M. R. Hoffmann, J. O. Edwards, *J. Phys. Chem.* **79** (1975) 2096–2098.
21. L. R. Martin, D. E. Damschen, *Atmos. Environ.* **15** (1981) 1615–1621.
22. J. V. McArdle, M. R. Hoffmann, *J. Phys. Chem.* **87** (1983) 5425–5429.

23. C. Drexler, H. Elias, B. Fecher, K. J. Wannowius, *Fresenius J. Anal. Chem.* **340** (1991) 605–615.
24. C. Drexler, H. Elias, B. Fecher, K. J. Wannowius, *Ber. Bunsenges. Phys. Chem.* **96** (1992) 481–485.
25. J. Lagrange, C. Pallarès, G. Wenger, P. Lagrange, *Atmos. Environ.* **27A** (1993) 129–137.
26. L. Breytenbach, W. van Pareen, J. J. Pienaar, R. van Eldik, *Atmos. Environ.* **28** (1994) 2451–2459.
27. B. Fecher, Kinetik und Mechanismus der S(IV) Oxidation mit Wasserstoffperoxid und organischen Peroxiden in wässriger Lösung, Dissertation Technische Hochschule Darmstadt (D17) 1995.
28. A. Lind, A. Lazrus, G.L. Kok, *J. Geophys. Res.* **92** (1987) 4171–4177.
29. H. Elias, K. J. Wannowius, Sonderforschungsbereich 233: *"Dynamik und Chemie der Hydrometeore"*; Arbeitsbericht für den Zeitraum 1991–1993 (1994) 481–504.
30. E. A. Betterton, M. R. Hoffmann, *J. Phys. Chem.* **92** (1988) 5962–5965.
31. U. Deister, W. J. McElroy, P. Warneck, in: P. Borrell, P. M. Borrell, W. Seiler (eds), *Proc. EUROTRAC Symp. '90*, SPB Academic Publishing bv, The Hague 1991, pp. 315–316.
32. R. E. Connick, S. Lee, R. Adamic, *Inorg. Chem.* **32** (1993) 565–571.
33. H. Elias, U. Götz, K.J. Wannowius, *Atmos. Environ.* **28** (1994) 439–448.
34. C. Drexler, Kinetik und Mechanismus der S(IV) Oxidation mit Peroxosäuren des Stickstoffs und salpetriger Säure, Dissertation Technische Hochschule Darmstadt (D17), 1993.
35. J. O. Edwards, R. C. Plumb, *Progress Inorg. Chem.* **41** (1994) 599–635.
36. R. A. Kenley, P. L. Trevor, B. Y. Lan, *J. Am. chem. Soc.* **103** (1981) 2203–2206.
37. E. H. Appelman and D. J. Gosztola, *Inorg. Chem.* **34** (1995) 787–789.
38. G. Lammel, D. Perner, P. Warneck, *J. Phys. Chem.* **94** (1990) 6141–6144.
39. T. Loegager, K. Sehested, *J. Phys. Chem.* **97** (1993) 10047–10052.
40. J. O.Edwards, *Peroxide Reaction Mechanism*, Interscience, New York 1962.
41. J. Lagrange, C. Pallarès, P. Lagrange, in: J. Peeters (ed) CEC *Air Pollution Research Report 45*, Guyot, Brussels 1993, pp. 265–268.
42. K. D. Fogelman, D. M. Walker, D. W. Margerum, *Inorg. Chem.* **28** (1989) 986–993.
43. B. S. Yiin, D. W. Margerum, *Inorg. Chem.* **27** (1988) 1670–1672.
44. K. Kumar, D. W. Margerum, *Inorg. Chem.* **26** (1987) 2706–2711.
45. E. Hellpointer, S. Gäb, *Nature* **337** (1989) 631–634.
46. C. P. Rinsland, R. Zander, C. B. Farmer, R. H. Norton, L. R. Brown, J. M. Russel III, J. H. Park, *Geophys. Res. Lett.* **13** (1986) 761–764.
47. A. Goldman, F. G. Murcray, R. D. Blatherwick, J. J. Kosters, F. H. Murcray, D. G. Murcray, *J. Geophys. Res.* **94** (1989) 14945–14955.
48. D. J. Jacob, *J. Geophys. Res.* **91** (1986) 9807–9826.

2.3 Transition Metal Ions and Their Role in Atmospheric Waters

2.3.1 Introduction

A number of HALIPP research groups have concentrated in their studies on the possible role of transition metal ions and complexes in the aqueous phase chemistry in atmospheric waters, i.e. cloud and rain water. The work has mainly focussed on the catalytic role of metal ions and complexes in the autoxidation of sulfur(IV) oxides, during which sulfuric acid and sulfates are produced. In general it is accepted that such reactions contribute to the acidification of cloud and fog water in the atmosphere [1]. Although the overall chemical processes involved in such oxidation processes are quite complex, it is in principle possible to focus on some elementary reaction steps that assist the elucidation of the autoxidation mechanism and the catalytic role of the metal ions. Numerous studies have dealt with the transition metal catalysed oxidation of sulfur(IV) oxides, and recent reviews treat atmospherically relevant processes and mechanisms in addition to the particular role of coordination chemistry in such processes [2, 3]. In this respect it is important to note that evidence for the redox cycling of iron in atmospheric water droplets and for a link between the iron and sulfur cycles has been reported [4, 5]. Other metals that are considered as potential catalysts are manganese, cobalt and copper [1, 6]. Their presence in atmospheric aqueous and particulate samples has been demonstrated [7–9]. Of all these systems, the catalytic role of iron and manganese has been studied in much detail within the context of the HALIPP project.

2.3.2 Iron-catalysed autoxidation of sulfur(IV) oxides

This system has been studied in detail by 4 participating groups: Grgic *et al.* [10, 11], Pasiuk-Bronikowska *et al.* [12, 13], van Eldik *et al.* [14, 15], and Warneck *et al.* [17, 18] (see also their individual reports). Direct kinetic evidence for the redox cycling of Fe(II/III) in the presence of bisulfite and oxygen has been found [14, 15]. Based on all the observations, the overall mechanism can be represented by the following reactions [14–18].

$$FeOH^{2+} + HSO_3^- \rightarrow Fe(OH)HSO_3]^+ \quad (1)$$

$$[Fe(OH)HSO_3]^+ \rightarrow Fe^{2+} + SO_3^- + H_2O \quad (2)$$

$$SO_3^- + O_2 \rightarrow SO_5^- \quad (3)$$

$$Fe^{2+} + SO_5^- + H_2O \rightarrow FeOH^{2+} + HSO_5^- \quad (4)$$

$$Fe^{2+} + HSO_5^- \rightarrow FeOH^{2+} + SO_4^- \quad (5)$$

$$Fe^{2+} + SO_4^- \rightarrow Fe^{3+} + SO_4^{2-} \quad (6)$$

$$SO_5^- + HSO_3^- \rightarrow HSO_5^- + SO_3^- \quad (7)$$

$$SO_5^- + HSO_3^- \rightarrow SO_4^{2-} + SO_4^- + H^+ \quad (8)$$
$$SO_4^- + HSO_3^- \rightarrow SO_4^{2-} + SO_3^- + H^+ \quad (9)$$
$$SO_5^- + SO_5^- \rightarrow S_2O_8^{2-} + O_2 \quad (10)$$
$$HSO_5^- + HSO_3^- + H^+ \rightarrow 2SO_4^{2-} + 3H^+ \quad (11)$$

The initiation reactions (1) and (2) are suggested to involve an ironsulfito complex, its exact nature still being unresolved [16, 19]. Reaction (4) is a crucial step since it accounts for the reoxidation of the reduced Fe(II) species with SO_5^- being a strong oxidant. A contribution from the oxidation of Fe(II) by molecular oxygen has also been suggested [12, 20]. Warneck *et al.* [17, 18] provided evidence for the participation of SO_4^- radicals during the Fe(III) catalysed autoxidation of S(IV) oxides. No evidence for the participation of OH radicals could be found. Simulations of this mechanism based on available rate constants, indicated that the reoxidation routes via SO_4^- and HSO_5^- are too slow to contribute meaning-fully [17, 18] under conditions of concentrations $[HSO_3^-] \gg [Fe]$. The available experimental results largely support the proposed mechanism. Wohlgemuth and Jaeschke [21] have studied the Fe(III)-catalysed reaction in small droplets in a specially designed flow reactor. They found the oxidation rate to vary linearly with the Fe(III) concentration and observed an increase in rate on going from pH 3 to 3.7. Their reported rate constants are higher than those found in the literature.

In terms of the possible role of soot and aerosol particles in the iron-catalysed autoxidation reactions, Grgic *et al.* [22] found that especially the iron leached from soot particles exhibited a meaningful catalytic effect [16]. In the case of aerosol particles collected in Ljubljana (size 0.4–1.6 μm) it was found that the reaction became faster at lower pH in the range 2.8 to 6.1.
The thermal reactions outlined above may exhibit photochemical components during daytime processes. Benkelberg and Warneck [23] showed that the photodecomposition of $FeOH^{2+}$ and $FeSO_4^+$ resulted in the formation of OH and SO_4^- radicals, respectively, the latter being a less important process. Hoigne and co-workers [24–27] also reported evidence for the formation of OH and HO_2/O_2^- radicals during the photolysis of aquated Fe(III) and oxalato complexes of Fe(III).

2.3.3 Manganese-catalysed autoxidation of sulfur(IV) oxides

The catalytic effect displayed by manganese(II/III) was studied by Pasiuk-Bronikowska [28] and in considerable detail by Elding and co-workers [29]. The reaction was shown to be autocatalytic with an induction period. The latter could be shown to be due to the formation of manganese(III), which is the actual catalyst similar to the case of iron(III). In the absence of initially added Mn(III) it was proposed that low concentrations of iron(III), present as unavoidable impurities at levels of 10^{-8} mol dm^{-3}, are responsible for the initiation of the reaction. A simplified reaction mechanism, rather similar to the one given above for the iron–catalysed reaction, can be written as follows [29]:

$$Mn^{2+} + HSO_3^- \rightarrow MnHSO_3^+ \quad (12)$$
$$Mn(III) + MnHSO_3^+ \rightarrow 2Mn(II) + SO_3^- + H^+ \quad (13)$$
$$SO_3^- + O_2 \rightarrow SO_5^- \quad (3)$$
$$Mn(II) + SO_5^- + H^+ \rightarrow Mn(III) + HSO_5^- \quad (14)$$
$$Mn(II) + HSO_5^- \rightarrow Mn(III) + SO_4^{2-} + OH \quad (15)$$
$$MnHSO_3^+ + HSO_5^- \rightarrow Mn(II) + 2HSO_4^- \quad (16)$$
$$MnHSO_3^+ + OH \rightarrow Mn(II) + SO_3^- + H_2O \quad (17)$$

These steps are followed by reactions (7) to (11) as given above. The redox cycling of manganese is an essential component of the catalytic process. Such a redox cycling could be clearly observed in the case of an azide complex of manganese in the presence of sulfite and oxygen [30]. The $MnHSO_3^+$ species was suggested to be the reactive complex in the above outlined mechanism. Pulse-radiolysis studies demonstrated that the rate constant for reaction (14) is in the range 10^8 to 10^9 dm^3 mol^{-1} s^{-1}, which is significantly higher than the rate constant for reaction (4), *viz.* 6×10^5 dm^3 mol^{-1} s^{-1} [15, 18, 31]. A detailed comparison with previous literature has been published [32]. The authors conclude that the manganese catalysed reaction can be described by the simple rate law $-d[S(IV)]/dt = k[Mn(II)][S(IV)]$, where $k = 1.4 \times 10^3$ dm^3 mol^{-1} s^{-1} over the range $2.5 < pH < 5$.

2.3.4 Other metals and synergistic effects

Other metals that have been shown to exhibit catalytic activity are cobalt and copper [33–36]. Their catalytic behaviour exhibits similar trends as those outlined above for iron and manganese and it is tempting to conclude that all one-electron metal oxidants behave rather similarly, an unproven hypothesis. It is clear that a metal ion or complex in a higher oxidation state is needed to initiate the formation of the SO_3^- radical. A laboratory demonstration experiment was developed on the basis of the general observations for one-electron metal oxidants [37]. A general mechanism based on the reactions outlined above is presented in Fig. 1 and illustrates the important role of redox cycling of the metal ion.

Mechanistic studies of redox reactions between sulfur(IV) oxides and two-electron metal oxidants, *viz.* platinum(IV), thallium(IV) and gold(III), in the presence of dissolved oxygen have shown that only a stoichiometric redox reaction is observed and no evidence for catalytic effects could be found [38–40]. It was concluded that a two-electron transfer takes place within the encounter complex formed between sulfite and these metal ions. Thus no sulfite radical is released into the solution and the chain mechanism is not initiated.

A number of reports on synergistic effects of different metal ions during the catalysed autoxidation of sulfur(IV) oxides have appeared. However, the systems are in general so complex that it is difficult to predict the contribution from such effects [41]. In some cases even negative synergism for cobalt-iron and cobalt-manganese has been reported [42]. The iron-manganese system has a high potential for synergistic effects based on the effect of iron impurities observed by Elding and co-workers [29, 32]. Further work is needed in this area.

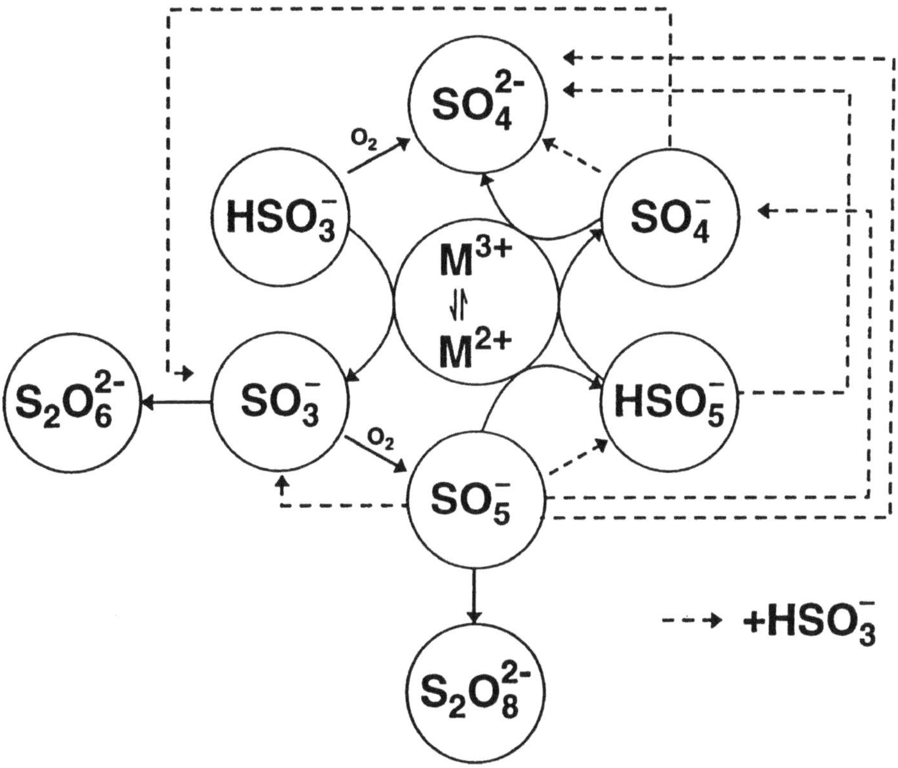

Fig. 1: Generalised reaction scheme, showing the role of one electron metal oxidants in the oxidation of sulfur(IV) in aqueous solution.

2.3.5 Other metal-catalysed reactions

Reactions of metal ions such as iron and copper with hydroperoxy radicals are considered important in atmospheric water drops [6, 43, 44]. The peroxy radicals enter the drops from the gas phase where they are produced by photochemical reactions. Copper is more important than iron in the scavenging of HO_2 and O_2^- radicals. Under such conditions the iron(III)-catalysed oxidation of sulfur(IV) may be largely inhibited, because copper(I) produced in the reaction of copper(II) with HO_2/O_2^-, may control the reduction of iron(III) to iron(II). In addition, the copper catalysed disproportionation of HO_2/O_2^- yields an enhanced formation of H_2O_2, which acts as main oxidant for aqueous SO_2 [44].

Nitrogen oxides, as well as mixed sulfur-nitrogen oxides, produced during the spontaneous or metal catalysed interaction of $NO/HONO/NO_2^-$ with $SO_2/HSO_3^-/SO_3^{2-}$, can interact with metal ions and complexes and affect their oxidation state.

In general, iron(II) complexes can rapidly bind NO to produce species of the type Fe(III)–NO^-, which decomposes to iron(III) and NO^-, *i.e.* N_2O. Coordinated NO can interact with bisulfite to produce hydroxylamine-disulfonate(HADS), N_2O and iron(III) [45, 46]. It has also been reported that the reaction between nitrous acid and bisulfite is a potential source of free radicals in natural environments [47]. Manganese(III) and iron(III) ions and complexes can affect the hydrolysis reactions of such mixed sulfur-nitrogen oxides [48]. For instance, HADS rapidly hydrolyses under such conditions to hydroxylaminemonosulfonate, during which the metal ions are reduced to the +II oxidation state. In one case evidence for the redox cycling of manganese(II/III) in the presence of a sulfur-nitrogen oxide was observed [49]. The role of these reactions in atmospheric oxidation processes is presently unknown despite their potential importance, especially in terms of controlling the oxidation state of the catalytic metal ions.

2.3.6 Conclusions

Transition metal ions can play an important role in the chemistry of atmospheric waters. The results reported here clearly demonstrate that the participating HALIPP groups have made good progress in the elucidation of the metalcatalysed autoxidation mechanisms. A certain degree of consistency within the different systems exists and predictions can be made to some degree. A major difficulty concerns the extrapolation of the data to atmospheric conditions, especially in terms of concentration ranges and pH. In the case of the manganese-catalysed autoxidation of sulfur(IV), the effect of pH did not seem to be that important. However, in the case of the iron-catalysed reaction, the system is extremely pH sensitive and it is almost impossible to obtain reliable data at pH > 3 due to solubility limitations and the formation of polymeric iron-hydroxo species under such conditions.

2.3.7 References

1. P. Warneck, *Fresenius J. Anal. Chem.* 340 (1991) 585–590.
2. Ch. Brandt, R. van Eldik, *Chem. Rev.* 95 (1995) 119–190.
3. N. Coichev, R. van Eldik, *New J. Chem.* 18 (1994) 123–131.
4. P. Behra, L. Sigg, *Nature* 344 (1990) 419–421.
5. G. Zhuang, Z. Yi, R. A. Duce, P. R. Brown, *Nature* 355 (1992) 537–539.
6. D. L. Sedlak, J. Hoigne, *Atmos. Environ.* 27A (1993) 2173–2185.
7. A. Molnar, E. Meszaros, L. Bozo, I. Borbely–Kiss, E. Koltay, Gy. Szabo, *Atmos. Environ.* 27A (1993) 2457–2461.
8. P. Hoffmann, T. Sinner, A. N. Dedik, V. K. Karandashev, A. A. Malyshev, S. Weber, H. M. Ortner, *Fresenius J. Anal. Chem.* 350 (1994) 34–37.
9. Ch. Brandt, V. Lepentsiotis, A. Paul, H. Hohmann, R. van Eldik, *GIT Fachz. Lab.* 38 (1994) 770–775.
10. I. Grgic, M. Bizjak, V. Hudnik, J. Levec, *Atmos. Environ.* 25A (1991) 1591–1597.

11. I. Grgic, M. Kocutar, M. Novic, M. Bizjak, V. Hudnik,in: G. Angeletti and G. Restelli (eds) *Physico–Chemical Behaviour of Atmospheric Pollutants, Proc. 6th Symp.*, EC, Luxembourg 1994, pp. 907–911.
12. W. Pasiuk–Bronikowska, K. J. Ruzinski, T. Bronikowski, J. Ziajka, in: *EUROTRAC Annual Reports Part 6*, EUROTRAC ISS Garmisch-Partenkirchen 1993, pp. 42–47.
13. P. Warneck, J. Ziajka, W. Pasiuk–Bronikowska, in: G. Angeletti and G. Restelli (eds) *Physico–Chemical Behaviour of Atmospheric Pollutants, Proc. 6th Symp.*, EC, Luxembourg 1994, pp. 901–906.
14. K. Bal Reddy, R. van Eldik, *Atmos. Environ.* 26A (1992) 661–665.
15. Ch. Brandt, I. Fabian, R. van Eldik, *Inorg. Chem.* 33 (1994) 687–701.
16. J. Kraft, R. van Eldik, *Inorg. Chem.* 28 (1989) 2297–2305, 2306–2312.
17. J. Ziajka, F. Beer, P. Warneck, *Atmos. Environ.* 28 (1994) 2549–2552.
18. P. Warneck, J. Ziajka, *Ber. Bunsenges. Phys. Chem.* 99 (1995) 59–65.
19. E. A. Betterton, *J. Atmos. Chem.* 17 (1993) 307–324.
20. W. Pasiuk-Bronikowska, J. Ziajka, T. Bronikowski, *Autoxidation of Sulphur Compounds*, PWN Polish Scientific Publ., Warszawa, Horwood, New York.
21. J. Wohlgemuth, W. Jaeschke, *Air Pollution Research Report 54*, European Commission, Brussels 1995, pp. 433–438.
22. L. Grgic, V. Hudnik, M. Bizjak, J. Levec, *Atmos. Environ.* 27A (1993) 1409–1416.
23. H.–J. Benkelberg, P. Warneck, *J. Phys. Chem.* 99 (1995) 5214–5221.
24. B. C. Faust, J. Hoigne, *Atmos. Environ.* 24A (1990) 79–89.
25. Y. Zuo, J. Hoigne, *Environ. Sci. Technol.* 26 (1992) 1014–1022.
26. R. G. Zepp, B. C. Faust, J. Hoigne, *Environ. Sci. Technol.* 26 (1992) 313–319.
27. Y. Zuo, J. Hoigne, *Atmos. Environ.* 28 (1994) 1231–1239.
28. Pasiuk-Bronikowska, T. Bronikowski, M. Ulejczyk, in: P. M. Borrell, P. Borrell, T. Cvitaš, W. Seiler (eds), *Proc. EUROTRAC Symp. '94*, SPB Academic Publishing bv, The Hague 1994, pp. 1030–1034.
29. J. Berglund, S. Fronaeus, L. I. Elding, *Inorg. Chem.* 32 (1993) 4527–4538.
30. N. Coichev, R. van Eldik, *Inorg. Chim. Acta* 185 (1991) 69–73.
31. J. Berglund, G. V. Buxton, L. I. Elding, S. McGowan, G. A. Salmon, *J. Chem. Soc., Faraday Trans.* 90 (1994) 3309–3313.
32. J. Berglund, L. I. Elding, *Atmos. Environ.* 29 (1995) in press.
33. N. Coichev, R. van Eldik, *Inorg. Chem.* 30 (1991) 2375–2380.
34. W. Pasuik-Bronikowska, K. J. Rudzinski, in: P. M. Borrell, P. Borrell, T. Cvitaš, W. Seiler (eds), *Proc. EUROTRAC Symp. '92*, SPB Academic Publishing bv, The Hague 1993, pp. 577–580.
35. W. Pasiuk-Bronikowska, K. J. Rudzinski, in: P. M. Borrell, P. Borrell, T. Cvitaš, W. Seiler (eds), *Proc. EUROTRAC Symp. '94*, SPB Academic Publishing bv, The Hague 1994, pp. 1026–1029.
36. W. Pasiuk-Bronikowska, K. J. Rudzinski, in: G. Angeletti and G. Restelli (eds) *Physico-Chemical Behaviour of Atmospheric Pollutants, Proc. 6th European Symp.*, European Commission, Luxembourg 1994, pp. 912–917.
37. N. Coichev, R. van Eldik, *J. Chem. Educ.* 71 (1994) 767–768.
38. J. Berglund, R. Voigt, S. Fronaeus, L. I. Elding, *Inorg. Chem.* 33 (1994) 3346–3353.
39. J. Berglund, P. Werndrup, L. I. Elding, *J. Chem. Soc., Dalton Trans.* (1994) 1435–1439.
40. J. Berglund, L. I. Elding, *Inorg. Chem.* 34 (1995) 513–519.

41. I. Grgic, V. Hudnik, M. Bizjak, J. Levec, *Atmos. Environ.* 26A (1993) 571–577.
42. W. Pasiuk–Bronikowska, A. Sokolowski, *Chem. Eng. Commun.*, in press.
43. M. von Piechowski, T. Nauser, J. Hoigne, R. E. Bühler, *Ber. Bunsenges. Phys. Chem.* 97 (1993) 762–771.
44. D. Sedlak, J. Hoigne, *Environ. Sci. Technol.* 28 (1994) 1898–1906.
45. V. Zang, R. van Eldik, *Inorg. Chem.* 29 (1990) 4462–4468.
46. V. Zang, R. van Eldik, *J. Chem. Soc., Dalton Trans.* (1993) 111–118.
47. S. N. Mendiara, E. Ghibaudi, L. J. Perissinotti, A. J. Colussi, *J. Phys. Chem.* 96 (1992) 8089–8091.
48. F. F. Prinsloo, J. J. Pienaar, R. van Eldik, H. Gutberlet, *J. Chem Soc., Dalton Trans.* (1994) 2373–2379.
49. F. F. Prinsloo, J. J. Pienaar, R. van Eldik, *J. Chem. Soc., Dalton Trans.* (1995) 293–299.

2.4 Photochemical Sources of Radicals in Atmospheric Water Drops

2.4.1 Introduction

Table 1: Summary of photochemical processes and quantum yield data.

Process Photodissociation Coef.[a]	Wavelength or Range (nm)	Radical Quantum Yield	References
$H_2O_2 + h\nu \rightarrow OH + OH$	254	1.0	1–3
$j = 1.0 \times 10^{-5}$ [22]	308, 351	0.98 ± 0.03, 0.96 ± 0.04	4
	250–310	1.55 ± 0.16 [b]	5
	254	1.43 ± 0.15, 1.22 ± 0.04	5
$NO_3^- + h\nu \rightarrow NO_2 + O^-$	305	$(9.2 \pm 0.4) \times 10^{-3}$	6
$j = 5.6 \times 10^{-7}$	308	0.017 ± 0.003	4
	313	0.013–0.017	7
$NO_2^- + h\nu \rightarrow NO + O^-$	280–390	see Fig. 1	4, 8–10
$j = 3.8 \times 10^{-5}$			8
$HNO_2 + h\nu \rightarrow NO + OH$	254, 365	0.46 ± 0.1, 0.45 ± 0.1	9
$j = 6.7 \times 10^{-4}$	280 – 385	0.35 ± 0.03 [b]	8
$FeOH^{2+} + h\nu \rightarrow Fe^{2+} + OH$	280 – 370	see Fig. 2	13
$j = 5.2 \times 10^{-3}$			12
$FeSO_4^+ + h\nu \rightarrow Fe^{2+} + SO_4^-$	280 – 350	$(1.5 - 7.5) \times 10^{-3}$ [c]	13
$j = 1.0 \times 10^{-4}$			13
$FeSO_3^+ + h\nu \rightarrow Fe^{2+} + SO_3^-$	280 – 350	insufficient data	15
$Fe(C_2O_4)_2^- + h\nu \rightarrow Fe^{2+} + C_2O_4^{2-} + C_2O_4^-$	300 – 420	0.62 [b]	17
$j = 6.2 \times 10^{-2}$			17

[a] Reference gives source of absorption cross sections used;
[b] independent of wavelength;
[c] quantum yield rises with decreasing wavelength.

The chemistry in tropospheric liquid water drops is driven at least in part by radicals produced by photochemical reactions in the aqueous phase. Among the compounds that occur in the atmosphere, undergo photodecomposition in sunlight, and that may serve as sources of radicals, are hydrogen peroxide, nitrate and nitrite anion, undissociated nitrous acid, and several complexes of trivalent iron, of which the hydroxo-, sulfito-, sulfato- and oxalato-complexes are potentially the most important. In order to assess the effectiveness of such photochemical processes it is necessary to compare for each individual compound its abundance in the atmospheric aqueous system and the photodissociation frequency associated with the photochemical process. Calculation of the rate coefficient for radical generation requires a knowledge of absorption coefficients and photochemical quantum yields as a function of wavelength. Both quantities must be measured in the laboratory.

Where as absorption coefficients were known in most cases, the quantum yield data available at the beginning of HALIPP referred to isolated wavelengths, and the full wavelength dependence needed to estimate photodissociation coefficients was not known. The radical scavenger technique was applied throughout to derive quantum yield data. Table 1 provides an overview on photochemical processes of interest and associated quantum yields obtained. These processes will be discussed below.

2.4.2 Hydrogen Peroxide

Prior to HALIPP, the photolysis of H_2O_2 in aqueous solution had been studied only at 253.7 nm (the wavelength of the mercury resonance line). A quantum yield of 0.5 for H_2O_2 photodecomposition, or 1.0 for OH production, had been inferred [1–3]. It has also been recognised that HO_2 radicals formed in the reaction of OH with H_2O_2 or added scavenger may trigger a chain reaction, in which OH is regenerated so that the observed quantum yield may be greater. Studies of H_2O_2 photolysis conducted within HALIPP have extended the wavelength range toward 351 nm, well into the wavelength region overlapping the solar spectrum reaching the troposphere.

Zellner *et al.* [4] used excimer laser photolysis at 308 nm and 351 nm wavelength and worked with thiocyanate ion, SCN^-, as scavenger for OH radicals. At the fairly high concentrations of scavenger employed (0.3 M), the $(SCN)_2^-$ radical is rapidly formed and its concentration determined by optical absorption measured at 475 nm. OH quantum yields obtained at 298 K were 0.98 ± 0.03 at 308 nm and 0.96 ± 0.04 at 351 nm. The temperature dependence of OH formation was studied, and activation energies 5.5 ± 1.6 and 4.8 ± 1.3 kJ mol^{-1}, respectively, were derived. They are the same within experimental error. If the former measurements of the temperature dependence at 254 nm [1, 2] are included, the activation energy calculated from the sum of all data is 6.29 ± 0.33 kJ mol^{-1}.

Schäfer [5] used 2-propanol as scavenger for OH radicals and followed the evolution of product acetone in the photodecomposition of H_2O_2 with light in the wavelength range 250–310 nm, wavelength selected in 10 nm steps by means of a monochromator. He confirmed that the OH quantum yield shows no significant wavelength dependence. However, the absolute values, which averaged 1.55 ± 0.16, were higher than those found by Zellner *et al.* [4]. The OH quantum yield derived from additional measurements with Hg 254 nm light from a Penray lamp was 1.43 ± 0.15. Here it was necessary to restrict irradiation times to less than 8 min, because oxygen was consumed in the oxidation of 2-propanol; complete consumption had to be avoided because it gave rise to a chain reaction involving H_2O_2 and the 2-propanolyl radical. The dependence of OH quantum yield on H_2O_2 concentration was investigated in all these experiments; none was found. The Penray lamp also was used to follow the decay of H_2O_2 by optical absorption in the absence of scavenger. In this case the OH quantum yield was 1.22 ± 0.04. All the values obtained were significantly greater than unity, but reasons for the disparity with previous data are not apparent.

2.4.3 Nitrate

The spectrum of the nitrate anion reaches out to about 355 nm, with a first maximum at or near 305 nm. In this wavelength region NO_3^- photodissociation produces $NO_2 + O^-$ in the main product channel and $NO_2^- + O$ in a minor one. The relative probabilities at 305 nm are 90 % and 10 %, respectively [6]. The pK_A value of the conjugate acid/base pair OH/O^- is about 12, so that at pH values usually encountered the O^- radical becomes rapidly protonated. Zellner *et al.* [4] have used SCN^- as scavenger for OH and determined the quantum yield and its temperature dependence by laser photolysis at 308 nm. At 298 K the value was 0.017 ± 0.003 with an apparent activation energy of 15 ± 4 kJ mol^{-1}. There was no dependence on pH in the range 4–9. The 298 K quantum yield is in good agreement with the result of Zepp *et al.* [7] obtained at 313 nm, who found quantum yields 0.013–0.017 for the 293–303 K temperature range. These authors used anisole or nitrobenzene as OH scavengers. Warneck and Wurzinger [6], who studied the NO_3^- photolysis at 305 nm, used 2-propanol as OH scavenger, determined product acetone and corrected for the presence of oxygen atoms. They reported $(9.2 \pm 0.4) \times 10^{-3}$ but noted that the yield increased to 0.0123 ± 0.007 when the scavenger concentration was increased from 0.0013 to 0.13 M. It should be noted that the entire data set refers to the wavelength region 305–313 nm. It would be of interest to determine quantum yields at other wavelengths.

2.4.4 Nitrite

The absorption spectrum of the nitrite anion extends toward 400 nm with a first maximum at or near 355 nm. Photodecomposition of NO_2^- leads primarily to the formation of NO and O^- (followed by rapid conversion of O^- to OH, except at very high pH). By using N_2O as scavenger for hydrated electrons, Fischer [8] explored the possibility that $NO_2 + e_{aq}^-$ are formed as products. He confirmed that this channel contributes less than 7 % to that producing OH radicals. Experimental data for OH quantum yields in NO_2^- photolysis at or near room temperature are summarised in Fig. 1, which demonstrates that the data are reasonably consistent. Two data points, one at 337 nm given by Zafiriou and Bonneau [10], and one at 351 nm reported by Zellner *et al.* [4] lie outside the common error range. If these two points are disregarded, the data can be represented by a function of the form $\Phi(OH) = a/(1 + \exp(\lambda-\lambda_0)/b) + c$. A best fit to this function with $a = (4.7 \pm 0.3)\ 10^{-2}$, $b = 10.0 \pm 1.6$, $c = (2.2 \pm 0.1)\ 10^{-2}$ and $\lambda_0 = 331 \pm 2$, is shown in Fig. 1 by the solid line. Zellner *et al.* [4] have explored the temperature dependence of the OH quantum yields at 308 and 351 nm and reported activation energies 13 ± 3 and 15 ± 3 kJ mol^{-1}, respectively, indicating that the temperature dependence does not depend on wavelength. These values are similar to that obtained for the OH quantum yield from the photodecomposition of nitrate, but markedly higher than that for the photodissociation of H_2O_2. This may indicate that protonation of O^- occurs after the radical leaves the solvent cage.

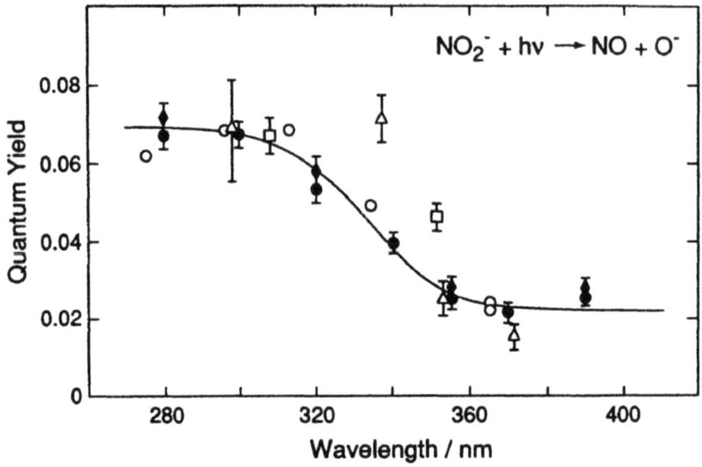

Fig. 1: Photodecomposition of nitrite, quantum yield of O^-/OH as a function of wavelength (T = 295 K). Filled circles: [8], open circles: [9], open diamonds: [10], open squares: [4], filled diamonds: in the presence of N_2O as scavenger for hydrated electrons, [8].

2.4.5 Nitrous Acid

The equilibrium constant for the dissociation of nitrous acid in dilute solution is about 6.3×10^{-4} M at 293 K, and its dependence on temperature and ionic strength is known. A significant fraction of nitrite thus exists in the form undissociated acid at pH ≤ 4. The absorption spectrum of undissociated nitrous acid in aqueous solution is similar to that in the gas phase, where photodissociation results in the production of NO and OH. Alif and Boule [9] have explored the photodissociation of 2×10^{-2} M HNO_2 in the aqueous phase at pH 1.4 at two wavelengths, 254 and 365 nm, with the same experimental technique as that used in studying NO_2^- photodecomposition. The quantum yields were 0.46 ± 0.1 and 0.45 ± 0.1, respectively. These values are surprisingly similar, although they refer to entirely different absorption regimes. At room temperature, the thermal decomposition of nitrous acid cannot be ignored. Fischer [8] therefore studied the photodecomposition of HNO_2 at 274 K with benzene as OH scavenger and product phenol as indicator (pH = 2). Measurements made at 8 wavelengths between 280 nm and 390 nm confirmed that the OH quantum yield is essentially constant in this wavelength region. The average value was 0.347 ± 0.025. Appreciable amounts of nitrate also were produced with an average quantum yield 0.42 ± 0.03 obtained after correction for thermal HNO_2 decomposition. The chemistry underlying the formation of nitrate is still not fully understood, as it involves reactions between HO_2/O_2^- and NO or NO_2, which are poorly known. Nevertheless, it appears that more HNO_2 is converted to nitrate than OH (and NO) is formed. Inasmuch as Alif and Boule [9] determined quantum yields from the loss of HNO_2, there exists good agreement between both data sets.

2.4.6 Fe(III)-hydroxo complex

Aqueous solutions of Fe(III) at pH ≤ 4 contain at least four ionic species in equilibrium with each other: Fe^{3+}, $FeOH^{2+}$, $Fe(OH)_2^+$, and the $Fe_2(OH)_2^{4+}$ dimer. Equilibrium constants are well known [11] except for the pair of species $FeOH^{2+}/Fe(OH)_2^+$, whose spectra appear to be indistinguishable. The concentration of the dimer is negligible for total Fe(III) concentrations less than about 10^{-4} M. Numerical values for absorption coefficients of Fe^{3+} and $FeOH^{2+}$ were given by Faust and Hoigné [12] and confirmed by Benkelberg and Warneck [13]. $FeOH^{2+}$ is the dominant absorber at wavelengths above 300 nm, whereas absorption by Fe^{3+} becomes more important at shorter wavelengths. Photodecomposition of $FeOH^{2+}$ generates OH + Fe^{2+} as products.

Fig. 2, taken from reference [13], shows the behaviour of the OH quantum yield in the wavelength region 280–370 nm. The quantum yield at 313 nm wavelength had been determined several times previously (reviewed by Faust and Hoigné [12]). David and David [14] obtained a value 0.105 at 340 nm, and Faust and Hoigné [12] found 0.017 at 360 nm. The latter value is lower than the average in Fig. 2. The data may be represented by the function

$$\Phi(\mathrm{OH}) = a/(1 + \exp(\lambda - \lambda_o/b)) + c.$$

The solid line in Fig. 2 was obtained as a best fit with

$$a = 0.359 \pm 0.08,\ b = 21.95 \pm 4.6,\ c = 0.054 \pm 0.01 \text{ and } \lambda_o = 300.7 \pm 9.1.$$

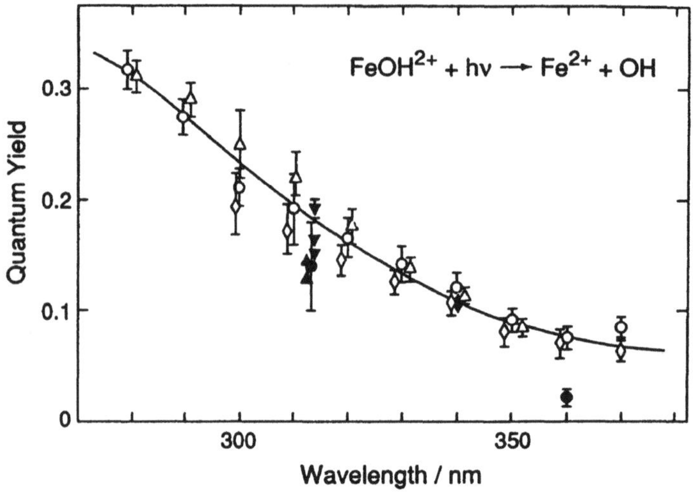

Fig. 2: Photodecomposition of iron(III)-hydroxo complex, OH quantum yield as a function of wavelength. Open circles and diamonds from measurements at pH 3, open triangles at pH 2 after correction for the contribution from the process
$Fe^{3+} + H_2O + h\nu \rightarrow Fe^{2+} + OH + H^+$, [13]; filled circles, [12]; filled triangles, as reviewed by Faust and Hoigné, [12]; filled diamond, [14].

2.4.7 Fe(III)-sulfato complex

$FeSO_4^+$ is formed by association of sulfate anion with Fe^{3+}. The equilibrium constant and its dependence on temperature and ionic strength is known. The products expected to result from photodecomposition are Fe^{2+} and SO_4^-. Benkelberg and Warneck [13] have determined the quantum yield of sulfate radicals using benzene as scavenger, phenol as indicator, and 2-methyl-2-propanol to discriminate against OH radicals formed from $FeOH^{2+}$ also present in the system. The wavelength dependence of the SO_4^- quantum yield was similar to that of the OH quantum yield from $FeOH^{2+}$, but the absolute values were about 50 times smaller.

2.4.8 Fe(III)-sulfito complexes

Up to three SO_3^{2-} (or HSO_3^-) anions can be added to Fe^{3+} or $FeOH^{2+}$ to form the corresponding Fe(III)-sulfito complexes. Only the first addition product is expected to be important in the atmosphere because of the low concentrations of both HSO_3^- and $FeOH^{2+}$ in atmospheric water drops. The stability constant and the absorption spectrum of $FeSO_3^+$ are approximately known [15]. Thermal dissociation of the complex to form Fe^{2+} and SO_3^- is thought to be responsible for the catalytic effect of iron in sulfite oxidation. Photolysis is expected to lead to the same products. Experimental difficulties have prevented the determination of quantum yields up to now.

2.4.9 Fe(III)-oxalato complexes

Three ferrioxalate complexes are formed by the successive addition of oxalate-anion to Fe^{3+} (or $FeOH^{2+}$). These complexes are very stable, and the equilibrium constants are quite well known [11]. Photodecomposition of $Fe(C_2O_4)_3^{3-}$ in sulfuric acid solutions has been extensively studied, because the system is widely used as an actinometer in the near ultraviolet spectral region. The quantum yield for Fe^{2+} formation from $Fe(C_2O_4)_3^{3-}$ photodecomposition is 1.24, independent of wavelength in the range 250–400 nm; at longer wavelengths the yield declines [16]. A quantum yield greater than unity indicates the influence of secondary reactions. The mechanism is thought to involve an oxalate radical $C_2O_4^-$, which reacts with iron-oxalate to produce Fe^{2+} and 2 CO_2. The primary quantum yield thus should be 1.24/2 = 0.62. Zuo and Hoigné [17] have shown that at lower iron–oxalate concentrations oxygen becomes competitive as a reactant for $C_2O_4^-$, and this reaction produces an O_2^- or HO_2 radical, which subsequently combine to form hydrogen peroxide. In atmospheric water drops, because of the low concentrations of both Fe(III) and oxalalic acid, the prevalent species would be the monooxalato and the bioxalato complexes, $FeC_2O_4^+$ and $Fe(C_2O_4)_2^-$. For the latter species, Zuo and Hoigné [17] have shown that at 313 nm wavelength the Fe^{2+} quantum yield is 1.2 in deaerated solution. This value is similar to that for the $Fe(C_2O_4)_3^{3-}$ complex, so that one may assume identical quantum yields (independent of wavelength).

2.4.10 Photodissociation coefficients

The photodissociation coefficient is defined by

$$j = \int \phi(\lambda)\, \sigma(\lambda)\, I_o(\lambda)\, d\lambda \ s^{-1}$$

where $\phi(\lambda)$ is the quantum yield for the photochemical process, $\sigma(\lambda)$ is the absorption cross section of the species undergoing photodecomposition, $I_o(\lambda)$ is the actinic solar flux in the atmosphere, and the integration must be performed over the entire photochemically active wavelength region. The absorption cross sections (m^2 molecule^{-1}) can be calculated from the decadic absorption coefficient $\varepsilon(\lambda)$ usually determined in the laboratory: $\sigma(\lambda) = (\varepsilon(\lambda)/N_A)$ ln 10, where $N_A = 6.02 \times 10^{23}$ is Avogrado's number, and $\varepsilon(\lambda)$ has the units m^2 mol^{-1}. Photodissociation coefficients were calculated with the same actinic flux so as to provide a comparison of data. Solar fluxes 0.8 km above sea level were calculated by the two-stream method of Brühl and Crutzen [18], which includes back-scattered light with ground albedo 0.25. A standard vertical ozone profile was adopted, and the zenith angle was set to 25° corresponding to noon time summer conditions at 48° latitude. The results are included in Table 1. In atmospheric water drops, the solar intensity is increased by nearly a factor of two due to multiple internal reflection [19]. This effect is not included in the data. The relative values in Table 1 suggest that Fe(III)-hydroxo and -oxalato complexes are most effective as radical sources. The $FeOH^{2+}$ complex produces OH radicals, which under continental conditions would react predominantly with hydrogensulfite or hydroxymetheanesulfonate, whereas $[Fe(C_2O_4)_n]^{3-2n}$ complexes ultimately produce HO_2/O_2^--radicals. They are scavenged by ions of transition metals [21], producing H_2O_2, which then contributes to the oxidation of hydrogensulfite.

2.4.11 References

1. J. P. Hunt, H. Taube, *J. Am. Chem. Soc.* 74 (1952) 5999–6002.
2. J. H. Baxendale, J. A. Wilson, *Trans. Faraday Soc.* 53 (1957) 344–356.
3. G. V. Buxton, W. K. Wilmarth, *J. Phys. Chem.* 67 (1963) 2835–2841.
4. R. Zellner, M. Exner, H. Herrmann, *J. Atmos. Chem.* 10 (1990) 411–425.
5. A. Schäfer, Photolyse von Nitrat, Bromat und H_2O_2 in wässrigen Lösungen: Bestimmung von Quantenausbeuten mit Hilfe von Radikalfängern, Dissertation, University Mainz, 1991.
6. P. Warneck, C. Wurzinger, *J. Phys. Chem.* 92 (1988) 6278–6283.
7. R. G. Zepp, J. Hoigné, H. Bader, *Environ. Sci. Technol.* 21 (1987) 443–450.
8. M. Fischer, Photolyse neutraler und acider wässriger Lösungen von Nitrit and Sulfit im ultravioletten Spektralbereich, Dissertation, University Mainz, 1994.
9. A. Alif and P. Boule, *J. Photochem. Photobiol.* A 59 (1991) 357–367.
10. O. C. Zafiriou, R. Bonneau, *J. Photochem. Photobiol.* 45 (1987) 723–727.
11. R. M. Smith, A. E. Martell, *Critical Stability Constants*, Plenum Press, New York 1976.
12. B. C. Faust, J. Hoigné, *Atmos. Environ.* 24A (1990) 79–89.
13. H.-J. Benkelberg, P. Warneck, *J. Phys., Chem.* 99 (1995) 5214–5221.
14. F. David, P. G. David, *J. Phys. Chem.* 80 (1976) 579–583.
15. Kraft, R. van Eldik, *Inorg. Chem.* 28 (1989) 2297–2305, 2306–2312.
16. C. G. Hatchard, C. A. Parker, *Proc. Roy. Soc. (London)* A235 (1956) 518–536.
17. Y. Zuo, J. Hoigné, *Environ. Sci. Technol.* 26 (1992) 1014–1022.
18. C. Brühl, P. J. Crutzen, *J. Geophys. Lett.* 16 (1989) 703–706.
19. J. Madronich, *J. Geophys. Res.* 92 (1987) 9740–9752.
20. D. L. Sedlak, J. Hoigné, *Atmos. Environ.* 27A (1993) 2173–2185.
21. D. L. Sedlak, J. Hoigné, *Atmos. Environ.* 28A (1994) 1898–1906.
22. R. Atkinson, D. L. Baulch, R. A. Cox, R. F. Hampson Jr., J. A. Kerr, J. Troe, *J. Phys. Chem. Ref. Data* 18 (1989) 881–1097.

2.5 Free Radical Reactions in the Aqueous Phase

2.5.1 Introduction

In 1982 Chameides and Davis [1] suggested that the free-radical-induced chain oxidation of S(IV) could be a major pathway for generating acidity in cloud and rainwater. They proposed that the chain reaction could be initiated either by ·OH or HO_2 scavenged from the gas phase or by radicals generated either chemically or photochemically in the aqueous phase. It was also proposed [2] that $NO_3\cdot$ could be a significant night-time source of radicals capable of initiating the chain oxidation. Thus, the quantification of the elementary steps in the free-radical oxidation of S(IV) and of the reactions of NO_3 became a main focus of HALIPP.

In aqueous solution, SO_2 behaves as a weak dibasic acid according to equilibria (1), (2) and (3)

$$SO_2 + H_2O \rightleftharpoons SO_2(aq) \quad (1)$$
$$SO_2(aq) \rightleftharpoons H^+ + HSO_3^- \quad pK = 1.89 \quad (2)$$
$$HSO_3^- \rightleftharpoons H^+ + SO_3^{2-} \quad pK = 7.2 \quad (3)$$

and at the pH pertaining in clouds is, therefore, present largely as HSO_3^- and to a lesser degree SO_3^{2-}. The main reactions involved in the chain reaction are:

initiation

$$\cdot OH + HSO_3^- \rightarrow H_2O + SO_3^-\cdot \quad (4)$$
$$\cdot OH + SO_3^{2-} \rightarrow OH^- + SO_3^-\cdot \quad (5)$$
$$SO_3^-\cdot + O_2 \rightarrow SO_5^-\cdot \quad (6)$$

propagation

$$SO_5^-\cdot + HSO_3^- \rightarrow HSO_5^- + SO_3^-\cdot \quad (7)$$
$$SO_5^-\cdot + SO_3^{2-}(+H^+) \rightarrow HSO_5^- + SO_3^-\cdot \quad (8)$$
$$SO_5^-\cdot + HSO_3^- \rightarrow SO_4^-\cdot + SO_4^{2-} + H^+ \quad (9)$$
$$SO_5^-\cdot + SO_3^{2-} \rightarrow SO_4^-\cdot + SO_4^{2-} \quad (10)$$
$$SO_4^-\cdot + HSO_3^- \rightarrow SO_4^{2-} + SO_3^-\cdot + H^+ \quad (11)$$
$$SO_4^-\cdot + SO_3^{2-} \rightarrow SO_4^{2-} + SO_3^-\cdot \quad (12)$$

chain transfer

$$SO_5^-\cdot + SO_5^- \rightarrow 2SO_4^-\cdot + O_2 \quad (13)$$

termination

$$SO_5^-\cdot + SO_5^- \rightarrow S_2O_8^{2-} + O_2 \quad (14)$$
$$SO_4^-\cdot + SO_4^- \rightarrow S_2O_8^{2-} \quad (15)$$

2.5.2 Experimental

The techniques employed by HALIPP investigators to generate and study the free radicals involved include laser flash photolysis [3, 4], pulse radiolysis [5, 6], steady-state radiolysis [6, 7] and steady-state photolysis coupled with scavenger methods [8, 9]. Notable innovations to the techniques include: (i) combining a rapid-mix system with pulse radiolysis apparatus so that reactants which undergo thermal reactions, *e.g.* $S(IV)/O_2$ mixtures, can be prepared immediately (~ 1 s) before delivery of the radiation pulse [5, 6], and (ii) laser flash photolysis with a long-path white cell [4].

2.5.3 Reactions of ·OH Radicals

·OH radicals have been generated by laser flash and steady-state photolysis of hydrogen peroxide solution [10], reaction (16)

$$H_2O_2 + h\nu \rightarrow Ø{\cdot}OH \qquad (16)$$

and by pulse radiolysis of N_2O saturated aqueous solutions [5, 6], reactions (17) and (18),

$$(4.8 \times 10^{-7})\, H_2O \rightsquigarrow (2.8 \times 10^{-7}) e_{aq}^{-} + (2.8 \times 10^{-7})\, {\cdot}OH + (2.8 \times 10^{-7})\, H^{+} + (6.2 \times 10^{-8})\, H{\cdot} + (7.3 \times 10^{-8})\, H_2O_2 \qquad (17)$$

$$e_{aq}^{-} + N_2O\,(+H_2O) \rightarrow N_2 + {\cdot}OH + OH^{-} \qquad (18)$$

$$k_{18} = 9.1 \times 10^{9}\ dm^3\ mol^{-1}\ s^{-1} \qquad [11]$$

where the numbers in parentheses in equation (17) are the radiolytic yields in mol J^{-1}.

The rate constants evaluated in the HALIPP project are listed in Table 1 where they are compared with values available in the literature.

Table 1: Rate constants of reactions of OH

No.	Reaction	$k/dm^3\ mol^{-1}\ s^{-1}$ [a]	Ref.	Method/Comments
4	$\cdot OH + HSO_3^{-}$	$(2.7 \pm 0.2) \times 10^9$	6, 12, ‡	PR, direct observation of SO_3^{-}
	$\rightarrow H_2O + SO_3^{-}\cdot$	4.5×10^9	13	PR, competition with pNBA
5	$\cdot OH + SO_3^{2-}$	$(4.6 \pm 0.2) \times 10^9$	6, 12 ‡	PR, direct observation of SO_3^{-}
	$\rightarrow OH^{-} + SO_3^{-}\cdot$	5.2×10^9	13	PR, competition with pNBA
20	$\cdot OH + CH_2(OH)SO_3^{-}$	$(3.0 \pm 0.3) \times 10^8$	14, 15 ‡	PR, competition with $Fe(CN)_6^{4-}$
	$\rightarrow CH(OH)SO_3^{-} + H_2O$			and SCN^{-}
		2.6×10^8	16	PR, competition with SCN^{-}

[a] Ambient temperature ‡ = HALIPP Project, PR = Pulse Radiolysis

McElroy and Waygood [10] have studied the mechanism of the oxidation of formaldehyde by OH radical and have concluded that in the presence of oxygen the hydrated formyl radical, $CH(OH)_2$, reacts with oxygen to produce formic acid and $HO_2\cdot$radical, reaction (19), with $k_{19} = (3.5 \pm 1.6) \times 10^9$ dm^3 mol^{-1} s^{-1}.

$$CH(OH)_2 + O_2 \rightarrow HO_2\cdot + HCOOH \tag{19}$$

In their study of the OH-induced oxidation of hydroxymethane sulfonate (HMS), Buxton *et al* [15, 17] have shown that the radical formed by the attack of OH on HMS undergoes acid-base dissociation of the –OH group with a $pK_A = 5.9$. Also, at pH 4.0 the radical reacts with O_2 to yield a peroxy radical, reaction (21), with $k_{2_1} = (2.6 \pm 0.3) \times 10^9$ dm^3 mol^{-1} s^{-1}, whereas at pH 7, the $CH(O^-)SO_3^-$-radical reacts with oxygen to form the superoxide radical ion, reaction (22), with $k_{22} = (1.64 \pm 0.08) \times 10^9$ dm^3 mol^{-1} s^{-1}.

$$CH(OH)SO_3^- + O_2 \rightarrow O_2CH(OH)SO_3^- \tag{21}$$
$$CH(O^-)SO_3^- + O_2 \rightarrow CHOSO_3^- + O_2^- \tag{22}$$

2.5.4 Reactions of $SO_x^-\cdot$Radicals

$SO_3^-\cdot$radicals have been generated by pulse radiolysis of N_2O-saturated S(IV) solutions, reactions (17), (18), (4) and (5) or laser photolysis of dithionate solutions with 193 nm light, reaction (23). However, it has been shown [18] that in addition, there is a second photolytic channel, reaction (24).

$$S_2O_6^{2-} + h\nu \rightarrow 2SO_3^-\cdot \tag{23}$$
$$S_2O_6^{2-} + h\nu \rightarrow SO_2 + SO_4^{2-} \tag{24}$$

$SO_5^-\cdot$radicals are generated from $SO_3^-\cdot$ by their reaction with O_2, reaction (6), and $SO_4^-\cdot$radicals can be produced either by photolysis of solutions containing peroxodisulfate, reaction (25), or by pulse radiolysis of oxygen-free solutions of peroxodisulfate, reaction (26).

$$S_2O_8^{2-} + h\nu \rightarrow 2SO_4^-\cdot \tag{25}$$
$$e_{aq}^- + S_2O_8^{2-} \rightarrow SO_4^-\cdot + SO_4^{2-} \tag{26}$$

In the latter case it is usual also to add 2-methylpropan-2-ol to convert the reactive OH-radical, which is also generated in the radiolysis, to the unreactive *tert*-butanol radical. To study the reactivities of SO_3^-, SO_4^- and SO_5^- by pulse radiolysis or flash photolysis, a knowledge of the spectral characteristics of the radicals is needed; these are listed in Table 2, where they are compared with literature values. In general the spectral shapes do not differ greatly from those observed previously, but the spectrum of SO_5^- determined by Zellner *et al.* [20, 21] is somewhat narrower than that reported by Buxton *et al.* [24]. Since their earlier work [24], the Leeds group have remeasured [6] the molar absorptivities of SO_3^- and SO_5^- at lower radiation doses and have used a recently determined radiation chemical yield for the SCN^- dosimeter [19] in order to assess the molar absorptivities. They have also confirmed [12] that the observed spectrum was independent of the bandwidth of the monochromator in the range 2 to 10 nm.

Table 2: Spectral parameters of $SO_3^{-}\cdot$, $SO_4^{-}\cdot$ and $SO_5^{-}\cdot$.

Radical	λ_{max}/nm	ε_{max}/dm^3 mol^{-1} cm^{-2}	Ref.	Method/Comments
$SO_3^{-}\cdot$	260	1090 ± 20	12 ‡	PR, dose 12 Gy, pH 4 and 9.0
		1140 ± 20	6 ‡	Revised from the value in 12 due to the recent re-evaluation of the radiation chemical yield of the dosimeter [19]
$SO_3^{-}\cdot$	260	–	18 ‡	LFP
$SO_3^{-}\cdot$	255	–	20, 21 ‡	LFP /diode array
$SO_3^{-}\cdot$	260	1380	22	PR, O_2^- dosimetry
$SO_3^{-}\cdot$	260	1300	23	PR, 1M NaOH
$SO_4^{-}\cdot$	450	1630 ± 50	6, 24 ‡	PR, data from [24] revised to recent value of radiation chemical yield of dosimeter. [19]
$SO_4^{-}\cdot$	450	1600 ± 100	3 ‡	LFP /diode array, comparison with Cl_2^-
$SO_4^{-}\cdot$	445	1650	20, 21 ‡	LFP /diode array
$SO_4^{-}\cdot$	450	1100	25	PR
$SO_5^{-}\cdot$	260	790 ± 20	6, 12 ‡	PR, dose 25 Gy, pH 8.9. Revised from data in[12] due to recent re-evaluation of the radiation chemical yield of the dosimeter. [19]
$SO_5^{-}\cdot$	265	–	20, 21 ‡	LFP /diode array
$SO_5^{-}\cdot$	260	1030	22	PR, O_2^- dosimetry

LFP = laser flash photolysis, PR = pulse radiolysis, ‡ = HALIPP project

The rate constants for the reactions of $SO_x^{-}\cdot$ radicals determined during the project are listed in Table 3.

Buxton *et al.* [6, 7] using ^{60}Co γ-radiolysis to study the formation of SO_4^{2-} in the chain reaction found $(2k_7 + k_9)/(2k_{14})^{1/2} = 1.76 \pm 0.07$
and $(2k_8 + k_{10})/(2k_{14})^{1/2} = 97.5 \pm 6.0$
and hence $(2k_7 + k_9) = 1.76 \times 10^4$ dm^3 mol^{-1} s^{-1}
and $(2k_8 + k_{10}) = (9.75 \pm 0.3) \times 10^5$ dm^3 mol^{-1} s^{-1}, respectively. Huie and Neta [13] using a method involving competition between ascorbate and S(IV) for SO_5^- radicals concluded that at pH 8.7 the rate constant of the propagation steps was ~ 1.3×10^7 dm^3 mol^{-1} s^{-1}, i.e. ~ 13 × the value of Buxton *et al.* At pH 4.9 they estimated an upper limit for the rate constant of 3×10^5 dm^3 mol^{-1} s^{-1}, which although higher than the value of 1.76×10^4 dm^3 mol^{-1} s^{-1} obtained by Buxton *et al.* is not in disagreement with it.

Values given for the termination rate constant, k_{14}, differ by a factor of almost 3, but this discrepancy may lie in the assumptions made about the value of the rate constant for reaction (13).

Buxton *et al.* [6] have evaluated their data assuming the occurrence of reaction (13), which was incorporated in the computer modelling. The evidence for the occurrence of reaction (13) was based on the observation of SO_4^- when SO_5^- was generated by pulse radiolysis. Similar observations have been made by Huie *et al.* [22] and Waygood [39], both of whom generated SO_5^- by flash photolysis. This latter worker, however, interpreted the formation of SO_4^- as being due to the propagation reaction (9), but this proposal seems unlikely to be correct in view of the low value of k_9 (see Table 3) and the very small concentration of hydrogen sulfite which can be generated in the dithionate system (reaction (24)). Herrmann *et al.* [21] did not observe SO_4^- in the flash photolysis of oxygenated dithionate solutions and therefore concluded that $k_{13}/(k_{13}+k_{14}) < 0.04$.

This result, however, is not in agreement with the work of Yermakov *et al.* [40] who measured the rate of the radiation induced oxidation under high dose rate conditions where the kinetic chain length becomes unity and hence gives a direct measure of k_{13}/k_{14}, which they deduced to be 7 ± 1. Under atmospheric conditions reactions (37) and (38) may be significant chain–terminating reactions and their rate constants have been evaluated by Buxton *et al.* [41, 6] (see Table 3).

Table 3: Rate Constants of Reactions of SO_x^- .

No.	Reaction	$k/dm^3\ mol^{-1}\ s^{-1\ a}$	Ref.	Method/Comments
6	$SO_3^{-}\cdot + O_2 \rightarrow SO_5^{-}\cdot$	$(2.5 \pm 0.1) \times 10^9$	6, 12 ‡	PR/Rapid–mix, decay at 340 nm.
		1.1×10^9	22	LFP, decay at 245 nm.
27	$SO_3^{-}\cdot + SO_3^{-}\cdot \rightarrow S_2O_6^{2-}$	$(3.11 \pm 0.06) \times 10^8$	6, 5 ‡	PR, I→0, $k(T) = (2.5 \pm 0.3) \times 10^{10} \exp\left[\frac{(-1240 \pm 48)K}{T}\right]$
		5.7×10^8	22	LFP
15	$SO_4^{-}\cdot + SO_4^{-}\cdot \rightarrow S_2O_8^{2-}$	$(4.45 \pm 0.15) \times 10^8$	26 ‡	LFP, I e 0, based on e = 1600 $dm^3\ mol^{-1}\ cm^{-1}$
	" " "	$(1.6 \pm 0.4) \times 10^8$	21 ‡	LFP, ε = 1650 $dm^3\ mol^{-1}\ cm^{-1}$
	" " "	1.9×10^8	22	LFP, ε = 1180 $dm^3\ mol^{-1}\ cm^{-1}$
	" " "	2.7×10^8	27	LFP, I ε 0, e = 1380 $dm^3\ mol^{-1}\ cm^{-1}$
	" " "	7.6×10^8	28	In conc. H_2SO_4, e = 1600 $dm^3\ mol^{-1}\ cm^{-1}$
18	$SO_4^{-}\cdot + S_2O_8^{2-} \rightarrow S_2O_8^{-}\cdot + SO_4^{2-}$	$(6.1 \pm 0.6) \times 10^5$	26 ‡	LFP
	" " "	$(6.3 \pm 1.5) \times 10^5$	21 ‡	LFP
29	$SO_4^{-}\cdot + H_2O \rightarrow HSO_4^- + \cdot OH$	$500 \pm 60\ s^{-1}$	26 ‡	LFP
	" " "	$660 \pm 40\ s^{-1}$	21 ‡	LFP, $k(T) = (4.7 \pm 0.2) \times 10^2 \exp\left[\frac{(-1100 \pm 40)K}{T}\right]$
	" " "	$360\ s^{-1}$	27	LFP
30	$SO_4^{-}\cdot + \cdot OH \rightarrow HSO_5^-$	$(9.0 \pm 0.4) \times 10^9$	5, 6 ‡	PR/Computer fit
		1×10^{10}	29	PR/ " "

31	$SO_4^{-}\cdot + H\cdot \rightarrow HSO_4^{-}$	$(6.8 \pm 0.4) \times 10^{9}$	5, 6 ‡	PR/ Computer fit
	" " "	1×10^{10}	28	PR, in 5 mol dm^{-3} H_2SO_4
11	$SO_4^{-}\cdot + HSO_3^{-}\cdot$ $\rightarrow SO_3^{-}\cdot + H^+ + SO_4^{2-}$	$(6.8 \pm 0.1) \times 10^{8}$	5,6 ‡	PR,I $\rightarrow$ 0, pH 4, decay of SO_4^- at $\lambda = 450$ nm
		7.5×10^{8}	30	LFP, I $\rightarrow$ 0, decay of SO_4^- at $\lambda = 453$ nm
12	$SO_4^{-}\cdot + SO_3^{2-} \rightarrow SO_4^{2-} + SO_3^{-}\cdot$	$(3.1 \pm 0.1) \times 10^{8}$	5, 6 ‡	PR, I $\rightarrow$ 0, pH 9, decay of SO_4^- at λ= 450 nm
	" " " "	5.7×10^{8}	8 ‡	SSP, competition with 2-PrOH
	" " " "	3.7×10^{8}	8 ‡	SSP, competition with EtOH
32	$SO_4^{-}\cdot + CH_2(OH)_2 \rightarrow P$	$(1.26 \pm 0.05) \times 10^{7}$	31 ‡	PR, decay of $SO_4^{-}\cdot$ at $\lambda = 450$ nm, product probably not $\cdot CH(OH)_2$
		1.1×10^{7}	32	LFP, decay of SO_4^- at $\lambda = 450$ nm
33	$SO_4^{-}\cdot + CH_2(OH)SO_3^{-} \rightarrow$ $HCHO + SO_3^{-}\cdot + H^+ + SO_4^{2-}$	$(2.8 \pm 0.2) \times 10^{6}$	33, 15 ‡	PR, decay of $SO_4^{-}\cdot$
34	$SO_4^{-}\cdot + Fe^{2+} \rightarrow Fe^{3+} + SO_4^{2-}$	$(3.0 \pm 1.9) \times 10^{8}$	3 ‡	LFP, rapid pre-equilibrium, ET in complex, $k_{ET} = (6.5 \pm 2.1) \times 10^3\ s^{-1}$, $k_{-34} = (4.8 \pm 3.0) \times 10^4\ s^{-1}$
		$(4.8 \pm 0.2) \times 10^{8}$	34 ‡	PR, pH 4, independent of I
		9.9×10^{8}	35	PR
35	$SO_4^{-}\cdot + Cu(II) \rightarrow Cu(III) + SO_4^{2-}$	$(8.1 \pm 2.6) \times 10^{7}$	36 ‡	LFP, rapid pre–equilibrium, ET in complex $k_{ET} = (5.1 \pm 3.3) \times 10^3\ s^{-1}$, $k_{-35} = (2.2 \pm 1.4) \times 10^4\ s^{-1}$
36	$SO_4^{-}\cdot + Mn^{2+} \rightarrow Mn(III) + SO_4^{2-}$	$(8.3 \pm 1.9) \times 10^{7}$	36 ‡	LFP, rapid pre–equilibrium, ET in complex, $k_{ET} = (4.5 \pm 2.5) \times 10^4\ s^{-1}$, $k_{-36} = (4.1 \pm 2.6) \times 10^4\ s^{-1}$
		2×10^{7}	37	PR, 6 mol dm^{-3} H_2SO_4
		2×10^{7}	38	PR, conc. H_2SO_4
14	$SO_5^{-}\cdot + SO_5^{-}\cdot \rightarrow S_2O_8^{2-} + O_2$	$(5.0 \pm 0.3) \times 10^{7}$	6 ‡	PR, decay of $SO_5^{-}\cdot$, computer fitting
	" " "	$(1.3 \pm 0.3) \times 10^{8}$	21 ‡	LFP, decay of $SO_5^{-}\cdot$, $k(T) = (1.1 \pm 0.1) \times 10^{12}\ \exp\left[\frac{(-1240 \pm 48)K}{T}\right]$
	" " "	1×10^{8}	22	LFP, decay of $SO_5^{-}\cdot$
	$SO_5^{-}\cdot + SO_5^{-}\cdot \rightarrow X$	$(9.3 \pm 0.1) \times 10^{7}$	39 ‡	LFP, decay of $SO_5^{-}\cdot$

13	$SO_5^{-}\cdot + SO_5^{-}\cdot \rightarrow 2SO_4^{-} + O_2$	$\sim 3 \times k_{14}$	6 ‡	PR, observation of SO_4^- at 450 nm
	" " " "	$< 0.04 \times k_{14}$	21 ‡	LFP, SO_4^- not observed
	" " " "	~ 0	39 ‡	LFP, SO_4^- observed but ascribed to reaction (9)
	" " " "	$(7 \pm 1) \times k_{14}$	40	SSR, removal of S(IV)
	" " " "	$\sim k_{14}$	22	LFP, observation of SO_4^- at 450 nm.
37	$SO_5^{-}\cdot + HO_2\cdot \rightarrow X$	$(1.7 \pm 0.1) \times 10^9$	41 ‡	PR, observation of X
38	$SO_5^{-}\cdot + O_2^- \xrightarrow{H^+} HSO_5^- + O_2$	$(2.67 \pm 0.16) \times 10^8$	6 ‡	PR, decay of SO_5^-/O_2^-, computer fit
7	$SO_5^{-}\cdot + HSO_3^- \rightarrow HSO_5^- + SO_3^{-}\cdot$	$(8.1 \pm 0.3) \times 10^3$	6, 7 ‡	SSR, formation of SO_4^{2-}, pH 4, data yields $(2k_7 + k_9)/(2k_{14})^{½} = 1.76 \pm 0.07$, uses $\frac{k_9}{k_7+k_9} = 0.04$ [42]
	" " " "	$< 3 \times 10^5$	13	PR, competition with ascorbate
9	$SO_5^{-}\cdot + HSO_3^- \rightarrow SO_4^{-}\cdot + SO_4^{2-} + H^+$	$(3.4 \pm 0.13) \times 10^2$	6, 7 ‡	SSR, formation of SO_4^{2-}, pH 4, uses $(2k_7 + k_9)/(2k_{14})^{½} = 1.76 \pm 0.07$, and $\frac{k_9}{k_7+k_9} = 0.04$ [42]
8	$SO_5^{-}\cdot + SO_3^{2-} \xrightarrow{H^+} HSO_5^- + SO_3^{-}\cdot$	$(3.43 \pm 0.33) \times 10^5$	6, 7 ‡	SSR, formation of SO_4^{2-}, pH 9, uses $(2k_8 + k_{10})/(2k_{14})^{½} = 97.5 \pm 6.0$ and $\frac{k_{10}}{k_8+k_{10}} = (0.72 \pm 0.1)$[8]
	" " "	1.3×10^7	13	PR, competition with ascorbate, complex scheme
10	$SO_5^{-}\cdot + SO_3^{2-} \rightarrow SO_4^{-}\cdot + SO_4^{2-}$	$(1.3 \pm 0.8) \times 10^5$	6, 7 ‡	SSR, formation of SO_4^{2-}, pH 9, uses $(2k_8 + k_{10})/(2k_{14})^{½} = 97.5 \pm 6.0$ and $\frac{k_{10}}{k_8+k_{10}} = (0.72 \pm 0.1)$ [8]
39	$SO_5^{-}\cdot + Fe^{2+} \rightarrow HSO_5^- + Fe^{3+}$	$(4 \pm 2) \times 10^7$	21 ‡	LFP, decay of SO_5^-
		3.2×10^6	43 ‡	PR, pH 4, $I = 71$ mmol dm^{-3}, formation of Fe(III)
40	$SO_5^{-}\cdot + Mn^{2+} \rightarrow HSO_5^- + Mn(III)$	$2 \times 10^8 - 2 \times 10^{10}$	44 ‡	PR, pH 3, $I = 10$ mmol dm^{-3}, formation of product, k depends on Mn^{2+} complexation

[a] Ambient temperature ‡ = HALIPP project, PR = pulse radiolysis, LFP = laser flash photolysis, SSP = steady-state photolysis, SSR = steady-state radiolysis.

A feature of reaction (37) was that the product is an adduct which is probably a tetroxide species. This species give rise to optical absorption in the wavelength region 230–300 nm with $\lambda_{max} = 260$ nm and $\varepsilon_{max} = 2700$ $dm^3\ mol^{-1}\ cm^{-1}$.

The role of transition metal ions (TMI) in atmospheric waters is discussed in section 2.3. of this report. Reactions between SO_4^- and SO_5^- and the lower oxidation state of the metal are key reactions in the catalytic oxidation of S(IV) by TMI. Several of these reactions have been studied by HALIPP investigators and the results are summarised in Table 3.

In studies where S(IV) is present, there is evidence that the oxidation of the metal ion can be influenced by complexation between the lower valency state of the metal and HSO_3^-/SO_3^{2-} [44, 45]. It is probable that similar effects will occur with other anions that are found in significant concentrations in cloud waters, such as sulfate and chloride.

2.5.5 Reactions of $NO_3\cdot$

Extensive studies of the reactivity of $NO_3\cdot$ in aqueous solution have been carried out by Zellner and co-workers [46–49] as part of HALIPP. $NO_3\cdot$ radicals were generated by laser flash photolysis of peroxodisulfate to form $SO_4^-\cdot$ which then reacted with NO_3^- (5–120 mmol dm^{-3}), reaction (41), for which
$k_{41}\ (I \rightarrow 0) = (5.0 \pm 0.5) \times 10^4\ dm^3\ mol^{-1}\ s^{-1}$.

$$SO_4^-\cdot + NO_3^- \rightarrow SO_4^{2-} + NO_3\cdot \qquad (41)$$

The main reactions pertinent to cloud chemistry are summarised in Table 4.

NO_3 reacts with HSO_3^- and SO_3^{2-} with rate constants of 1.4×10^9 and 3×10^8 dm^3 $mol^{-1}\ s^{-1}$ and, therefore, will be an effective initiator of the free-radical chain oxidation of S(IV). The reaction between NO_3 and Cl^- has important implications because of the high concentrations of Cl^- found in maritime cloud water. However, Buxton *et al.* [56] have reported that the reverse reaction, reaction (50), is fast.

$$Cl\cdot + NO_3^- \rightarrow Cl^- + NO_3\cdot \qquad (50)$$

2.5.6 Conclusions

Most of the reactions of $\cdot OH$, $SO_x^-\cdot$ and $NO_3\cdot$ radicals which are expected to be significant to the free-radical chemistry occurring in cloud and fog droplets have been characterised and their rate constants evaluated as part of the HALIPP subproject. These data thus provide the requisite data-set for the effective modelling of the effects of cloud on tropospheric chemistry In a few instances, however, substantial discrepancies still exist in the reported values, *e.g.* on the reported rate constants for the two pathways for the mutual reaction of $SO_5^-\cdot$ radicals, and in these cases further work in needed.

Table 4: Rate Constants of Reactions of $NO_3^-\cdot$ relevant to cloud chemistry .

Reaction	$k/dm^3\ mol^{-1}\ s^{-1\ a}$	Ref.	Comments
$NO_3\cdot + HSO_3^- \rightarrow SO_3^{-}\cdot + NO_3^- + H^+$	$(1.4 \pm 0.1) \times 10^9$	46 ‡	LFP, decay of NO3·, $I = 0.13$ mol dm^{-3}, $T = 298$ K, $k(T) = (1.1 \pm 0.2) \times 10^{12}\ \exp\left[\frac{(-2000 \pm 400)K}{T}\right]$
" " " " "	9.0×10^8	50 ‡	PR, decay of $NO_3\cdot$, $T = 293 \pm 2$ K, $I = 3\text{-}4$ mol dm^{-3}
$NO_3\cdot + HSO_3^-/SO_3^{2-}$	2.0×10^9	51 ‡	Decay of NO_3, $I = 5$ mol dm^{-3}, pH = 7.0, $T = 295$ K
$NO_3\cdot + SO_3^{2-} \rightarrow NO_3^{-}\cdot + SO_3^-$	$(3.0 \pm 0.4) \times 10^8$	46 ‡	LFP, decay of $NO_3\cdot$, $I = 0.13$ mol dm^{-3}, $T = 278$ K,
$NO_3\cdot + SO_2 + H_2O \rightarrow NO_3^- + HSO_3^{-}\cdot + H^+$	$(2.3 \pm 0.5) \times 10^8$	46 ‡	LFP, decay of NO_3, $I = 0.13$ mol dm^{-3}, $T = 298$ K, $k(T) = (5.6 \pm 0.7)$ x $10^{13}\ \exp\left[\frac{(-3700 \pm 1100)K}{T}\right]$
$NO_3\cdot + Cl^- \rightarrow NO_3^- + Cl$	$(1.0 \pm 0.2) \times 10^7$	46 ‡	LFP, decay of NO_3, I ® 0, $T = 298$ K,
" " " "	7.1×10^7	51 ‡	$k(T) = (1.9 \pm 0.2) \times 10^{13}\ \exp\left[\frac{(-4300 \pm 500)K}{T}\right]$
" " " "	1.0×10^8	52	PR, decay of $NO_3\cdot$ and formation of $Cl_2^-\cdot$, $I = 5$ mol dm^{-3} PR, decay of $NO_3\cdot$ and formation of $Cl_2^-\cdot$, $I = 2$ mol dm^{-3}
$NO_3\cdot + CH_2(OH)_2 \rightarrow NO_3^- + \cdot CH(OH)_2 + H^+$	$(7.8 \pm 1.2) \times 10^5$	47 ‡	LFP, decay of $NO_3\cdot$, $T = 298$ K,
$NO_3\cdot + HCHO \rightarrow HNO_3 + \cdot CHO$	4.9×10^5	53	LFP, decay of $NO_3\cdot$, $I = 2$ mol dm^{-3}, $T = 298$ K
" " " "	1.1×10^6	"	LFP, decay of $NO_3\cdot$, $I = 6$ mol dm^{-3}, $T = 298$ K
$NO_3\cdot + HCOOH \rightarrow H^+ + NO_3^- + \cdot COOH$	$(3.3 \pm 1.0) \times 10^5$	49 ‡	LFP, decay of $NO_3\cdot$, $I = 0.19$ mol dm^{-3}, $T = 298$ K, $k(T) = (3.4 \pm 0.3) \times 10^{10}\ \exp\left[\frac{(-3400 \pm 600)K}{T}\right]$
" " " " "	$< 1 \times 10^5$	51	PR, decay of NO_3, $T = 295$ K, 6 mol dm^{-3} $HNO_3\cdot$
" " " " "	$(2.1 \pm 0.1) \times 10^5$	54	FP, decay of $NO_3\cdot$, T = RT, 0.1 mol dm^{-3} $K_2Ce(NO_3)_6$, pH = 0.7
$NO_3\cdot + HCOO^- \rightarrow NO_3^- + H^+ + CO_2^-$	$(5.0 \pm 0.4) \times 10^7$	49 ‡	LFP, decay of $NO_3\cdot$, $T = 298$ K, $I = 0.19$ mol dm^{-3}, $k(T) = (8.2 \pm 0.8)$ x $10^{10}\ \exp\left[\frac{(-2200 \pm 700)K}{T}\right]$
$NO_3\cdot + CH_3COOH \rightarrow NO_3^- + \cdot CH_2COOH + H^+$	$(1.3 \pm 0.3) \times 10^4$	49 ‡	LFP, decay of $NO_3\cdot$, $T = 298$ K, $I = 0.19$ mol dm^{-3}
" " " "	4.6×10^4	54	FP, decay of $NO_3\cdot$, T = RT, 0.1 mol dm^{-3} $K_2Ce(NO_3)_6$, pH = 0.7
" " " "	230	55	FP, decay of $NO_3\cdot$, $T = 296$ K, 6 mol dm^{-3} HNO_3

‡ = HALIPP project

PR = pulse radiolysis

LFP = laser flash photolysis

FP = flash photolysis

2.5.7 References

1. W. L. Chameides, D. D. Davis, *J. Geophys. Res.*, **87** (1982) 4863–4877.
2. W. L. Chameides, *J. Geophys. Res.*, **91**, (1986) 5331–5337.
3. W. J. McElroy, S. J. Waygood, *J. Chem. Soc., Faraday Trans.*, **86** (1990) 2557–2564.
4. H. Herrmann, A. Reese, R. Zellner, CEC *Air Pollution Research Report 45*, J. Peeters (ed.) Guyot, Brussels 1993, pp. 253–258.
5. G. V. Buxton, G. A. Salmon, N. D. Wood, in: K. H. Becker (ed) *Air Pollution Research Report 33*, CEC, Guyot, Brussels, 1991, pp. 156–161.
6. G. V. Buxton, S. McGowan, G. A. Salmon, J. E. Williams, N. D. Wood, in preparation.
7. G. V. Buxton, S. MGowan, G. A. Salmon, in: P. M. Borrell, P. Borrell, T. Cvitaš, W. Seiler (eds), *Proc. EUROTRAC Symp. '92*, SPB Academic Publishing bv, The Hague 1993, pp. 599–604.
8. U. Deister, P. Warneck, *J. Phys. Chem.*, **94** (1990) 2191–2198.
9. U. Deister, P. Warneck, C. Wurzinger, *Ber. Bunsenges, Phys. Chem.*, **94** (1990) 594–599.
10. W. J. McElroy, S.. Waygood, *J. Chem. Soc. Faraday Trans.*, **87** (1991) 1513–1521.
11. G. V. Buxton, C. L. Greenstock, W. P. Helman, A. B. Ross, *J. Phys. Chem. Ref. Data*, **17** (1988) 513–886.
12. S. McGowan, The oxidation of sulfur dioxide in cloud water, Ph. D. Thesis, University of Leeds,1994.
13. R. E. Huie, P. Neta, *Atmos. Environ.*, **21** (1987) 1743–1747.
14. G. V. Buxton, S. Croft, G.A. Salmon, in: R. A. Cox (ed) *Air Pollution Research Report 42*, CEC, Guyot, Brussels 1992, pp. 19–22.
15. G. V. Buxton, S. Croft, S. A. Murray, G. A. Salmon, to be published.
16. T. M. Olson, R. W. Fessenden, *J. Phys. Chem.*, **96** (1992) 3317–3320.
17. S. Barlow, G. V. Buxton, S. A. Murray, G. A. Salmon, in: K. H. Becker (ed) *Air Pollution Research Report 54*, EC, Brussels 1995, pp. 451–456.
18. S. J. Waygood, W. J. McElroy, *J. Chem. Soc., Faraday Trans.*, **88** (1992) 1525–1530.
19. G. V. Buxton, C. R. Stuart, *J. Chem. Soc., Faraday Trans.*, **91** (1995) 279–281.
20. R. Zellner, H. Herrmann, M. Exner, A. Reese, *EUROTRAC Annual Report Part 6*, EUROTRAC ISS Garmisch-Partenkirchen 1992, pp. 70–81.
21. A. Reese, H. Herrmann, H. C. Weiss, R. Zellner, in: K. H. Becker (ed) *Air Pollution Research Report 54*, EC, Brussels 1994, pp. 61–66.
22. R. E. Huie, C. L. Clifton, N. Alstein, *Radiat. Phys. Chem.*, **33** (1989) 361–370.
23. Z. P. Zagorski, K. Sehested, S. O. Nielsen, *J. Phys. Chem.*, **75** (1971) 3510–3517.
24. G. V. Buxton, G. A. Salmon, N. D. Wood, in: P. Borrell, P. M. Borrell, W. Seiler (eds), *Proc. EUROTRAC Symp. '90*, SPB Academic Publishing bv, The Hague 1991, pp. 303–305.
25. W. Roebke, M. Renz, A. Henglein, *Int. J. Radiat. Phys. Chem.*, **1** (1969) 39–44.
26. W. J. McElroy, S. J. Waygood, *J. Chem. Soc., Faraday Trans.*, **86** (1990) 2557–2564.
27. Y. Tang, R. P. Thorn, R. L. Maudlin III, P. H. Wine, *Photochem. Photobiol.*, **44** (1988) 2–43.
28. P. Y. Jiang, Y. Katsumura, R. Nagaishi, M. Domae, K.Ishikawa, K. Ishigure, *J. Chem. Soc., Faraday Trans.*, **88** (1992) 1653–1658.
29. U. K. Klaning, K. Sehested, E. H. Apelman, *Inorg. Chem.*, **30** (1991) 3582–3584.
30. P. H. Wine, Y. Tang, R. P. Thorn, J. R. Wells, D. D. Davis, *J. Geophys. Res., D* **94** (1989) 1085–1094.

31. G. V. Buxton, G. A. Salmon and N. D. Wood, in: G. Restelli and G. Angeletti (eds) *Physico-Chemical Behaviour of Atmospheric Pollutants, 5th European Symp.*, Kluwer Academic Publishers, Dordrecht 1990, pp. 245–250.
32. C. L. Clifton and R. E. Huie, *Int. J. Chem. Kinet.*, **21** (1989) 677–687.
33. G. V. Buxton, S. Croft and G. A. Salmon, in: P.M. Borrell, P. Borrell, T. Cvitaš, W. Seiler (eds), *Proc. EUROTRAC Symp. '92*, SPB Academic Publishing bv, The Hague 1993, pp. 605–609.
34. G. V. Buxton, T. N. Malone and G. A. Salmon, to be published.
35. E. Heckel, A. Henglein and G. Beck, *Ber. Bunsenges, Phys. Chem.*, **70** (1966) 149–54.
36. S. J. Waygood, in: P. M. Borrell, P. Borrell, T. Cvitaš, W. Seiler (eds), *Proc. EUROTRAC Symp. '92*, SPB Academic Publishing bv, The Hague 1993, pp. 585–588.
37. P. Neta, R. E. Huie, *J. Phys. Chem.* **90** (1986) 4644–4648.
38. A. V. Gogolev, I. E. Makarev, A. M. Fedoseev, A. K. Pikaev, *High Energy Chem.*, **20** (1986) 229–233.
39. S. J. Waygood, in: R. A. Cox (ed.) *Air Pollution Research Report 42*, CEC, Guyot, Brussels 1992, pp. 23–26.
40. A. N. Yermakov, B. V. Zhitomirsky, G. A. Poskrebyshev, D. M. Sozuratov, *J. Phys. Chem.*, **97** (1993) 10712–10714.
41. G. V. Buxton, T. N. Malone, G. A. Salmon, in preparation.
42. P. Warneck, J. Ziajka, *Ber. Bunsenges, Phys. Chem.*, **99** (1995) 59–65.
43. G. V. Buxton, G. A. Salmon, J. E. Williams, in: K. H. Becker, (ed) *Air Pollution Research Report 54*, EC, Brussels 1994, pp. 439–443.
44. J. Berglund, L. I. Elding, G. V. Buxton, S. McGowan, G. A. Salmon, *J. Chem. Soc., Faraday Trans.*, **90** (1994) 3309–3313.
45. S. Barlow, G. V. Buxton, S. McGowan, G. A. Salmon, in: J. Peeters (ed) *Air Pollution Research Report 45*, CEC, Guyot, Brussels 1993, pp. 279–283.
46. M. Exner, H. Herrmann, R. Zellner, *Ber. Bunsenges. Phys. Chem.*, **96** (1992) 470–477.
47. H. Herrmann, M. Exner, R. Zellner, *Geochimica et Cosmochimica Acta.*, **58** (1994) 3239–3244.
48. G. Raabe, M. Exner, H. Herrmann, R. Zellner, in: K. H. Becker (ed) *Air Pollution Research Report 54*, EC, Brussels 1994, pp. 457–461.
49. M. Exner, H. Herrmann, R. Zellner, *J. Atmos. Chem.*, **18** (1994) 359–278.
50. S. Turner, Leeds University Project Report, 1995.
51. P. Neta, R. E. Huie, *J. Phys. Chem.*, **90** (1986) 4644–4648.
52. K. J. Kim, W. H. Hamill, *J. Phys. Chem.*, **80**, (1976) 2320–2325.
53. O. Ito, S. Akiho, M. Iino, *J. Phys. Chem.*, **93** (1989) 4079–4003.
54. L. Dogliotti, E. Hayon, *J. Phys. Chem.*, **71** (1967) 3802–3808.
55. T. W. Martin, R. E. Rummel, R. C. Gross, *J. Am. Chem. Soc.*, **86** (1964) 2595–2600.
56. G. V. Buxton, J. L. Eccles, G. A. Salmon, in: P. M. Borrell, P. Borrell, T. Cvitaš, W. Seiler (eds), *Proc. EUROTRAC Symp. '92*, SPB Academic Publishing bv, The Hague 1993, pp. 610–614.

2.6 Gas-Liquid Interactions

2.6.1 Introduction

It is now recognised that heterogeneous processes occurring in or on hydrometeors may change the chemistry of the lower atmosphere. For example, modelling calculations showed that in-cloud oxidation accounts for much of the oxidation of sulfur(IV) observed in the atmosphere (which cannot be described by homogeneous reactions alone) provided interfacial mass transport is efficient enough to enable reactants to enter the condensed phase [1]. Gas-liquid interactions are not only important for the aqueous phase chemistry (because they act as a source of reactants as stated above) but also for the gas phase chemistry because they can change the abundance of some trace gases (*e.g.* ozone and, in turn, No_x) [2], and they may initiate new pathways by activating radicals [3] (see also part 2.7.), an effect apparently occurring during Arctic spring.

In order to describe quantitatively such heterogeneous processes it is necessary to obtain a better understanding of the rate of uptake of gases by the condensed phase in addition to the partitioning of the reactants between the different phases. The partitioning can be described at equilibrium (if it exists), by Henry's law, which correlates the partial pressure P_x of a gas to its aqueous phase concentration $[C_x]$ where the proportionality factor is called Henry's law constant H_x, *i.e.*:

$$[C_x] = H_x\, P_x \qquad (1)$$

The lapse of time necessary to achieve equilibrium conditions depends on the rate at which a trace gas can be transferred to the condensed phase. It is important to accurately know this rate in order to model heterogeneous chemistry. If equilibrium exists the description appears to be straightforward whereas if not, interfacial mass transport must be explicitly treated.

The rate of uptake of trace gases by a liquid phase is a multi-step process which can be related to fundamental properties of the gas and of the interface. The uptake process can be summarised as follows:

1. Diffusion of the molecules in the gas phase toward the liquid interface,
2. Mass transfer across the vapour/liquid interface,
3. Hydrolysis, ionisation or chemical reactions in the liquid phase,
4. Diffusion of the species in the liquid phase.

The subprocesses 1 and 4 are governed by diffusion, in both gas and liquid phases. The second process, transport across the interface, is characterised by the mass accommodation coefficient α (previously named sticking coefficient). This coefficient which represents the fraction of molecular collisions of a gaseous species impinging on the interface followed by its incorporation in the liquid phase, can be defined as:

$$\alpha = \frac{\text{number of absorbed molecules}}{\text{number of molecular collisions on the interface}} \tag{2}$$

The flux of molecules Φ (cm^{-2} s^{-1}) across the air/water interface from the gas phase into the liquid equals the gas kinetic collision rate (1/4) c_g $<v>$ of the trace gas with the water surface, times the dimensionless mass accommodation coefficient $\alpha \leq 1$ (c_g = trace gas concentration adjacent to the air/water interface; $<v>$ = mean thermal velocity of the trace gas). The net flux Φ_{net} is obtained by multiplication with a factor κ_{sat} which corrects for surface saturation (c_{aq} = solute concentration in the aqueous phase):

$$\Phi_{net} = \frac{c_g}{4} < v > \alpha \kappa_{sat} = \frac{c_g}{4} < v > \alpha \left(1 - \frac{c_{aq}}{c_g HRT}\right)_{interface} \tag{3}$$

In most cases, the uptake process cannot be fully deconvoluted into its elementary steps and the overall kinetics are expressed in terms of uptake coefficients (γ_{obs}), which represent the fraction of gas/liquid collisions that are effective in the uptake and/or chemical transformation. In order to extract from the uptake coefficient further information on elementary steps, it is necessary to investigate the effect of mass transport (controlled by the working pressure and the volume of the condensed phase) or the effect of chemical reactions (controlled by the composition of the exposed surfaces or its nature, *i.e.* solid or liquid) [4]. However, the following definition pertains to the *true* mass accommodation coefficient α:

$$\alpha = \lim \left(\frac{4\Phi_{net}}{c_g < v >}\right) \qquad c_{aq} \to 0 \tag{4}$$

α may thus be deduced from measurements of the net flux Φ_{net} if c_g at the interface and the correction factor κ_{sat} can be computed. The net flux Φ_{net} across the phase boundary becomes small (a) if α is very small, and/or (b) if $c_{aq}/c_g H$ approaches unity. The latter happens if diffusion in the gas and/or liquid phases becomes flux limiting, and/or if H is very small. Speaking in terms of resistances in series [5], the transport resistances in the gas and liquid phases should be kept as small as possible in comparison with the interfacial resistance if α is to be deduced from measurements of Φ_{net}. This may be achieved by employing very short contact times, and by reducing transport resistance in the gas phase. The latter depends on the curvature of the exposed water surface [6] and on the binary diffusion coefficient. It is thus desirable to expose strongly curved surfaces at low total pressures or to use helium instead of air as carrier gas. The transport resistance in the liquid phase can be reduced to a limited extent by reactive scavenging, but this approach poses a minor problem at short contact times if the trace gas under study is sufficiently soluble.

These considerations have led to the development of:

(1) the droplet train technique,
(2) the liquid jet technique,
(3) the wetted-wall flow tube,
(4) the wind tunnel experiments, and
(5) the aerosol chamber technique for the determination of uptake rates.

These techniques are all working in different (but complementary) regimes and the treatment of the results is also different i.e. both numerical simulation and explicit mathematical solutions are used. The different characteristics of these techniques are summarised in Table 1. In (1) small water droplets are exposed to gases in a flow system at low total pressure. The flux Φ_{net} of a trace gas into the droplets is deduced from the amount of material taken up by collecting and analysing the exposed droplets. In (2) a thin water jet of variable length is exposed in a flow tube to a trace gas diluted in a carrier gas at ambient pressure. The net flux Φ_{net} is determined by analysing the collected jet water. In (3), the uptake rates are deduced from the wall loss rate as measured in vertically mounted flow tube the walls of which are covered by a solution. The net flux Φ_{net} is determined by monitoring the gas phase loss. In (4), suspended drops are analysed after a given exposure time in order to deduce the uptake rate from the chemical composition. Finally, in (5) the uptake rates are deduced from measured gas losses arising from the interaction with suspended aerosol particles under controlled conditions.

Table 1: Characteristics of the different experimental set-ups used.

	Droplet train technique	Jet technique	Flow tube technique	Wind tunnel experiment	Aerosol Chamber
Working pressure (Torr)	10–100	760	760	760	760
Typical exposure time (s)	$10^{-3} - 10^{-2}$	$10^{-4} - 10^{-2}$	1	> 10	> 100
Gas phase diffusion limitations	small limitations	limited	limited	limited	very small limitations
Analytical technique	MS – HPLC	DOAS – HPLC	FTIR – HPLC	HPLC	HPLC–gas phase monitoring
total surface area exposed (cm^2)	10^{-1}	$10^{-3} - 10^{-2}$ (a)	10^2	$10^{-4} - 2$	3×10^3 (b)
Nature of the surface	droplet	jet	film	droplet	droplet
Detection limit	$\gamma \geq 10^{-5}$	$10^{-5} \leq \gamma \leq 10^{-2}$	$10^{-7} \leq \gamma \leq 10^{-4}$		$10^{-7} \leq \gamma \leq 10^{-1}$

(a) not counting liquid jet flow rate, (b) total particle concentration 5×10^{11} m^{-3}.

2.6.2 The droplet train technique

The droplet train technique consists of a low pressure flow reactor in which monodisperse droplets are brought into contact with known concentrations of traces gases for a short period of time (5–20 ms) [11–14]. This technique involves the production of a micrometric droplet train (drop diameters ≤ 100 μm) produced by a vibrating orifice.

The apparatus consists of several different chambers: a droplet generating chamber, a buffer zone, the interaction chamber where gas and liquid phases interacts, a second buffer zone and the collection chamber where the droplets are collected for subsequent analysis by means of liquid chromatography. The gases flowing in and out of the interaction chamber can be monitored with high sensitivity, using an ion-trap mass spectrometer. The amount of substance taken up by the droplets is related to the kinetics of the uptake.

The design of this technique provides not only very good sensitivity for high uptake rate measurements (*i.e.*, $\gamma \leq 5 \times 10^{-3}$) but allows also the determination of uptake coefficients down to 10^{-5}. Because of short contact times it also allows the detection of highly reactive primary products.

2.6.3 The liquid jet technique

The liquid jet technique combines measurements of integrated absorption rates of a trace gas through the surface of a thin water jet with time-dependent numerical simulations of transport by molecular diffusion in the gas and liquid phases, and of chemical reactions affecting the solute concentration in the liquid phase, using the FACSIMILE/CHEKMAT computer package [7–10]. The mass accommodation coefficient is treated as a free parameter and deduced by fitting the numerical model to solute concentrations measured as a function of jet length.

The liquid jet features are due to very short contact times (0.1–0.8 ms) between the co-axially flowing gases and the jet, whose diameter is typically around 100 μm, emerging from a short quartz capillary. This liquid flow is neatly trapped in a counter capillary fixed in a tapered glass tube which can be moved in the x, y plane. Water from the trapping capillary is collected for analysis by ion chromatography, or (after mechanical decoupling) is pumped directly through a conductivity detector for on-line analysis, as explained in detail elsewhere [5]. The length of the exposed jet is varied in the range 0.3–4 mm by displacing the upper glass tube co-axially with 1μm precision.

2.6.4 The wetted-wall flow tube

The approach employed here is based on measuring the loss rate of gaseous compounds flowing along a vertically aligned flow tube, the walls of which are covered by a film of a slowly flowing aqueous solution. In addition to being very sensitive (*i.e.*, $10^{-7} \leq \gamma \leq 10^{-4}$ or 10^{-2} depending on the total pressure used), this method has also the advantage that the well established mass transport theory for cylindrical flow tube reactors applies [15]. The contact times in these experiments are in the order of a few seconds.

Schematically, the principle of operation relies on measuring the decrease (by FTIR) of the concentration of the trace gas, which is related to the first order rate constant of uptake by the liquid film at the flow tube's wall. The knowledge of this rate constant allows to determine the uptake coefficient γ either by straightforward calculation using gas kinetic theory or by numerical modelling. This latter option must be used when gas phase diffusion limits the rate of uptake.

Analysis (by means of ion chromatography) of the liquid flowing out of the reaction zone allows to establish a mass balance, which provides a further check of the experiments.

2.6.5 Wind tunnel technique

Another technique consists of a wind tunnel and the free suspension of drops of radii between 20 and 4000 μm under controlled conditions in the laminar vertical air stream of the tunnel [16, 17]. The long times available for drop suspension in this set-up allows exposure of drops to low concentrations (ppb range) of trace gases (*e.g.* SO_2). The amount of gas taken up by the drops is then measured *in-situ* by the pH change (for example during the oxidation of S(IV) to S(VI) by H_2O_2) using a calibrated pH colour indicator or by chemical analysis with HPLC after extraction of the drops.

The measured rate of uptake is compared to a gas phase diffusion model which was reformulated to take into account the dynamics of a falling drop. From this comparison, it becomes possible to obtain information about the accommodation process and the approach to saturation of the aqueous phase.

2.6.6 Aerosol chamber technique

In this set-up submicron aerosol particles (dry or deliquescent) are exposed to the gas phase in a chamber sufficiently large to ensure long residence times of the particles [15]. Gaseous reactants are added at low concentrations (at the ppb level). The uptake rates are determined either by monitoring gas phase losses of reactants by means of FTIR or HPLC (after trapping samples in aqueous solution) or by measuring the products formed after reactive uptake by the particles. The aerosol size distribution is monitored during the experiment. The small size of the particles used in this technique reduces corrections for gas phase diffusion so that they remain small even at atmospheric pressure. This technique requires a careful consideration of possible reactions on the walls of the chamber. Therefore, blank runs in the absence of aerosol particles are regularly performed in order to assess the influence of wall losses on uptake rates. The affinity of the gaseous reactant toward the wall must be smaller than that toward the aerosol. Information about the interfacial mass transport process is obtained by comparison of the observed concentration - time dependence with a theoretical model.

2.6.7 Results

Table 2 summarises the results obtained within the framework of HALIPP using the different techniques described above as well as data taken from the literature, as far as available, for comparison. In most cases the agreement between the different sets of data is good.

Table 2: Mass accommodation coefficients of different trace gases measured during the course of the project (at 298 K; if not available, the results are extrapolated to that temperature)

Trace gas	Schurath *et al.* [7–10]	Mirabel *et al.* [11–14, 19]	Zetzsch *et al.* [15]	Worsnop *et al.* [4]	Other authors	
SO_2	0.1	0.13		0.09	1 > 0.02	[16] [24]
H_2O_2	≥ 0.01	0.11		0.11 0.08	1	[16]
HCl	≥ 0.01			0.08		
N_2O_5	≥ 0.005	< 0.01	0.03	0.02	0.05	[21]
NO_2	only N_2O_4 was absorbed	> 0.0015			> 0.001 > 0.0006	[20] [23]
PAN	≥ 0.001					
NH_3	0.04	0.09				
HONO	0.05	0.05			$0.004 < \alpha < 0.04$	[20]
HCOOH	0.013			0.015		
CH_3COOH	0.0175			0.021		
CO_2	$1 - 2 \times 10^{-4}$					
O_3	0.0042	> 0.002		0.05	≥ 0.002 0.0005	[22] [24]
$ClNO_2$		> 0.01	> 0.01			
halogenated acetyls		> 0.001		≥ 0.01		

The study of these gas/liquid interactions can be divided into two cases, *i.e.* for well soluble and poorly soluble gases. Gases such as HNO_3, HCl, N_2O_5, CH_3COOH, HCOOH are very well soluble in water, and interfacial mass transport may represent a significant resistance to the uptake of these species into water during the time scale of laboratory experiments (in the order of milliseconds) [7–17]. Depending on the pH of the solution, some other gases may also behave like well soluble gases, *e.g.* NH_3, HONO and SO_2, due to the aqueous dissociation equilibria involved. Also in such cases the net flux across the interface contains information about the accommodation process. By comparison of the data with numerical modelling (taking into account mass transport micro-physical properties of the condensed phase) or analytical treatment (obtained from the solution of the differential equations of transport), it becomes possible to extract from the experiments values for the mass accommodation coefficient. The experiments also allow the direct measurement of the rate of heterogeneous oxidation of SO_2 by H_2O_2 (*e.g.* in wind tunnel experiments) and/or the modelling of this process. The results show that in this case the rate of oxidation deviates significantly from the behaviour expected from the assumption that bulk equilibrium and known oxidation kinetics apply [16, 18].

For less soluble gases, *i.e.* NO_2, CO_2, O_3 $ClNO_2$ or halogenated carbonyl compounds, the measurements of mass accommodation coefficients is much more difficult. In some cases the uptake rate is completely described by bulk properties such as solubility and reaction rates (for example CCl_3COCl [12]). However, by addition of chemical scavengers it becomes possible to speed up bulk processes and to obtain information about the accommodation step of the uptake process. Such a strategy has been successfully applied to gases such as O_3, $ClNO_2$ [10, 19] where iodide has been employed as an aqueous phase reactant to scavenge the gases after dissolution. For example, in the case of $ClNO_2$, the uptake coefficient was increased from 10^{-6} for pure water to nearly 10^{-2} for a 10^{-2} M NaI solution. This gave more insight into the chemistry of this process. In fact, using different scavengers and varying the composition of the condensed phase provides an opportunity to study the bulk chemistry of some dissolved gases. Again for $ClNO_2$, it was possible to develop a reaction scheme for its formation from N_2O_5 and for its reaction in salt containing solutions [15]. A discussion will be given in part 2.7.

The subproject HALIPP has provided an ideal framework for collaboration at the community level. In particular, joint experiments were done between the Universities of Strasbourg and Bonn, and between the University of Strasbourg and the Fraunhofer Institute at Hannover. In both cases, the droplet train apparatus was moved to another laboratory in order to perform joint measurements. Such experiments were performed for HONO, N_2O_5 and $ClNO_2$ in order to allow comparison of the different techniques [5, 19]. For example, the good agreement for HONO between the jet and droplet train techniques, which differ considerably with respect to exposure and shape of the exposed water surface, lends credibility to the measured mass accommodation coefficient. For another gas, *i.e.* nitryl chloride, this collaboration has enabled the identification of a new pathway for the uptake of trace gas into an aqueous solution. In fact, it was shown that the uptake rate of this compound was not completely described by bulk properties and a surface reaction was postulated. The latter possibility may become important since this reaction appears to be "facile and fast" at the interface compared to bulk chemical transformations. In the case of NO_2, the results reported in the literature differ considerably and cannot be explained by bulk properties. A reason for the difficulty with NO_2 could be due to the fact that its uptake proceeds via the dimer N_2O_4. Another reason may be attributed to the occurrence of a surface reaction (similar to the experiments with $ClNO_2$) as identified by a group in Jülich [20].

As can be seen from Table 2, most of the mass accommodation coefficients reported here are larger than approximately 10^{-2} with the exception of CO_2 and possibly O_3 (in the case of NO_2, the values reported are inconsistent and therefore should be handled carefully). According to the theoretical treatment of Schwartz [6] for such large values, the interfacial transport should not limit the uptake. In contrast, gas phase diffusion or bulk properties (*i.e.*, solubility and reactivity) are probably the main limiting processes. Note however, that for values close to 10^{-2}, a limitation due to the interfacial mass transport may become apparent [18].

Another aspect of the study performed during the course of this project concerns the determination of Henry's law constant in approximately neutral ion–free water, the data for which are summarised in Table 3. While all alkyl mononitrates exhibit low solubilities, and do not hydrolyse under ambient conditions, peroxyacyl nitrates are nearly two orders of magnitude more soluble and hydrolyse slowly with typical activation energies of 54.5 kJ mol^{-1}. Although the dinitrates and hydroxynitrates are up to three orders of magnitude more soluble, most of them are stable or exhibit very low hydrolysis rates in neutral aqueous solutions.

Using a simulated multiphase system, it was possible to study the phase partitioning of gases such as SO_2 and HNO_3 [26]. The results obtained agree well with data found in the literature and the phase partitioning appears to be described Henry's law (at least for SO_2 in water and for HNO_3 in H_2O/H_2SO_4 mixtures).
In conclusion, different sophisticated and innovative techniques devoted to the study of gas/liquid water interactions were developed which are the jet, the droplet train, the wetted-wall flow tube, the wind tunnel and the aerosol chamber techniques. The experimental set-ups are applicable to a wide range of conditions allowing the measurement of reliable uptake rates. The techniques have been employed for different gases (see Table 2) leading to the determination of mass accommodation coefficients larger than 10^{-2}, meaning that in most cases interfacial mass transport does not represent a limiting factor for the uptake by cloud droplets. It should be noted that the uptake of radicals such as OH and HO_2 was not studied because of the intrinsic difficulty to handle these species when in contact with water.

Table 3: Henry's law constants, enthalpies and entropies of solution for organic nitrates [25].

Compound	*H* at 298K M atm–1	ΔH^0_{soln} kJ mol^{-1}	ΔS^0_{soln} kJ mol^{-1}
CH_3ONO_2	2.64	–39.4 ± 0.6	–126.4 ± 3
$C_2H_5ONO_2$	2.18	–44.5 ± 1.5	–145.5 ± 5
1-$C_3H_7ONO_2$	1.55	–45.6 ± 0.4	–152.2 ± 2
2-$C_3H_7ONO_2$	1.08	–44.5 ± 0.7	–151.3 ± 3
1-$C_4H_9ONO_2$	1.42	–48.1 ± 0.9	–161.3 ± 3
2-$C_4H_9ONO_2$	0.89	–45.0 ± 0.6	–154.6 ± 3
2-methyl-propyl-1-nitrate	0.96	–43.6 ± 1.2	–149.2 ± 4
1-$C_5H_{11}ONO_2$	1.2		
2-$C_5H_{11}ONO_2$	0.53	–52.3 ± 2.2	–187.7 ± 4
1,2-ethane-dinitrate	640 ± 8		
2-nitrooxy-ethanol	39900 ± 900		
2-nitrooxy-1-propanol *)	7300 ± 300		
1-nitrooxy-2-propanol *)	6700 ± 300		
1,2-propane dinitrate	175 ± 20		
nitrooxyacetone	1010 ± 40		
perooxyacetyl nitrate	97.9 ± 2.0	–47.4 ± 0.8	–150 ± 2.5
peroxyproponyl nitrate	70.8 ± 1.5	–49.4 ± 0.9	–160 ± 3
peroxy-*n*-butyl nitrate	54.8 ± 1.5		
peroxymethacryloyl nitrate	41.0 ± 3.8		
peroxy-*i*-butyryl nitrate	24.0 ±0 .7		

2.6.8 References

1. S. E. Schwartz, in: J. G. Calvert (ed) *SO_2, NO and NO_2 Oxidation Mechanisms: Atmospheric Considerations*, Butterworth, Boston, MA 1984, pp. 173–208.
2. J. Lelieveld, P. J. Crutzen, *Nature*, **343** (1990) 227–233.
3. B. J. Finlayson-Pitts, M. J. Ezell, J. N. Pitts Jr., *Nature*, **337** (1989) 241–244.
4. C. E. Kolb, D. R. Worsnop, M. S. Zahniser, P. Davidovits, D. R. Hanson, A. R. Ravishankara, L.F. Keyser, M. T. Leu, L. R. Williams, M. J. Molina, M. A. Tolbert, in: J. R. Barker (ed) *Current Problems of Atmospheric Chemistry*, Advan. Phys. Chem. Ser. C, in press.
5. A. Bongartz, J. Kames, U. Schurath, Ch. George, J. L. Ponche, Ph. Mirabel, *J. Atmos. Chem.*, **18** (1994) 149–169.
6. S. E. Schwartz, in: W. Jaeschke (ed) *Chemistry of Multiphase Atmospheric Systems*, NATO ASI Ser. G6, Springer-Verlag, Berlin Heidelberg 1986, pp. 415–471.
7. W. Kirchner, F. Welter, A. Bongartz, J. Kames, S. Schweighoefer, U. Schurath, *J. Atmos. Chem.*, **10** (1990) 427–449.
8. A. Bongartz, S. Schweighoefer, C. Roose, U. Schurath, *J. Atmos. Chem*, (1995) in press.
9. A. Bongartz, Massen–Akkommodationskoeffienten schwacher atmosphärischer Säuren und Basen, Dissertation, Bonn, 1993.

10. C. Wunderlich, A. Schlemm, K. Hesse, D. Hupperich, U. Schurath, in: K. H. Becker (ed) *Air Pollution Research Report 54*, EC, Brussels 1995, pp. 45–50.
11. J. L Ponche, Ch. George, Ph. Mirabel, *J. Atmos. Chem.*, **16** (1993) 1–21.
12. Ch. George, J. Lagrange, Ph. Lagrange, Ph. Mirabel, C. Pallares, J. L. Ponche, *J. Geophys. Res.*, **99** (1994) 1255–1262.
13. Ch. George, J. Y. Saison, J. L. Ponche, Ph. Mirabel, *J. Phys. Chem.*, **98** (1994) 10857–10862.
14. Ch. George, J. L. Ponche, Ph. Mirabel, W. Behnke, V. Scheer, C. Zetzsch, *J. Phys. Chem.*, **98** (1994) 10857–10862.
15. W. Behnke, Ch. George, V. Scheer, C. Zetzsch, in preparation.
16. A. Waltrop, S. K. Mitra, A. I. Flossmann, H. R. Pruppacher, *J. Atmos. Chem*, **12** (1991) 1–17.
17. S. K. Mitra, A.U. Hannemann, *J. Atmos. Chem.* **16** (1993) 201–218.
18. Ph. Mirabel, Ch. George, N. Miloshev, in: G. Angeletti and G. Restelli (eds) *Physico-Chemical Behaviour of Atmospheric Pollutants, Proc. 6th European Symp.*, EC, Luxembourg 1994, pp. 864–871.
19. Ch. George, W. Behnke, V. Scheer, C. Zetzsch, L. Magi, J. L. Ponche, Ph. Mirabel, *Geophys. Res. Lett.* (1995) in press.
20. D. Becker, A. Wahner, *Annales Geophysicae*, **13(SII)** (1995) C376.
21. M. Mozurkewich, J. G. Calvert, *J. Geophys. Res.*, **93** (1988) 15889–15896.
22. R. G. Utter, J. B. Burkholder, C. J. Howard , A. R. Ravishankara, *J. Phys. Chem.*, **96**, (1992) 4973–4979.
23. J. H. Lee, I. N.Tang, *Atmos. Environ.* 22 (1988) 1147.
24. J. H. Lee, I. N.Tang, in: R. W. Johnson, G. E. Gordon (eds), The Chemistry of Acid Rain, *Am. Chem. Soc. Symp.* 1987, p.109.
25. K. Kames, U. Schrath, J. Atmos Chem. (1995) in press.
26. W. Jaeschke, J. F. Diersen, S. Volkwein, J. Wolgemuth, this volume.

2.7 Aerosol Surface Reactions and Photocatalysis

2.7.1 Mineral dusts and organic compounds.

Heterogeneous kinetics at aerosol surfaces is a field in atmospheric chemistry which covers a broad range of interactions between gases ↔ solids, gases ↔ liquids and solids ↔ liquids [1]. During the last 10 years a multitude of publications appeared in the literature on the use of photocatalytic processes for treatment of polluted air and waste waters but a direct link to atmospheric chemistry can seldom be made.

Photocatalytic activity of coal fly ashes and metal oxides has been shown to consume atmospheric pollutants [2]. TiO_2 aerosol under irradiation by simulated sunlight has been observed to increase the level of OH in an aerosol smog chamber experiment by an order of magnitude [3]. Volatile hydrocarbons, such as aromatics and alkanes, were significantly degraded by an additional oxidation process on the surface of the aerosol. The efficiency was observed to be proportional to the mean residence time of the hydrocarbons in the adsorbed state [3] and inversely proportional to their vapour pressures [3, 4]. Fe_2O_3 (haematite) was shown to be much less photoactive, and an additional, heterogeneous photodegradation was found to be significant for alkylated benzenes alone [5]. Compounds adsorbed permanently on the aerosol can be rapidly degraded by photocatalytic oxidation, and half-lives of a few hours have been reported for di-2-ethylhexyl-phthalate (adsorbed on Fe_2O_3) and shorter than 20 min (adsorbed on TiO_2), whereas a fly ash from a coal power plant had no significant effect [5].

The attention in this section is mostly focused on the conversion of organic compounds at solid surfaces or at suspended particles in aqueous solutions. Especially the photocatalytic activity of inorganic solids such as muscovite, titanium dioxide, silicon dioxide, aluminium dioxide, ferric oxide and fly ash has been investigated. The photocatalytic activity has been looked at from different angles using various experimental techniques.

In smog chamber experiments the enhanced degradation rate of organic compounds due to the presence of aerosols can be compared with the reaction rate in the homogeneous gas phase. In surface coated flow reactors or in smog chamber experiments with coated aerosols [6] more specific gas ↔ solid interactions initiated by irradiation can be investigated. In aqueous solutions both the effect of heterogeneous interactions at suspended particles and of dissolved ionic species can be distinguished. Only few quantitative kinetic data have been derived in these reacting systems and most of the experiments yielded only qualitative and mechanistic information on the processes occurring. A summary of the experimental conditions and results is given in Table 1. Here a list of organic substances is shown with the type of interactions considered and the experimental technique used.

Table 1: Heterogeneous reactions of organic compounds.

Organic compound	Interaction medium liquid solid	Technique	Effect*	Products	Ref.	
enzaldehyde	muscovite	smog chamber	L	benzoic acid	7	$2 m^2 g^{-1}$
	fly ash	smog chamber	L	benzoic acid		$1 m^2 g^{-1}$
naphtalene	muscovite	smog chamber	M	} 23 chemical substances	8	
	fly ash	smog chamber	M			
	TiO_2	smog chamber	M			
	muscovite	static/flow reactor	L	1,4 naphtoquinone	9	$\lambda > 340$ nm
	Fe_2O_3	static/flow reactor	L	1,4 naphtoquinone		}various surfaces/weights
	fly ash	static/flow reactor	M	1,4 naphtoquinone		
	TiO_2	static/flow reactor	H	1,4 naphtoquinone, 2 naphtol, phthalide, phthaldialdehyde, phthalic acid, acetophenone, benzaldehyde, benzoic acid		$50 m^2 g^{-1}$
α-pinene	SiO_2	flow reactor	L	-	12	$\gamma \leq 1.4 \times 10^{-4}$
	NaCl	flow reactor	L	-	13	$\gamma \leq 1.4 \times 10^{-4}$
	SiO_2	solution	L	limone, *trans*-verbenol, borneol, isopinocampheol, *trans*-carveol perillylic alcohol, *p*-menth-3-ene-1-ol, pinene oxide, limonene oxide, *trans-p*-menth-6-ene-2,8-diol, 6-endo-hydroxyborneol, 2,3 bornanediol, 2,10 bornanediol, pinoaldehyde, methylethyl-2-methyl-2-cyclo-hexen-1-one, campholene aldehyde, pino-camphone	16	$\lambda > 220$ nm
	SiO_2	solution + H_2O_2	H			
β-pinene	fly ash	smog chamber	M	-	18	
	TiO_2	smog chamber	H	-		
xalic acid	Fe_2O_3	solution	M	CO_2	19 20,21	$\lambda > 290$ nm
etaldehyde	TiO_2	solution	H	CO_2, acetic acid	19	$\lambda > 290$ nm
	Fe_2O_3	solution	L	CO_2, acetic acid		
4 dichloro-phen-oxyacetic acid	laponite	static flow reactor	L	} 2,4 dichlorophenol, chlorohydroxyquinone, chloro-*p*-benzoquinone, 4-chlorocatechol	21	λ:220-290 nm
	SiO_2	static flow reactor	H			
	laponite	solution	L			
	SiO_2	solution	L			
	SiO_2	solution + H_2O_2	H			
rbonaceous materials	presence of O_2			CO, CO_2	24,25	$\lambda > 325$ nm $100 m^2 g^{-1}$

* Catalytic Effect: L = low; M = medium; H = high.

When a high, medium or low catalytic effect on the degradation rate was observed, it is indicated as H, M or L in the column 'catalytic effect'. Reaction products identified as a result of the heterogeneous interactions are also given in the column 'products'. A negative sign (–) indicates that no reaction products could be detected.

The photocatalytic conversion of benzaldehyde was somewhat faster on muscovite than on fly ash and in both cases benzoic acid was the main reaction product [7]. Naphthalene is from an atmospheric viewpoint more interesting since it is a model compound for the class of noxious polyaromatic hydrocarbons (PAH). Its photodegradation on muscovite, fly ash and TiO_2 in a chlorine containing atmosphere showed the formation of a multitude of products both in the gas phase and adsorbed on the catalysts [8]. More elaborate results were obtained for the photocatalytic reactions of naphthalene in oxygen. TiO_2 and fly ash are efficient catalysts while the degradation on muscovite and Fe_2O_3 is slower [9]. On TiO_2 a degradation rate of 0.4 molecule nm^{-2} h^{-1} was measured at $\lambda > 340$ nm and with a radiant flux of *ca.* 22 m W cm^{-2}. Water vapour sharply increases the photodegradation rate. The major reaction product is 1,4-naphthoquinone together with a large number of oxidation products which are listed in Table 1. Most of these products have also been detected in smog chamber studies [10]. Depending on their volatility these products are detected either in the gas phase or adsorbed on the surface of the catalyst (9). The only kinetic study on the naphthalene degradation on TiO_2 has shown that the degradation constant is a function of the initial naphthalene concentration which could be explained in terms of the Langmuir adsorption isotherms [11].

α-Pinene and β-pinene are major hydrocarbons emitted in the atmosphere from biogenic sources. Sticking of α-pinene on SiO_2 and NaCl surfaces seems to be a slow process so that the heterogeneous destruction of α-pinene on pre-existing particles is unimportant under atmospheric conditions [12, 13]. However it was seen that α-pinene in its reaction with hydroxyl radicals in the homogeneous gas phase forms the condensable products pinonaldehyde, pinol, pinocamphone, campholene aldehyde, carvotanacetone, transcarveol and possibly an NO adduct [14]. It is interesting to note that pinonic acid was not observed. These condensables may undergo homogeneous nucleation processes leading to the formation of aerosols [15]. Many of these products were also detected with silicon dioxide as carrier where H_2O_2 strongly increased the α-pinene degradation rate [16]. Not much information on the photocatalytic degradation of biogenic hydrocarbons is available in the literature. Only recently has the role of photocatalysis been investigated [17]. Products observed and a comparison with the homogeneous destruction with hydroxyl radicals showed that the surface of natural aerosols can be an important sink for these organic compounds under atmospheric conditions.

The effect of both suspended Fe_2O_3 particles and dissolved Fe^{3+} ions on the photodegradation rate of oxalic acid has been studied quite in detail [19–21]. The major activity must be attributed to the ferric ions in solution while the photocatalytic activity of the solid is only of minor importance.

Acetaldehyde which plays a very important role in atmospheric chemistry as a precursor of peroxyacetyl nitrate (PAN) is converted to acetic acid and CO_2 in solutions in the presence of suspended TiO_2 and Fe_2O_3 particles [19]. Chloride ions markedly decrease the carbon dioxide formation rate.

Somewhat outside the field of atmospheric chemistry, an extensive study on the photodegradation of the herbicide 2,4 dichlorophenoxyethanoic acid on SiO_2 and the clay mineral laponite has been undertaken [21]. A detailed degradation scheme involving H_2O_2 reactions has been established.

All the experiments demonstrated the efficiency of TiO_2 as an efficient photocatalyst for the degradation of organic compounds. The mechanistic aspects of this catalytic activity are well documented in the literature and have been covered in a recent review [22].

At a more fundamental level it was thought that the formation of Lewis acid centres (LAC) at solid surfaces under UV illumination could be important for their photocatalytic activity. Only for Al_2O_3 and TiO_2 a noticeable production of LACs has been observed. For fly ash and SiO_2 these centres are destroyed quickly due to the photodesorption of water [23–25].

Finally the formation of CO and CO_2 from carbonaceous materials has been investigated as a function of the illumination and the water content [24]. Hydrogen peroxide has been observed as an intermediate which is of course an important photochemical oxidant in the atmosphere [25].

A major point related to the above mentioned results is the question whether photocatalytic reactions at aerosol surfaces are important under atmospheric conditions. This of course strongly depends on the composition and the concentration of the aerosols involved. One can make a crude estimation of the magnitude of the heterogeneous removal constant k_{het} of a gaseous component G with molar mass M:

$$k_{het} = \left(\frac{RT}{2\pi M}\right)^{0.5} \cdot S \cdot \alpha \tag{1}$$

where S stands for the average surface of the aerosol per unit volume and α for the accommodation coefficient [26]. For a gaseous compound with a molar mass $M = 40$ g mol^{-1}, an accommodation coefficient of 10^{-3} and an aerosol with $S = 1.5 \times 10^{-5}$ cm^2 cm^{-3}, one arrives at a value of $k_{het} = 1.7 \times 10^{-4}$ s^{-1}. One can now assume that G reacts with hydroxyl radicals with a rather high value of a rate constant $k_{OH} = 1 \times 10^{-11}$ molecule^{-1} cm^3 s^{-1}. Assuming a hydroxyl radical concentration in the atmosphere equal to 1×10^6 molecules cm^{-3} leads to a homogeneous destruction rate constant $k_{hom} = 1 \times 10^{-5}$ s^{-1} which is more than an order of magnitude lower than k_{het}. This calculation, which is of course a crude approximation, clearly shows the potential of heterogeneous processes. They can provide a realistic reaction path for the removal of organic compounds in the atmosphere.

The main task for future research efforts in this field should now be directed at not only gathering mechanistic information on the photocatalytic degradation paths of hydrocarbons but also on the determination of quantitative kinetic parameters. The latter are needed in order to incorporate heterogeneous chemistry in atmospheric chemistry models.

2.7.2 Sea salt and NO_x

Direct heterogeneous removal of NO_2 is slow (see also section on mass accommodation coefficients and uptake rates of gaseous species in water). The predominant night-time removal process of NO_x from the atmosphere is initiated by the oxidation of NO_2 to NO_3 radicals by ozone, followed (via reaction with another NO_2-molecule) by the formation of N_2O_5 [27]. Heterogeneous reactions with the aqueous content of aerosols in the troposphere are thought to convert N_2O_5 to nitric acid rapidly. Since the gas phase reaction of N_2O_5 with H_2O is very slow [28], with a rate constant lower than 2.7×10^{-22} cm^3s^{-1}, and N_2O_5 is thermally unstable [29] with a lifetime of 0.35 min at 298 K, N_2O_5 plays an important role in the tropospheric night-time chemistry as a reservoir for NO_3 radicals [30, 31]. The apparent lifetime of NO_3 radicals (in equilibrium with N_2O_5) has been observed to decrease from more than 1 h below 20 % relative humidity (r.h.) to less than 10 min above 50 % r.h. in field measurements in California during the night [32], and cannot be limited by reactions with terpenes or other gaseous air constituents. This apparent lifetime of NO_3 might instead be controlled by heterogeneous loss of N_2O_5, where the uptake of N_2O_5 by aerosols with high electrolyte concentration (close to the deliquescence point) may yield compounds other than HNO_3. $BrNO_2$ (a photolytic source of atomic Br) might be formed from the reaction of N_2O_5 with NaBr, contained in sea salt [33].

The intrusion of acids, such as sulfuric and nitric acid, into the marine sea-spray aerosol [34, 35] may release HCl and thereby (via the reaction with OH) lead to the formation of Cl atoms in the troposphere [36]. Tropospheric Cl atoms might also be formed via photolysis of NOCl, a product of the reaction of NO_2 with NaCl salt [37, 38]. The reaction of N_2O_5 with NaCl (the major component of sea-salt) was shown to lead to $ClNO_2$ [39], providing another potential photolytic source of Cl atoms in the troposphere. As already mentioned, Br atoms might be formed via the analogous reactions with bromides [33, 40].

The tropospheric relevance of these processes was demonstrated within HALIPP. The studies employed a Knudsen cell reactor, where the uptake of NO_2, N_2O_5, $ClONO_2$ and HNO_3 by powders of NaCl, NaBr, $CaCO_3$ and amorphous carbon was explored, and a smog chamber facility, where the reactions of NO_2 and N_2O_5 with dry and deliquescent NaCl aerosol were investigated These studies were supplemented by experiments with a wetted-wall flow reactor (see part 2.6.).

While earlier smog chamber experiments on the behaviour of NaCl aerosol in the presence of NO_x had suffered from contamination of the sodium chloride by a (high temperature) aerosol production process and from heterogeneous reactions on the (hydrophilic) glass walls, in later experiments during the preparatory phase of HALIPP such artefacts were eliminated by the use of (hydrophobic) FEP-200 Teflon bags (initially at 50 % r.h., later up to 94 % r.h.), for which the photochemical formation of Cl atoms from HCl and O_3 was shown to be low (in agreement with the expected reaction of OH with HCl), and no contribution of interfering wall reactions was detected [41, 42].

2.7.3 Almost dry NaCl aerosol (at 50 % r.h.)

The reaction of NO_2 with NaCl aerosol at 50 % relative humidity (where most of the aerosol is dry and only minor portions might be deliquescent) was found to be inefficient for the formation of Cl atoms with NOCl as intermediate at NO_2 levels up to 1 ppm. Accordingly, this process most probably is unimportant in the troposphere [42]. On the other hand, in the presence of NO_2 (200 ppm) and O_3 (100 ppm) under otherwise similar conditions, the formation of Cl atoms from $ClNO_2$ was observed to be very efficient. Chlorine atoms were shown to arise from the reaction sequence

$$NO_2 + O_3 \rightarrow NO_3 + O_2 \quad (2)$$

$$NO_3 + NO_2 \rightleftharpoons N_2O_5 \quad (3)$$

$$N_2O_5 + NaCl(s+l) \rightarrow ClNO_2 + NaNO_3(s+l) \quad (4)$$

$$ClNO_2 + h\nu \rightarrow Cl + NO_2 \quad (5)$$

The intermediate formation of $ClNO_2$ via uptake of N_2O_5 by the aerosol (with a high uptake coefficient, γ_{obs}, of at least 10^{-3}) was identified from the observed stoichiometry ($2NO_2$ consumed per Cl atom formed), from the efficiency of an initial dark phase (to avoid photolysis of the NO_3 required to produce N_2O_5) and from the expected photolysis rate of $ClNO_2$.

Experiments with synthetic sea-salt aerosol (instead of pure NaCl) under similar conditions and even more so with genuine sea water (which yields significantly higher aerosol mass densities) indicated an additional (not yet fully characterised) source of Cl atoms in the absence of NO_x. This source might be related to the observation that a formation of Cl_2 from HCl and O_3 can be detected in the dark in a glass chamber or in Teflon bags in the presence of the quartz aerosol Aerosil 200. The source strength of such active chlorine has been observed to be enhanced by the presence of humidity and light and even more so by semiconducting aerosols (Fe_2O_3 and TiO_2 or contaminated NaCl [43, 44]. Bromide, metal ions and the organics contained in genuine sea water might also play a role.

2.7.4 Deliquescent NaCl aerosol (at 70–94 % r.h.)

Techniques for the production of deliquescent NaCl aerosol (at humidities up to 92 %) with a long-lived, well characterised size distribution were developed to study its reactions with NO_2 and N_2O_5 in detail [45, 46]. The reaction of NO_2 was reinvestigated with wet NaCl aerosol with well characterised size at 291.5 K and 86 % r.h. and found to be slow, with an upper limit of the uptake coefficient of $\gamma_{obs} < 10^{-4}$. Instead of the expected NOCl, HONO was detected from the photolytic formation of OH [47, 48] according to the mechanism:

$$2NO_2(g) + NaCl(l) \rightarrow NOCl(l) + NaNO_3(l) \quad (6)$$
$$NOCl(l) + H_2O(l) \rightarrow HONO(g) + HCl(l) \quad (7)$$
$$HONO(g) + h\nu \rightarrow OH(g) + NO(g) \quad (8)$$

The reaction of N_2O_5 with wet NaCl aerosol was also reinvestigated between 70.9 and 92.2 % r.h. by injecting N_2O_5, prepared in a flow of NO_2 and O_3. Under consideration of the Fuchs equation, the first experiments [45, 46] with well characterised size distribution led to uptake coefficients, γ_{obs}, between 0.02 and 0.05 (with a mean value of 0.03) at 291.5 K and yields of $ClNO_2$ between 31 and 75 % of the N_2O_5 reacted.

By taking into account heterogeneous loss processes of N_2O_5 and NO_3 on the walls of the Teflon bags (determined in separate blank experiments) [47], the uptake coefficient of N_2O_5 was refined (5 runs) to $\gamma \pm \sigma = 0.032 \pm 0.003$ at 291 K and relative humidities between 77 and 94 % (corresponding to aqueous concentrations of NaCl in the range 5.8–1.8 mol/L) with yields of $ClNO_2$ up to 76 % of the N_2O_5 consumed by the deliquescent aerosol. Adding a mole fraction of 0.2 of Na_2SO_4 to the NaCl did not influence the uptake coefficient of N_2O_5, which under these conditions averaged (from 5 runs) to $\gamma = 0.032 \pm 0.002$. The yields of $ClNO_2$ were observed to be insignificantly smaller in the presence of Na_2SO_4 [49]. Nevertheless, the high yields of $ClNO_2$ observed in the wet system represent a surprising result as one expects primarily the fast hydrolysis of N_2O_5 to HNO_3 to take place at the expense of the exchange reaction leading to $ClNO_2$.

A wetted-wall flow tube method was subsequently developed, and the influence of pH on the yield of $ClNO_2$ was investigated [47]. Below pH = 2 the yield of $ClNO_2$ decreased significantly, and NOCl was formed instead, with a yield of almost 10 % at pH = 0. The yield of $ClNO_2$ from the reaction of N_2O_5 with NaCl solutions was significantly higher in the wetted-wall flow tube reactor, attaining a limiting value of almost 100 % at high NaCl concentration [50, 51]. The dependence of the uptake of $ClNO_2$ on the concentration of Cl^- (and the influence of NO_3^- and SO_4^{2-}) was investigated in detail and can be explained by a mechanism involving the nitronium ion, NO_2^+ (see also part 2.6.) [51]. The yields of $ClNO_2$, determined at various concentrations of chloride ion, Cl^-, in the wetted-wall flow tube reactor and various r.h. in the aerosol smog chamber experiments are compared in Fig. 1. The yields of $ClNO_2$ in the smog chamber show no obvious trend and their scatter is much larger. It is assumed that a heterogeneous loss of $ClNO_2$ at the Teflon walls is the reason for the lower yields in comparison to the wetted-wall flow tube results.

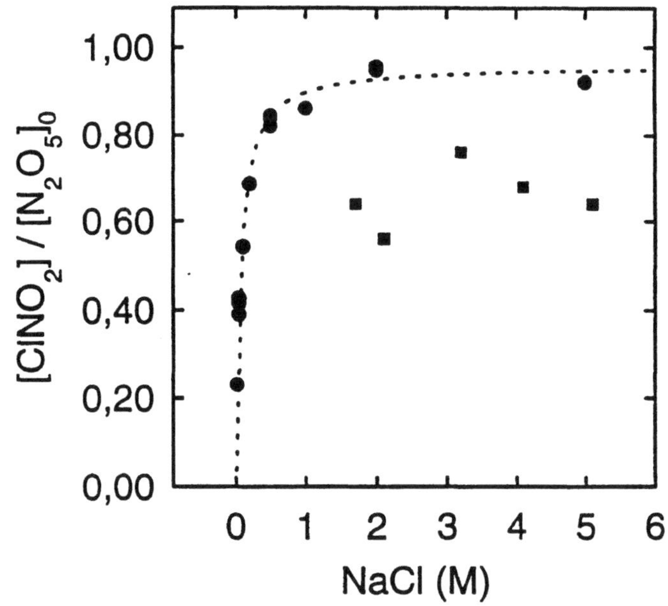

Fig. 1: Yields of $ClNO_2$ from the reaction of N_2O_5 with NaCl solutions as a function of Cl^- observed in wetted–wall flowtube experiments (points, determined by FTIR) and in aerosol smog chamber experiments (squares, determined from the rate of Cl atom formation by photolysis of $ClNO_2$ and the NO_x balance).

2.7.5 Mechanism for the uptake of nitrogen pentoxide N_2O_5

The uptake coefficient of $ClNO_2$ in pure water was determined by the wetted-wall flow tube technique (at 291K) to be $\gamma \pm 2\sigma = (5 \pm 0.6) \times 10^{-6}$ (much smaller than that for N_2O_5). The value decreased with increasing chloride concentration to 0.27×10^{-6} at $[Cl^-] = 4.6$ mol/L [50, 51]. This decrease of the uptake coefficient in the presence of Cl^- is much stronger than expected from a salting-out effect, indicating that a reversible dissociation of $ClNO_2$ to NO_2^+ and Cl^- occurs. The observations, which include the Cl^- dependent yield of $ClNO_2$ from N_2O_5, are consistent with the reaction mechanism shown in Fig. 2, which describes the uptake of N_2O_5 and the subsequent release and reabsorption of $ClNO_2$.

Assuming a steady state for NO_2^+, the differential equation describing the reaction mechanism and the diffusive transport of N_2O_5 and $ClNO_2$ in the liquid film can be solved [72], leading to estimates of the ratio k_4/k_5, the product $H_o k_6^{1/2}$ and the ratio $k_3\,[H_2O]/k_1$, where k_1 is large [69]. A value of $k_1 > 10^5\ s^{-1}$ follows from the uptake coefficient of N_2O_5 (determined by the droplet train technique at Hannover in collaboration with the University of Strasbourg) in pure water, in 1 M NaCl solution (see above) and on deliquescent NaCl aerosol, assuming a reasonable solubility and diffusion coefficient of N_2O_5 [63]. Numerical integration and fitting the rate constants by the software package FACSIMILE [51] leads to a final value of the order of 10^3 for the ratio k_4/k_5, in contrast to another determination about three orders of magnitude higher by Johnson and Margerum [74]. Furthermore, $H_o k_6^{1/2} \pm \sigma$ for $ClNO_2$ is obtained to be 0.44 ± 0.01 mol L^{-1} $s^{-1/2}$ with a reasonable salting-out constant associated with NaCl. Reaction 2 is too slow in comparison with reaction 4 to be significant in the mechanism, and reaction 3 can also be neglected if the observed yields of $ClNO_2$ are corrected for hydrolysis in the absorption cell [51]. The observed dependence on NO_3^- is found to be due to its influence on the ionic activities. The equilibrium of NO_2^+ with Cl^- (k_4/k_6) appears to control the concentration of NO_2^+ rather than its equilibrium with NO_3^- (k_2/k_1) [51]. The uptake of $ClNO_2$ by water is greatly enhanced in the presence of iodide as demonstrated by use of the droplet train technique (at Hannover, again in collaboration with the University of Strasbourg). The efficiency of I^- is too strong to be explained solely by scavenging of NO_2^+. A direct reaction of I^- with $ClNO_2$ leading to NO_2^- (detected by ion chromatography) and I_2 (detected by its colour), with an intermediate formation of ICl, is postulated [73]. Wetted-wall flow tube investigations of NaCl solutions (0.5 mol/L), mixed with NaBr (0.0009–0.03 mol/L) indicate that the uptake of N_2O_5 by sea water on account of its Br^- content may lead to a significant formation of $BrNO_2$ and other active volatile bromine compounds (Br_2 or BrCl) [69].

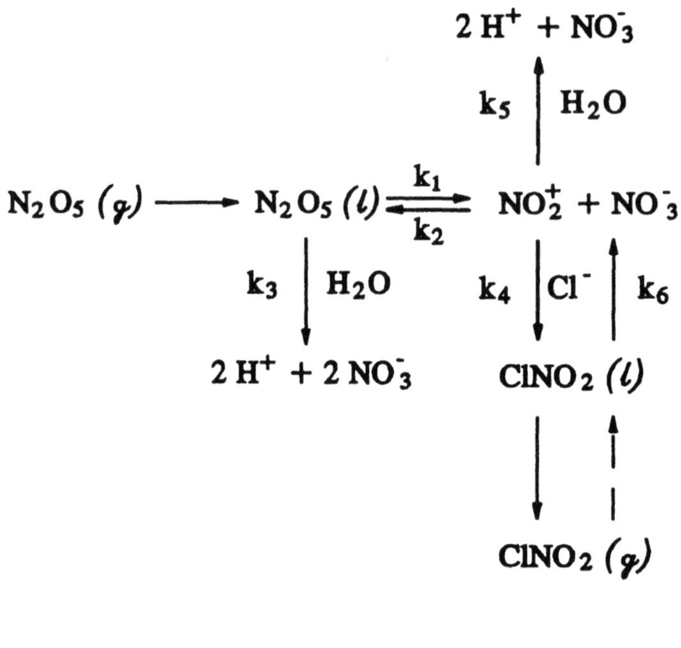

Fig. 2: Reaction scheme advanced to explain the formation of $ClNO_2$ by the interaction of N_2O_5 with aqueous sodium chloride solutions.

2.7.6 Powder samples

Uptake coefficients of nitrogen oxides and related trace gases on solid model compounds of atmospheric aerosol were determined by Rossi and co-workers [52] in a Knudsen cell reactor consisting of a two-chambered cell made of Teflon coated Pyrex or stainless steel, with mass spectrometric detection of the gas-phase reactants and products at ms time resolution. The goal was to study the displacement of chloride and bromide from sodium halides by their reactions with HNO_3(g) [53], N_2O_5(g) [54], $ClONO_2$(g) [55] and NO_2(g) [52] (which leads to an exchange of the anion in the first case and to a release of photolabile gaseous products in the three latter cases), the uptake and reactions of HNO_3 with $CaCO_3$ [56] (which neutralises atmospheric acidity) and of NO_2 [57] and HNO_3 [58] with amorphous carbon, a model compound for the particulate emissions of combustion engines (including those of aircraft).

These studies focused on the fundamental chemical kinetic aspects of the heterogeneous interaction and were conducted in the absence of significant water vapour such as encountered in the troposphere. The quantitative results on the uptake coefficients have therefore to be viewed as lower limits to the uptake coefficients valid for high relative humidities present in the atmosphere. It also became clear in the course of these studies that most heterogeneous reaction systems are complex in nature and do not merely correspond to elementary reactions or simple unimolecular rate laws. Therefore caution has to be applied in every case where extrapolations of laboratory results are required.

Salt samples with a varying degree of internal void and grain size were studied in detail [52], including powders with large a but poorly characterised external surface, monodisperse grains with a large and well characterised external surface, micron thick coatings corresponding to the geometrical surface and salt IR windows (single crystal). The uptake coefficients of HNO_3 and $ClONO_2$ were insensitive to the internal surface, but N_2O_5 was strongly dependent on it (in accord with a recently developed theory [59]). The uptake coefficients of NO_2 on NaCl(s) and NaBr(s) were found to be very low, about 10^{-7} [52], those of N_2O_5 were about three orders of magnitude higher [54], those of HNO_3 were equal to 3×10^{-2} [53], and those of $ClONO_2$ were by about an order of magnitude higher again, approaching unity [55] similar to the uptake of HNO_3 on $CaCO_3$ [56]. Uptake coefficients on bromide often were larger than those on chloride. This might indicate an influence of the halide on the reactivity. Amorphous carbon was found to be an efficient adsorber and a reactive reducing agent for NO_2 [57] and HNO_3 [58], releasing NO in the former and NO_2 in the latter case. In addition, HONO formation was observed when NO_2 interacted with amorphous carbon in the presence of residual moisture.

The uptake coefficients and products of NO_x species on salts and amorphous carbon are summarised in Table 2 and compared with other data in the literature.

No nitrosyl halides from reactions of NO_2 were detected in HALIPP in contrast to the earlier studies by Finlayson-Pitts and co-workers [38, 40]. In another recent study, a fairly high uptake coefficient of 6×10^{-5} was obtained for N_2O_4 (assuming that this is the reacting species) using diffuse reflectance and FTIR [60]. In contrast, Rossi and co-workers [52] found the hydrolysis products NOCl (HCl and HONO) to be formed from NaCl. A similar observation was made on partly and fully deliquescent NaCl aerosol by Zetzsch and co-workers [47, 48, 51], and it may be questioned whether the formation of NOCl can occur in the tropospheric environment, where humidity levels are mostly above 40 %. NOBr may be even more susceptible to hydrolysis and other loss reactions, so that it has not been detected in spite of the measurable uptake of NO_2 on NaBr.

The uptake of N_2O_5 by NaCl leads to $ClNO_2$ as the main product from dry and deliquescent NaCl. The uptake coefficients on dry NaCl and NaBr salts are much lower than on deliquescent NaCl aerosol, although an earlier measurement [61] implied that the behaviour of liquid and solid phase might differ much less. Another recent work [62] confirms that the uptake coefficient for NaCl(s) is more than an order of magnitude lower than that for NaCl(l), and the higher uptake coefficient [46, 47] in the second case was confirmed by the droplet train technique [63]. From the interaction of N_2O_5 with NaBr, Br_2 and HONO (due to traces of moisture) were observed [54] instead of the expected $BrNO_2$ owing to a rapid secondary reaction with NaBr.

The uptake of $ClONO_2$ on halides leads to the release of photolabile halogens and interhalogens (Cl_2, Br_2 and BrCl) [55]. This is an interesting feature of $ClONO_2$ (although its role in tropospheric chemistry is not established). Uptake probabilities of $ClONO_2$ on NaCl have also been determined in a flow reactor between 220 and 300 K [64]. After considering the total external surface area, an uptake probability of $\gamma_{obs} = (4.6 \pm 3.0) \times 10^{-3}$ was obtained at 296 K from both, the decay rates of $ClONO_2$ and the growth rate of Cl_2 (much lower than those obtained within HALIPP [55] on both salts). Traces of water did not alter the uptake probability, but traces of HOCl were observed as an additional product. A discrepancy of close to a factor of 100 exists between the uptake coefficients reported by Timonen *et al.* [64] and by Caloz *et al.* [55]. This is due to a correction by Caloz *et al.* [55] of the raw data given by the first group of authors. The correction is based on the pore diffusion model [64]. It has been shown conclusively in [55] that this correction does not apply to heterogeneous interaction of "sticky" molecules such as HNO_3 and $ClONO_2$, which brings the two data sets closer to agreement. The remainder of the discrepancy may be due to the absence of a correction for radial diffusion in the flow of Timonen *et al.* [64].

The uptake of HNO_3 on a single crystal of NaCl has recently been studied in an ultra high vacuum reactor by X-ray photoelectron spectroscopy of the surface [65], leading to an uptake coefficient almost two orders of magnitude lower than those obtained within HALIPP for NaCl and NaBr [53] where results for the uptake of HNO_3 on grains and on single crystals were identical. To what extent surface saturation by the trace gas or accumulating $NaNO_3$ may influence the uptake probability needs further investigation.

Table 2: Uptake coefficients, γ_{obs}, and products of reactions of NO_2, N_2O_5, HNO_3, and $ClONO_2$ determined on powders of NaCl, NaBr, $CaCO_3$ and C_{amorph}, single crystals of NaCl, coatings of solid and dissolved NaCl, on humidified, deliquescent NaCl aerosol and on NaCl solution at room temperature.

Trace Gas	Substrate	γ_{obs}	Products and Yields	Technique*	Ref.
NO_2	NaCl(s)	$< 10^{-7}$	HCl and HONO observed as products	KR	52
"	NaCl(s)	6×10^{-5}	uptake of N_2O_4; second order in NO_2	FR	60
"	NaCl(l) [a]	$< 10^{-5}$	HONO detected via OH from photolysis	AC	48
"	NaBr(s)	$\approx 2 \times 10^{-7}$	No products observed	KR	52
"	amorph. C(s)	4.0×10^{-2}	NO yield varies with soot source	KR	57
N_2O_5	NaCl(s)	5.0×10^{-4}	$ClNO_2$ main product, 60 % yield	KR	54
"	NaCl(s)	$> 2.5 \times 10^{-3}$	$ClNO_2$ main product	SR	61
"	NaCl(s)	0.1×10^{-2}	NO_3^- uptake by NaCl coating	FR	62
"	NaCl(l) [b]	3.2×10^{-2}	$ClNO_2$ main product, ≈ 66 % yield	AC	47
"	NaCl(l) [c]	3.2×10^{-2}	$ClNO_2$ main product, ≈ 57 % yield	AC	49
"	NaCl(l)	1.5×10^{-2}	NO_3^- uptake into solution	FR	62
"	NaCl(l) [d]	1.4×10^{-2}	NO_3^- uptake into solution	DT	63
"	NaBr(s)	4.0×10^{-3}	Br_2 and HONO products with 35 % yield	KR	54
$ClONO_2$	NaCl(s)	2.0×10^{-1}	Cl_2 produced with near 100 % yield	KR	55
"	NaCl(s)	4.6×10^{-3}	Cl_2 produced with high yield,	FR	64
"	NaCl(s) [e]	4.1×10^{-3}	Cl_2 and traces of HOCl	FR	64
"	NaBr(s)	3.0×10^{-1}	Br_2, BrCl and Cl_2 observed	KR	55
HNO_3	NaCl(s)	3.0×10^{-2}	HCl produced with 100 % yield	KR	53
"	NaCl(s)	4.0×10^{-4}	Cl^- consumed and NO_3^- formed on surface	SR	65
"	NaBr(s)	3.0×10^{-2}	HBr produced with 100 % yield	KR	53
"	$CaCO_3$(s)	1.0×10^{-1}	H_2O and CO_2 (100 % yield) produced	KR	56
"	amorph. C(s)	$> 1.0 \times 10^{-2}$	NO_2 yield on the order of 1 %	KR	58

* Technique: KR: Knudsen cell reactor, AC: Aerosol chamber, FR: Flow reactor, SR: Static salt reactor, DT: Droplet train

[a] NaCl aerosol (at 50 % r.h., possibly partly deliquescent)

[b] NaCl aerosol (above 75 % r.h. fully deliquescent)

[c] NaCl aerosol with a mole fraction of 0.2 Na_2SO_4 (above 75 % r.h., fully deliquescent)

[d] NaCl solution (1 mol/l)

[e] NaCl treated with water vapour

2.7.7 Tropospheric implications

The formation of $ClNO_2$ from NaCl has been estimated [46] to lead to levels of Cl atoms up to 4.3×10^3 cm^{-3} in the tropospheric marine boundary layer at 60° Northern Latitude. Typical wind speeds for January have been assumed together with a level of 0.5 ppb NO_2. This can be updated to [Cl] = 1×10^3 cm^{-3} at a typical level of 0.1 ppb NO_2. Such a source of atomic Cl may double the source strength arising from the well known reaction of OH with HCl [36] and other (still highly speculative) heterogeneous sources of Cl atoms in the troposphere [43, 44], which possibly are initiated by the reaction of O_3 with Cl^-. Some evidence for the existence of photoactive precursors of atomic Cl in the marine boundary layer comes from recent field measurements [66, 67]. The presence of chlorine atoms has been inferred from measurements of nonmethane hydrocarbons and their consumption pattern in the Arctic during polar sunrise [68]. The authors found also evidence (from the behaviour of acetylene and ethene) for the presence of Br atoms, which presumably are responsible for a sudden springtime ozone depletion in the tropospheric boundary layer of the Arctic. The occurence of atomic Br can be explained by the formation of $BrNO_2$ from the reaction of N_2O_5 with the relatively high Br^- content of the arctic aerosol [33, 69]. Halogen activation in the marine boundary layer has been considered in a recent box model [70], where a residence time of the aerosol particles of two days was assumed and the uptake of NO_3 leads to an oxidation of Br^- in the aerosol and releases Br_2 during night-time. Photolysis of Br_2 and the subsequent reaction cycle of atomic Br, BrO and HOBr (involving HO_2) leads to a significant destruction of maritime O_3. Other mechanisms, involving sulfur species, may similarly activate the halogens [71].

2.7.8 References

1. D. R. Shryer, *Heterogeneous Atmospheric Chemistry*, Geophysical Monograph 26, American Geophysical Union, Washington DC 1982.
2. H. Güsten, in: W. Jaeschke (ed) *Chemistry of Multiphase Atmospheric Systems*, NATO ASI Ser. G 6, Springer Verlag, Berlin Heidelberg 1986, pp. 567–594.
3. W. Behnke, W. Holländer, W. Koch, F. Nolting, C. Zetzsch, *Atmos. Environ.* **6** (1988) 1113–1120.
4. W. Behnke, F. Nolting, C. Zetzsch, *J. Aerosol Sci.* **18** (1987) 65–71.
5. C. Zetzsch, in: R. Zellner (ed) *Formation, Distribution and Chemical Transformation of Air Pollutants*, Dechema Monograph **104** VCH Verlagsges. Weinheim, Germany 1987, pp. 187–212.
6. W. Behnke, F. Nolting and C. Zetzsch, *J. Aerosol Sci.* **18** (1987) 849–852.
7. P. Foster, I. Denis, V. Jacob, M. Laffond, in: *EUROTRC Annual Report Part 6*, ISS Garmisch-Partenkirchen 1992, pp. 81–85.
8. P. Foster, V. Jacob, M. Laffond, P. Baussand, I. Dennis, C. Ferronato, *EUROTRAC Annual Report Part 6*, EUROTRAC ISS Garmisch-Partenkirchen 1992, pp. 95–100.
9. C. Guillard, H. Delprat, C. Hoang-Van, P. Pichat, *J. Atmos. Chem.* **16** (1993) 47–59.
10. V. Jacobs, Etude de la dégradation d'hydrocarbones aromatiques polycyclics légers an présence d'aerosols de nature diverses, Thesis, Université Joseph Fourier Grenoble I (1989).

11. R. W. Mathews, *J. Phys. Chem.* **91** (1987) 3328–3333.
12. C. Vinckier, N. Van Hoof, in: P. M. Borrell, P. Borrell, T. Cvitas, W. Seiler (eds), *Proc. EUROTRAC Symp. 92*, SPB Academic Publishing bv, The Hague 1993, pp. 652–654.
13. C. Vinckier, N. Van Hoof, *EUROTRAC Annual Report* 6, EUROTRAC ISS Garmisch-Partenkirchen 1992, pp. 118–120.
14. C. Vinckier, N. Van Hoof, F. Compernolle, S. Ashty, in: P. M. Borrell, P. Borrell, T. Cvitas, W. Seiler (eds), *Proc. EUROTRAC Symp. 94*, SPB Academic. Publishing bv, The Hague 1994, pp.120–123.
15. S. H. Zang, M. Shaw, J. H. Seinfeld, R. C. Flagan, *J. Geophys. Res.* **97 D** (1992) 20717–20729.
16. P. Pichat, C. Guillard, C. Hoang–Van, H. Delprat, D. Mas, F. Marme, T. Bouvier, B. Servajean, this volume.
17. V. A. Isidorov, E. M. Klokova, E. B. Prilepski, in: G. Angeletti and G. Restelli (eds) *Physico–Chemical Behaviour of Atmospheric Pollutants, Proc. 6th European Symp.*, EC, Luxembourg (1994) 41–46.
18. P. Foster, I. Dennis, V. Jacob, *EUROTRAC Annual Report Part 6,* EUROTRAC ISS Garmisch-Partenkirchen 1993, pp.87–92.
19. P. Pichat, C. Guillard, H. Delprat, C. Hoang-Van, *EUROTRAC Annual Report Part 6,* EUROTRAC ISS Garmisch-Partenkirchen 1991, pp. 80–85.
20. P. Pichat, C. Guillard, F. Marme, C. Hoang-Van, *EUROTRAC Annual Report Part* 6 EUROTRAC ISS Garmisch-Partenkirchen 1992, pp. 104–109.
21. P. Pichat, C. Guillard, C. Hoang-Van, J. Distier, D. Mas, B. Servajean, *EUROTRAC Annual Report Part 6,* EUROTRAC ISS Garmisch-Partenkirchen 1993, pp. 99–105.
22. M. R. Hoffman, S. T. Martin, W. Cloi, D. W. Bahneman, *Chem. Rev.* **95** (1995) 69–96.
23. J. C. Petit, J. Fatnassi, *EUROTRAC Annual Report Part 6,* EUROTRAC ISS Garmisch-Partenkirchen 1991, pp. 74–78.
24. J. C. Petit, J. Fatnassi, *EUROTRAC Annual Report Part 6,* EUROTRAC ISS Garmisch-Partenkirchen 1992, pp. 101–103.
25. J. C. Petit, J. Thlili, *EUROTRAC Annual Report Part 6,* EUROTRAC ISS Garmisch-Partenkirchen 1993, pp. 93–98.
26. T. E. Graedel, C. J. Weschler, *Rev. Geophys. Space Phys.* **19** (1981) 505–539.
27. P. Warneck, *Chemistry of the Natural Atmosphere*, Academic Press, San Diego(1988)
28. G. M. Sverdrup, C. W. Spicer, G. F. Ward, *Int. J. Chem. Kinet.* **19** (1987) 191–205.
29. R. P. Wayne, I. Barnes, P. Biggs, J. P. Burrows, C. E: Canosa–Mas, J. Hjorth, G. Le Bras, G. K. Moortgat, D. Perner, G. Poulet, G. Restelli, H. Sidebottom, *Atmos. Environ.* **25A** (1991) 1–203.
30. U. Platt, D. Perner, J. Schroeder, C. Kessler, A. Toennissen, *J. Geophys. Res.* **86** (1981) 11965–11970.
31. W. L. Chameides, *J. Geophys. Res.* **91** (1986) 5331–5337.
32. U. Platt, D. Perner, A. M. Winer, H. Biermann, R. Atkinson, J. N. Pitts, Jr., *Env. Sci. Technol.* **18** (1984) 365–369.
33. B. J. Finlayson-Pitts, F. E. Livingston, H. N. Berko, *Nature* **343** (1990) 622–625.
34. D. R. Hitchcock, L. L. Spiller, W. E. Wilson, *Atmos. Environ.* **14** (1980) 165–182.
35. C. S. Martens, J. J. Weosolwski, R. C. Harris, R. Kaifer, *J. Geophys. Res.* **78** (1973) 8778–8792.
36. H. B. Singh, J.F. Kasting, *Atmos. Chem.* **7** (1988) 261–285.

37. W. H. Schroeder, P. Urone, *Environ. Sci. Technol.* **8** (1974) 756–758.
38. B. J. Finlayson-Pitts, *Nature*, **306** (1983) 676–677.
39. B. J. Finlayson-Pitts, M. J.Ezell, J. N. Pitts, Jr., *Nature* **337** (1989) 241–244.
40. B. J. Finlayson-Pitts, S. N. Johnson, *Atmos. Environ.* **22** (1988) 1107–1112.
41. W. Behnke, C. Zetzsch, *J. Aerosol Sci.* **S21** (1990) 229–232.
42. C. Zetzsch, M. Elend, H.–U. Krüger, P. Kühn, V. Scheer, W. Behnke, in: K. H. Becker (ed) *Air Pollution Research Report 33*, CEC, Guyot, Brussels (1991) 167–170.
43. C. Zetzsch, W. Behnke, in: K.H. Becker (eds) *The Tropospheric Chemistry in the Polar Regions, NATO ASI Ser.*17, H. Niki, Springer, Berlin/Heidelberg (1993) 291–305.
44. W. Behnke, V. Scheer, C. Zetzsch, in: *Proc. Internat. Conf. on Naturally–Produced Organohalogens,* TNO, Delft ,in press.
45. W. Behnke, H.–U. Krüger, V. Scheer, C. Zetzsch, *J. Aerosol Sci.* **S22** (1991) 609–612.
46. C. Zetzsch, W. Behnke, *Ber. Bunsenges. Phys. Chem.* **96** (1992) 488–493.
47. W. Behnke, H.–U. Krüger, V. Scheer, C. Zetzsch, *J. Aerosol Sci.* **S23** (1992) 933–936.
48. W. Behnke, H.–U. Krüger, V. Scheer, C. Zetzsch, in: P. M. Borrell, P. Borrell, T. Cvitas, W. Seiler (eds), *Proc. EUROTRAC Symp. 92,* SPB Acad. Publishing, The Hague 1993, pp. 565–570.
49. W. Behnke, H.-U. Krüger, V. Scheer, C. Zetzsch, *J. Aerosol Sci.* **S24** (1993) 115–116.
50. W. Behnke, Ch. George, V. Scheer, C. Zetzsch, in: K. H. Becker (ed) *Air Pollution Research Report 54*, EC, Brussels (1994) 489–493.
51. W. Behnke, M. Elend, A. Frenzel, Ch. George, H.–U. Krüger, V. Scheer, C. Zetzsch, this volume
52. M. J. Rossi, F. F.Fenter, K. Tabor, F. Caloz, L. Gutzwiller, this volume
53. F. F. Fenter, F. Caloz, M. J. Rossi, *J. Phys. Chem.* **98** (1994) 9801–9810.
54. F. F. Fenter, F. Caloz, M. J. Rossi, in: P. M. Borrell, P. Borrell, T. Cvitas, W. Seiler (eds), *Proc. EUROTRAC Symp. 94,* SPB Academic. Publishing bv, The Hague 1994, pp. 1069–1073.
55. F. Caloz, F. F. Fenter, M. J. Rossi, *J. Phys. Chem.*, submitted.
56. F. F. Fenter, F. Caloz, M. J. Rossi, *Atmos. Environ.*, submitted
57. K. Tabor, L. Gutzwiller, M. J. Rossi. *Geophys. Res. Lett.* **20** (1993) 1431–1434.
58. K. Tabor, L. Gutzwiller, M. J. Rossi, *J. Phys. Chem.* **98** (1994) 6172–6186.
59. L. F. Keyser, S. B. Moore, M. T. Leu. *J. Phys. Chem.* **95** (1991) 5496–5502.
60. R. Vogt, B.J. Finlayson-Pitts, *J. Phys. Chem* .**98** (1994) 3747–3755.
61. F. E. Livingston, B.J. Finlayson-Pitts, *Geophys. Res. Lett.* **18** (1991) 17–20.
62. I. M. Msibi, Y.Li, J. P. Shi, R. M. Harrison, *J. Atmos. Chem.* **18** (1994) 291–300.
63. Ch. George, J. L. Ponche, Ph. Mirabel W. Behnke, V. Scheer, C. Zetzsch, *J. Phys. Chem.* **98** (1994) 8780–8784.
64. R. S. Timonen, L. T. Chu, M.-T. Leu, L. F. Keyser, *J. Phys. Chem.* **98** (1994) 9509–9517.
65. J. M. Laux, J. C. Hemminger, B. J. Finlayson–Pitts, *Geophys. Res. Lett.* **21** (1994) 1623–1626.
66. W. C. Keene, J. R. Maben, A. A. P. Pszenny, J. N. Galloway, *Environ. Sci. Technol.* **27** (1993) 866–874.
67. A. A. P. Pszenny, W. C. Keene, D. J. Jacob, S. Fan, J. R. Maben, M. P. Zetwo, M. Springer–Young, J. N. Galloway, *Geophys. Res. Lett.* **20** (1993) 699–702.
68. B. T. Jobson, H. Hiki, Y. Yokouchi, J. Bottenheim, F. Hopper, R. Leaitch, *J. Geophys. Res.* **99** (1994) 25355–25368.
69. W. Behnke, V. Scheer, C. Zetzsch, *J. Aerosol Sci.* **S25** (1994) 277–278.

70. R. Sander, P. J. Crutzen, *J. Geophys. Res,* submitted
71. M. Mozurkewich, *J. Geophys. Res.* **99** (1994) 330–335.
72. W. Behnke, Ch. George, V. Scheer, C. Zetzsch, in: P. M. Borrell, P. Borrell, T. Cvitas, W. Seiler (eds), *Proc. EUROTRAC Symp. 92,* SPB Academic. Publishing bv, The Hague 1993, pp. 1064–1068.
73. Ch. George, W. Behnke, V. Scheer, C. Zetzsch, L. Magi, J. L. Ponche, Ph. Mirabel *Geophys. Res. Lett.* (1995) in press.
74. D. W. Johnson, D. W. Margerum, *Inorg. Chem.* **30** (1991) 4845–4851.

2.8 General Conclusions

Eight years of HALIPP research have witnessed significant advances in the knowledge and understanding of heterogeneous and liquid phase chemical processes in the atmosphere. This section provides a general assessment of the progress achieved and gives some recommendations for the future.

Many of the uncertainties that had existed with regard to sulfur(IV) oxidation mechanisms in the aqueous phase have been removed. In particular, the mechanism of the chain oxidation by sulfuroxy radicals has been thoroughly studied and rate coefficients for the reactions involved have been specified. The role of iron as a catalyst in such mechanisms has been fully delineated. Manganese and cobalt also exhibit autocatalytic features, and a similar mechanism probably applies. The generation of radicals from photolabile precursors has been quantified, and the dominant role of iron complexes as photochemical sources of radicals has been established. Ions of transition metals, especially copper, are scavengers for the HO_2-radical and its conjugate anion, O_2^-, which are converted to H_2O_2. A common mechanism has been established for the reactions of S(IV) with H_2O_2 and other peroxides, specifically peroxomonosulfate, HSO_5^-, which is an important intermediate in the chain oxidation of hydrogensulfite. The oxidation of formaldehyde by OH radicals in the presence of O_2-forms exclusively formic acid and HO_2 as products. Rate coefficients for the reactions of OH and SO_4^--radicals with hydroxymethanesulfonate and major parts of the reaction mechanisms have been determined. One of the consequences of sulfur(IV) oxidation under atmospheric conditions is that SO_4^--radicals interact with chloride to produce chlorine atoms and Cl_2^--radicals. Their reactions have not yet been sufficiently explored to establish the fraction that reacts with water to regenerate OH radicals. The latter reaction determines to some extent the chain length of the radical chain oxidation of hydrogensulfite in clouds. An assessment of its significance in comparison to other oxidation pathways requires further study.

A number of reactions involving the NO_3-radical, which can enter cloud drops from the gas phase, have been studied. Most reactions with organic compounds proceed by hydrogen abstraction, but the rate coefficients have low values. The reactions with sulfur(IV) species and chloride are rapid. Reactions with transition metal ions have not yet been explored, and this remains a task for the future.

Several new techniques for the study of gas-liquid mass transfer have been developed and applied to determine mass accommodation coefficients for a greater number of trace gases participating in the chemistry of clouds. Some of the techniques will be suitable to study the uptake of radicals by the aqueous phase. However, this type of investigation has not been performed within HALIPP. It is hoped that appropriate measurements can be carried out in the future. Some trace gases, *e.g.* NO_2, may react on the surface of water drops without becoming fully dissolved. Such processes will need greater attention in the future, and the wetted-wall flow reactor might provide a suitable experimental arrangement for this purpose.

Regarding the investigation of heterogeneous reactions on the surface of materials present in aerosol particles it was recognised from the outset that they required a largely exploratory approach. Therefore, few quantitative results were expected. Various studies have shown that minerals can increase the degradation of adsorbed organic compounds under illumination, and that the reaction products often are different from those evolving from gas phase reactions. This result can have significant consequences, especially when product compounds turn out to be more toxic than reagent materials. Another critical result obtained concerns the formation of Lewis acid centres under illumination: Lewis acid centres turned out to be relatively unimportant as active sites for photocatalysis. The presence of water on minerals associated with atmospheric particles can have a significant influence on reaction rates, particularly with regard to dissolution of transition metals by the formation of organic complexes. This feature should be given more attention in the future. Unfortunately, the concepts advanced so far for surface reactions involving mineral components of the atmospheric aerosol have not been substantiated in the laboratory. Acceptable reaction mechanisms and associated rate data are still unavailable. The persistent lack of understanding regarding these processes thus precludes any critical assessment of their impact on atmospheric chemistry. This contrasts with the progress achieved in the field of reactive interactions between trace gases and components of sea salt.

The reaction of nitrogen pentoxide, N_2O_5, with deliquescent sodium chloride particles has been shown to produce $ClNO_2$. A chemical mechanism has been advanced to explain this unexpected result. $ClNO_2$ is a volatile product, which in the gas phase undergoes photodissoiation resulting in the liberation of chlorine atoms. The consequences to atmospheric chemistry are twofold: NO_2 is recovered rather than converted to nitrate, and the process represents a source of chlorine atoms additional to that arising from the reaction of OH radicals with HCl. The displacement of chloride and bromide from dry sodium halides by gaseous HNO_3, NO_2, $ClONO_2$, and N_2O_5 have been studied, and the associated uptake coefficients have been determined. The reactions cause an exchange of the anion in the first case and the release of photolabile gaseous products in the other three cases. Reactions of ozone or radicals with materials contained in sea salt, in the dry or wet state, will have to be explored in the future.

This report demonstrates that most of the original aims of HALIPP have been reached. Remarkable progress has been made in advancing the knowledge and understanding of the aqueous chemistry and gas-liquid mass transfer in clouds to the point where the new data can be successfully incorporated in simulation models. Good progress has also been made with respect to the interaction of gaseous nitrogen oxides with sodium halides present in sea salt particles, whereas the complexities experienced with surface reactions on minerals associated with atmospheric aerosol particles have defied explanations in terms of mechanistic concepts. Some new knowledge has been obtained nevertheless.

On the whole, HALIPP has been a successful project. It should be emphasised that the achievements of HALIPP are based on several critical prerequisites. One is the diversity of scientific expertise of the research groups assembled within HALIPP, another the implementation of experimental techniques that were available or were developed during the programme, a third is the generous exchange of know-how between the contributing scientists, and a fourth the close collaboration, which was achieved early in the programme. All these factors were essential in making HALIPP a success.

Individual Reports from HALIPP Contributors

Chapter 3

Sulfur(IV) Oxidation

3.1 Kinetic Investigation of the Stability of Peroxonitric Acid and of its Reaction with Sulfur(IV) in Aqueous Solution

P. Amels, H. Elias, U. Götz, U. Steingens and K. J. Wannowius

Institut für Anorganische Chemie, Technische Hochschule Darmstadt, Petersenstraße 18, D-64287 Darmstadt

Summary

The kinetics of the decomposition of aqueous peroxonitric acid, O_2NOOH, have been studied as a function of pH in the range pH = 0–9. In addition to literature studies on the stability of peroxonitric acid solutions two more reactions have been measured. They are assigned to the formation and the decay of a short-lived intermediate. The composition of the intermediate is discussed. A peroxo-species containing two nitrogens in different oxidation states (+III and +V) is favoured.

The kinetics of the oxidation of S(IV) by peroxonitric acid as well as by the intermediate have been studied as a function of pH in the range pH = 0–8.2 (T = 285 K, I = 1.0 M $NaClO_4$).

Aims of the research

The research aims at the detailed understanding of the oxidation mechanisms of S(IV) in the aqueous phase. Cloud models will be reliable only when the chemical reactions occurring are mechanistically understood. The work of the group concentrates on the laboratory investigation of those reactions which are presumably of importance for the liquid phase in the atmosphere. The studies were extended to aqueous redox processes involving nitrogen species such as HNO_2 and HNO_4.

Activities during the year

The kinetics of the decomposition of peroxonitric acid as well as of its reaction with sulfur(IV) have been studied in buffered aqueous solutions in the pH range 0–12. Some more light could be shed on the reactions occurring in aqueous solutions of peroxonitric acid, but they are still not fully understood.

Principal experimental results

***Synthesis of* HNO_4**

Stock solutions of peroxonitric acid were obtained by reacting solid $NO_2^+BF_4^-$ with excess H_2O_2 (99 %) according to a literature procedure [1, 2]. The reaction mixture could be stored at –20° C over weeks. Dilute solutions of peroxonitric acid were obtained by bubbling nitrogen through the stock solution and passing the gas stream into a 0.1M $HClO_4$ solution.

***Analysis of* HNO_4**

The concentration of peroxonitric acid was determined spectrophotometrically from the yield in Br_3^- in its reaction with excess bromide ions according to a literature procedure [2]. The molar absorptivity of Br_3^- was determined by studying the kinetics of its formation from BrO_3^- and Br^- with a rapid-scan stopped-flow spectrophotometer. Because the formation of the peak at 266 nm is followed by a decay, ε_{max} for Br_3^- could be underestimated.
From the kinetic analysis $\varepsilon_{max} = (2.3 \pm 0.11)\cdot 10^4$ M^{-1} cm^{-1} has been determined. In the literature deviating values have been reported [1, 2].

***Decomposition of* HNO_4**

The decomposition of peroxonitric acid is strongly pH dependent. At pH = 1 and T = 285 K it decomposes with a half-time of approx. $t_{½}$ = 15 min. As final products nitrite (pH = 12.4) and a mixture of nitrite and nitrate (pH = 4.8) were detected by ion chromatography. From the yield of nitrite as a function of pH. Lammel *et al.* [2] estimated the pK_a = 5.0 for HNO_4. Loegager and Sehested [3] and Appleman *et al.* [4] found that during the HNO_4 decay per mole of HNO_4 1 mol [3] (0.6 mol [4]) of hydrogen peroxide (present in the solutions due to the synthesis) is consumed at low pH. In that pH range reaction (1) is fast and may be responsible for the consumption of H_2O_2 and for the major part of the NO_3^- production [3,4].

$$HNO_2 + H_2O_2 \longrightarrow HOONO + H_2O \longrightarrow NO_3^- + H^+ + H_2O \quad (1)$$

In a preliminary study of the rate of the oxidation of HNO_2 by HNO_4 it is found that at pH = 0 the rate is surprisingly low, *i.e.* an interference of the decomposition of HNO_4 with its decomposition product NO_2^- or HNO_2, respectively, should not be of importance. This finding is in line with the literature [3].
According to the literature [1–4] reactions (2)–(4) are believed to describe the decomposition of peroxonitric acid.

$$HNO_4 \longrightarrow NO_2 + HO_2 \quad (2)$$
$$HNO_4 \longrightarrow HNO_2 + O_2 \quad (3)$$
$$NO_4^- \longrightarrow NO_2^- + O_2 \quad (4)$$

On the one hand Loegager and Sehested [3] questioned, however, the formation of radicals assumed by Lammel *et al.* [2] according to (2). On the other hand Appleman *et al.* [4] doubt the significance of reaction (3).
In the present study pH-jump experiments were done with the dilute HNO_4 solutions of pH 1 with appropriate buffer solutions to follow the rate of decomposition as a function of pH with conventional UV spectroscopy (pH < 3).
With increasing pH the rate increases. Above pH 3 the experiments were performed in a stopped-flow apparatus equipped with a multi-wavelength UV diode array spectrophotometer. Up to pH 5 three reaction steps (I, II, and III) were observed. Above pH 5 a single step (III) is found. Here the first two steps were too fast for the stopped-flow technique, or have reaction amplitudes which are small compared to step III.

Above pH 5 the anion of peroxonitric acid ($pK_a = 5.0$ [2]) should instantaneously be formed in the pH-jump experiment. The first spectrum measured immediately after mixing (delay: 5 ms) shows a distinct absorption peak at 285 nm. Loegager *et al.* [3] observed the same spectrum. They made an assignment to the anion NO_4^-. If the pH-jump experiment is done in the presence of increasing sulfur(IV) held in excess over peroxonitric acid, the absorbance of the first spectrum after the mixing time drops (see Fig. 1 a–d).

The anion NO_4^- turned out to be a highly reactive species. In the absence of S(IV) it reacts within less than 5 ms to an intermediate. Presumably "dimerisation" occurs with subsequent loss of a molecule of oxygen (5). In the present report this fundamental assumption is made, because an intermediate in the NO_4^- decay other than a "dimer" could not be imagined.

$$2\,NO_4^- \longrightarrow [N_2O_8^{2-}] \longrightarrow N_2O_6^{2-} + O_2 \quad (5)$$

Formally the nitrogens in the species $N_2O_6^{2-}$ are in the oxidation state IV. It seems to be more reasonable, however, that only one of the two nitrogens has been reduced to N(III) in the redox-reaction (5). In the presence of high concentrations of sulfite, reaction (6) can compete with (5), which reduces the yield of the intermediate observed immediately after mixing (see Fig. 1 a–d).

$$NO_4^- + SO_3^{2-} \longrightarrow NO_3^- + SO_4^{2-} \tag{6}$$

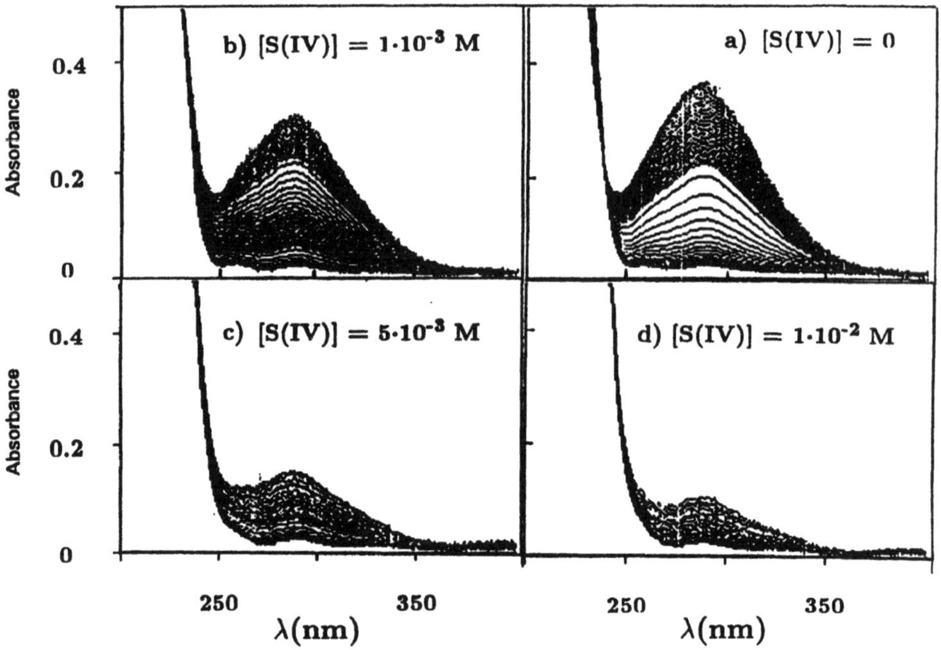

Fig. 1: Spectra obtained in the pH junp experiments (pH = 1 ⟶ 7.8)in the absence and the presence of S(IV) (T = 285 K; I = 1.0 M $NaClO_4$; [S(IV)] = 0 (a), 0.001 M (b), 0.005 M (c), 0.01 M (d)). The first (upper) spectrum is recorded after 5ms. Time intervals: a): 25 spectra at 3.5ms, 25 spectra at 35ms, 25 spectra at 350ms; b): 25 spectra at 3.5ms, 25 spectra at 24.5ms, 25 spectra at 175ms; c) and d): 25 spectra at 3.5ms, 25 spectra at 14ms, 25 spectra at 70ms.

If the first spectrum, which is measured after the mixing time, would be that of the anion of peroxonitric acid (as claimed by Loegager *et al.* [3]), and if reaction (6) is fast as compared to the stopped-flow time scale, the NO_4^- ions should have reacted completely with excess sulfite. Fig. 1 a–d definitively contradicts this consideration. That means that the first spectrum recorded after the mixing time could not be that of the anion of peroxonitric acid, NO_4^-, as postulated by Loegager *et al.* [3].

The intermediate $N_2O_6^{2-}$ is still unstable, but more stable than NO_4^-. Its decay leads to nitrite and oxygen (7).

$$N_2O_6^{2-} \longrightarrow 2\,NO_2^- + O_2 \tag{7}$$

In the presence of sulfite the rate of disappearance of the intermediate is enhanced via reaction (8).

$$N_2O_6^{2-} + SO_3^{2-} \longrightarrow NO_2^- + NO_3^- + SO_4^{2-} \quad (8)$$

The systematic variation of pH in the pH-jump experiments performed with the acidic HNO_4 solution of pH = 1 gave the pH dependence of the three observed reaction steps. The absorbance/time data for the decay of peroxonitric acid were fitted to a function describing the superposition of a second order reaction step with two first order steps. For the evaluation of the second order rate constant k^I (step I) the initial concentration of HNO_4 is needed, which was determined before each kinetic run. The pH profile is shown in Fig. 2 for the second order rate constant k^I as well as for the first order constants k^{II} and k^{III}. The rate constants for step I and II increase linearly with increasing pH. For step III a stepwise increase in the rate constants (going through three plateaus k_a, k_b, and k_c) is found.

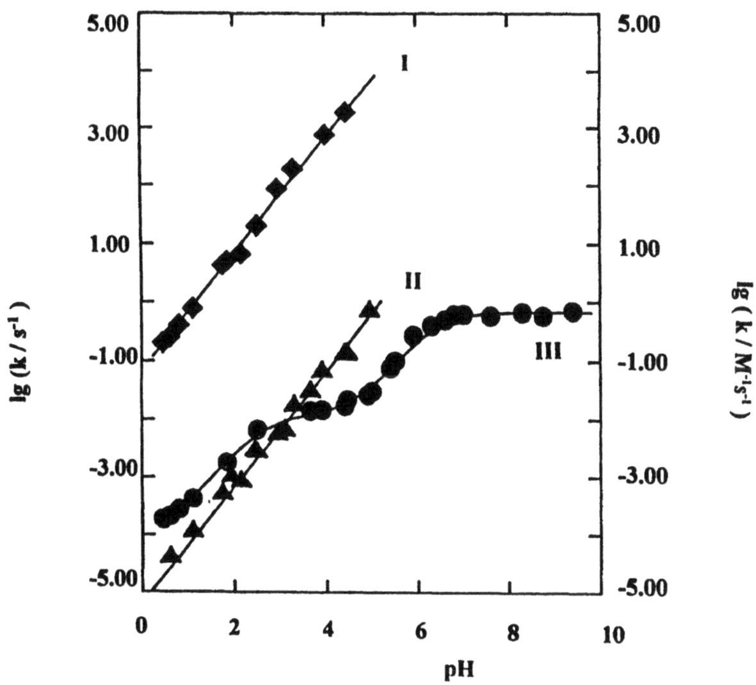

Fig. 2: pH profile for the second order rate constant k^I (step I; formation of the intermediate) as well as for the first order rate constants k^{II} for the decay of HNO_4 (step II), and k^{III} for the decay of the intermediate (step III). Conditions: $[HNO_4]_o = 2 \cdot 10^{-3}$ M; $T = 285$ K; $I = 1.0$ M $NaClO_4$.

The pH dependence of k^{III} (step III) is interpreted as being indicative for a protonation of a dibasic intermediate (9)–(10). The computer fit leads to the acid dissociation constants $pK_a(1) = 2.6 \pm 0.2$ and $pK_a(2) = 6.4 \pm 0.2$.

$$H_2N_2O_6 \rightleftharpoons HN_2O_6^- + H^+ \qquad (9)$$

$$HN_2O_6^- \rightleftharpoons N_2O_6^{2-} + H^+ \qquad (10)$$

Apparently, protonation increases the stability of the intermediate. The pH independant rate constants ($T = 285K$, $I = 1.0M$ $NaClO_4$) for the decay of its various protonated species are: $k_a <= 1 \cdot 10^{-4} s^{-1}$ ($H_2N_2O_6$);

$k_b = 0.018 \pm 0.002\ s^{-1}$ ($HN_2O_6^-$);

$k_c = 0.61 \pm 0.05\ s^{-1}$ ($N_2O_6^{2-}$).

Step I (described by the second order rate constant k^I) is assigned to the formation of the "dimer" from peroxonitric acid solutions of "high" concentrations, *i.e.* in the millimolar range. It could, however, not be determined, whether at pH < 5 the "dimer" is formed by (5) or by (11). The latter reaction is thought to be more plausible at pH < 5.

$$HNO_4 + NO_4^- \longrightarrow HN_2O_6^- + O_2 \qquad (11)$$

Step II is assigned to the known [2–4] decomposition of peroxonitric acid according to (4). Even small amounts of the extremely unstable anion NO_4^- being present due to (12) at pHs far below the pK_a of HNO_4 lead to a decomposition of HNO_4.

$$HNO_4 \rightleftharpoons NO_4^- + H^+ \qquad (12)$$

The above assignment is supported by the spectral information obtained with the diode array detection. Steps II and III lead to the formation of HNO_2 and/or NO_2^- ions (depending on the pH); step I yields an unknown UV absorption spectrum, which is also pH dependent. The product spectrum obtained in step I still contains the characteristics of the absorption expected for a peroxo-group.

Sulfur(IV) oxidation

With freshly prepared solutions of HNO_4 sulfur(IV) can be very rapidly converted to sulfate (13).

$$HNO_4 + HSO_3^- \longrightarrow NO_3^- + SO_4^{2-} + 2\,H^+ \qquad (13)$$

The products nitrate and sulfate have been identified by ion chromatography.
In the range pH < 3.5 the S(IV) oxidation by HNO_4 is found to proceed in a single step (A). Above pH 3.5 a second step (B) appears. With increasing pH its rate decreases with $[H^+]$ [2]. The decrease in rate of step B is accompanied by an increase of its spectral amplitudes. At the same time the amplitudes of step A vanish.

Step A is assigned to reaction (13). Its rate seems to be independent on pH in the range pH < 5 (14) (see Fig. 3). The variation of the concentration of the excess partner reveals, however, that the observed rate constant k_{obs} does not increase linearly with increasing concentration of S(IV) (15). The latter finding points to a more complex reaction mechanism.

rate $= k\cdot[HNO_4]\cdot[S(IV)]$ (14)

$k_{obs} = k\cdot[S(IV)] = k_1\cdot K\cdot[S(IV)] / (1+K\cdot[S(IV)])$ (15)

In terms of the concentration of bisulfite the rate law is given by (16). At $T = 285$ K and $I = 1.0$ M ($NaClO_4$) the rate constants $k_H = (1.08 \pm 0.15)\cdot 10^7\ M^{-2}s^{-1}$ and $k_P = (3.3 \pm 0.1)\cdot 10^5\ M^{-1}s^{-1}$ were determined.

rate $= (k_H[H^+] + k_P)\cdot[HSO_3^-]\cdot[HNO_4]$ (16)

In view of the reaction mechanism of related peroxo-compounds ROOH [5–7] the oxidation of HSO_3^- should be subject to general acid catalysis. The weak acid water is, however, a very powerful catalyst in the case of the strong peroxo-acid HNO_4. The influence of the buffers used in the experiments is of minor significance, only. The rate constants k_H and k_P (16) obtained for the system HNO_4/S(IV) are on the one hand in excellent agreement with the LFE relationship found for the systems ROOH/S(IV) which were previously studied (see Fig. 4). On the other hand the excellent fit of k_P may be taken as proof for the acid dissociation constant, which had been reported by Lammel *et al.* [2].

Step B is assumed to describe the reaction of sulfur(IV) with the intermediate in its various protonated forms. At pH > 7 the SO_3^{2-} ion and the $N_2O_6^{2-}$ ion are the dominant species of the reactants in solution. Eq.(8) seems to describe the main reaction at pH > 7 which could be followed with the stopped-flow technique (apart from the decay (7), k_{dec}), which is also observed. The rate of disappearance of the intermediate in the presence of sulfite is given by (17) pointing to a complex mechanism. Increasing concentrations of sulfite reduce, however, via (6) the yield of the intermediate (see Fig. 1 a–d).

$-d[IM]/dt = \{(k_{dec}+k_{SO3}(k'/k'')[SO_3^{2-}])[IM]\}/(1+(k'/k'')[SO_3^{2-}])$ (17)

A few runs have been done with dimethylsulfide (DMS) instead of SO_3^{2-} as reductant in the range pH > 7. Even high concentrations of DMS are not able to compete with the decay of NO_4^- given by (5). The oxidation of DMS by the intermediate is given by rate law (18). Sulfite turned out to be a better nucleophile than DMS: $k_{SO3}(k'/k'') > k_{DMS}$.

$-d[IM]/dt = (k_{dec} + k_{DMS}[DMS])\cdot[IM]$ (18)

Below pH 7 reactions (19)–(21) have to be considered in the system IM/S(IV).

$HN_2O_6^- + SO_3^{2-} \longrightarrow NO_2^- + NO_3^- + H^+ + SO_4^{2-}$ (19)

$HN_2O_6^- + HSO_3^- \longrightarrow NO_2^- + NO_3^- + 2H^+ + SO_4^{2-}$ (20)

$H_2N_2O_6 + HSO_3^- + H^+ \longrightarrow NO_2^- + NO_3^- + 3H^+ + SO_4^{2-}$ (21)

The dependence of k_{obs} on $[H^+]^2$, as observed for step B in the range pH < 5 (see Fig. 3), is thought to be indicative for the significance of reaction (21) below pH 5. It is however, not yet known which of the two reaction partners is protonated, or whether one of the protons is needed for an acid-catalysed reaction channel of the HSO_3^- ion.

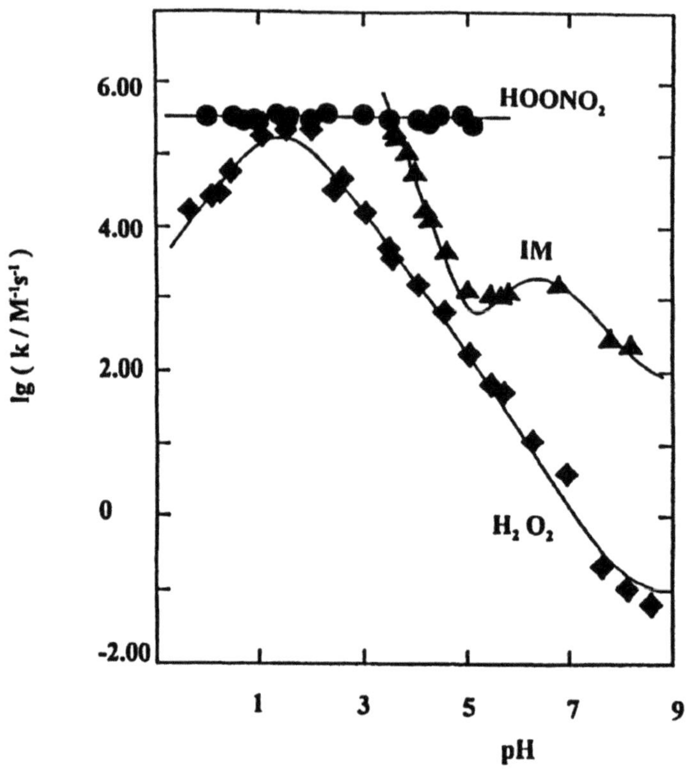

Fig. 3: pH profile for the second order rate constant (corrected for the contribution of buffers if necessary) in the systems HNO_4/S(IV) (circles), intermediate I M/S(IV) (triangles), and H_2O_2/S(IV) (diamonds) (T = 285 K; I = 1.0 M $NaClO_4$ (HNO_4 and I M), I = 0.5 M (H_2O_2)).

Reaction (20) is assumed to be of minor significance only. The "plateau" in the range pH = 5–6 is ascribed to reaction (19). This assignment is uncertain, however, because in the range pH > 6 only a few experimental data points are available. For comparison the system HSO_5^-/SO_3^{2-} [7] may be taken. At neutral pH a rate maximum is observed in the pH profile, which stems from the fact that the SO_3^{2-} ion is a better nucleophile than the HSO_3^- ion. In view of this observation reaction (19) is believed to contribute to the S(IV) oxidation in that pH range.

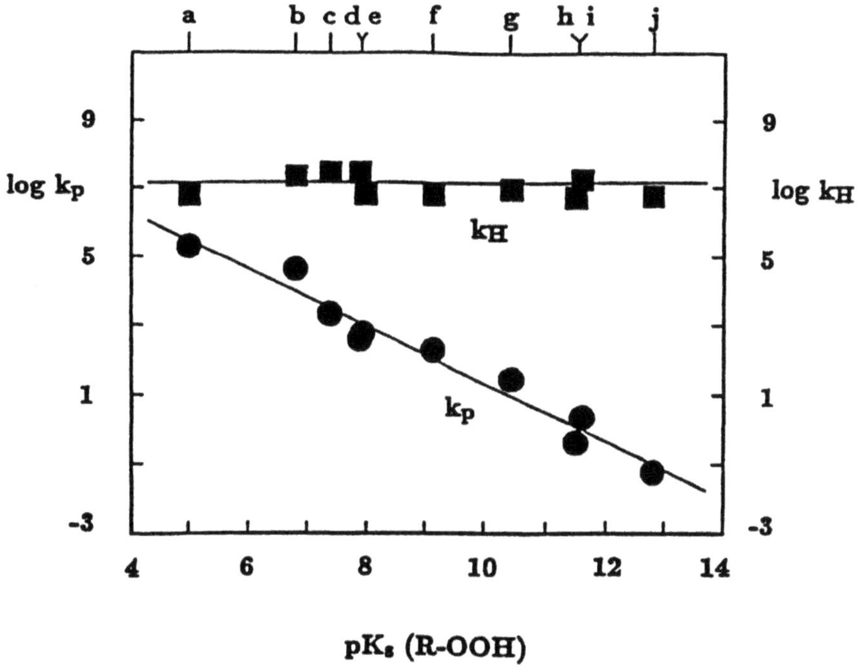

Fig. 4: A linear free energy relationship for the rate of the oxidation of sulfur(IV) in the systems HSO_3^-/ROOH via the proton (k_H) and water(k_P = $k_{HOH}[H_2O]$) catalyzed reaction channel (cf. (16); T = 285 K, I = 0.5 or 1.0 M $NaClO_4$). Oxidants: ROOH = H_2O_2(i), CH_3OOH(h), *t*-BuOOH(j), HC(O)OOH(c), $CH_3C(O)OOH$(e), $C_2H_5C(O)OOH$(d), $^-O_3SOOH$(f), $HOCH_2OOH$ (g), ONOOH (b, I = 1.0 M Na_2SO_4, T = 293 K), O_2NOOH(a).

Working on the above assumptions, the pH dependence of the second order rate constant for the oxidation of S(IV) by the intermediate is given by (22) and (23).

$$-d[IM]/dt = k_2[IM]_T[S(IV)]_T = k_A[N_2O_6^{2-}][SO_3^{2-}] + k_B[HN_2O_6^-][SO_3^{2-}] + k_C[H_2N_2O_6][HSO_3^-][H^+] \quad (22)$$

$$k_2 = \frac{k_A + k_B[H^+]/K_{I2} + k_C[H^+]^4/(K_{I1}K_{I2}K_{S2})}{\{1+[H^+]/K_{I2}+[H^+]^2/(K_{I1}K_{I2})\}\cdot\{1+[H^+]/K_{I2}+[H^+]^2/(K_{I1}K_{I2})\}} \quad (23)$$

The rate constants k_A = 60 ± 50 $M^{-1}s^{-1}$, k_B = (7.5 ± 1.2)·10^3 $M^{-1}s^{-1}$, and k_C = (1.2 ± 0.11)·10^{10} $M^{-2}s^{-1}$ (T = 285 K, I = 1.0 M $NaClO_4$) stand for the second order rate constants for the reactions (8), (19), and (21), respectively. The constants K_{I1}, K_{I2}, K_{S1}, and K_{S2} refer to the acid dissociation constants of the intermediate and of sulfur(IV), respectively. In Fig. 3 the pH profile of the second order rate constant for the oxidation of S(IV) by peroxonitric acid, by the intermediate, and by hydrogen peroxide are shown. Obviously, the rate of the S(IV) oxidation by the peroxo-species HNO_4 and the intermediate is at any pH higher than that with the oxidant H_2O_2.

Although the concentration of peroxonitric acid and that of its decomposition product might be low in the atmosphere, the high S(IV) oxidation rates of these compounds point to the necessity of considering these reactions in model calculations of clouds.

Composition of the intermediate observed in the HNO_4/NO_4^- ***decay***

The present study shows that in addition to the direct decomposition to nitrite according to (3) or (4) [2, 3] the formation of an intermediate and its decay is observed.

The findings obtained in the present study suggest that two peroxonitric acid species, *i.e.* HNO_4 or NO_4^-, "dimerise" via (5) or (11), respectively. Based on this assumption oxygen should be released in two steps (24). This is in line with the experimental observation of Kenley *et al.* [1].

$$2\,NO_4^- \longrightarrow N_2O_6^{2-} + O_2 \longrightarrow 2\,NO_2^- + 2\,O_2 \qquad (24)$$

If oxygen is liberated in the formation of the intermediate, one of the nitrogens in the "dimer" should have been reduced to N(III). From its ability to oxidize S(IV) (and DMS) it is concluded that the intermediate still contains a peroxo-group. A "double" pH-jump experiment (pH = 1 —> 7.8 —> 1) with a stopped-flow unit equipped with three syringes and two mixing jets did not show the decomposition kinetics of the starting solutions. After that treatment instead of three decomposition steps only a single step (III) is observed. That means that the liberation of oxygen (see (3)) is an irreversible process.

Interestingly enough, Barlow *et al.* [9] postulated the "dimeric" species N_2O_6 and $N_2O_6^-$ in their pulse-radiolytic study on NO_3 radicals in aqueous solution.
Because of the short life-time of the intermediate at pH > 7 usual techniques are not applicable to get further information on its composition. The UV spectrum of the intermediate (at pH > 7) resembles very much that of peroxonitrous acid [5], which also contains a peroxo group and nitrogen(III) as neighbouring chromophores.

With a "double" pH-jump experiment the intermediate can be rapidly formed (pH 1 —> >7) and stabilized (pH >7 —> 1). The life time of the intermediate at pH = 1 should be long enough to get more spectral information (*e.g.* IR, Raman spectroscopy). Experiments with ^{15}N enriched peroxonitric acid are scheduled for the future. The new synthetic route given by Appleman *et al.* [4] might be better suited for this purpose than the preparation used in the present study.

Henry's law constant of HNO_4

To our knowledge Henry's law constant is not known for peroxonitric acid. The method of preparation used in the present study allows to get a rough estimate for $H(HNO_4)$ from the partitioning of HNO_4 and H_2O_2 between the gas phase and the two solutions. A value of $H = 4 \cdot 10^3$ M·atm^{-1} (298 K) could be estimated. In that respect peroxonitric acid lies in between nitrous acid and hydrogen peroxide $H(HNO_3) = 2.5 \cdot 10^5$ M·atm^{-1}; $H(HNO_2) = 49$ M·atm^{-1}; $H(H_2O_2) = 6.9 \cdot 10^4$ M·atm^{-1}(298 K) [8].

Atmospheric relevance

Peroxonitric acid has been detected in the atmosphere [10–11]. The chemistry of HNO_4 is rather unknown. The few studies reported in the literature on HNO_4 reveal that it is a quite unstable compound. The present study shows that its anion is extremely unstable ($t_{½} < 5$ ms), but an intermediate is formed in the decomposition of HNO_4. Peroxonitric acid as well as the intermediate formed during its decomposition are both able to oxidize sulfur(IV). Both oxidants react faster than the commonly accepted atmospheric oxidant for S(IV) in the aqueous phase: hydrogen peroxide. Even low concentrations of HNO_4 or of its decomposition product may contribute significantly to the S(IV) > S(VI) conversion because of their high reaction rates towards S(IV). Low temperatures would increase the stability of HNO_4 and of the intermediate, which gives further support for their importance to atmospheric chemistry. The estimated Henry constant for HNO_4 implies that appreciable amounts of HNO_4, which is formed in the gas phase by the NO_2/HO_2 reaction, have to be expected in the aqueous phase.

Acknowledgements

The financial support by the Deutsche Forschungsgemeinschaft within the Sonderforschungsbereich 233: "Dynamik und Chemie der Hydrometeore" is gratefully acknowledged.

References

1. Kenley, R. A.; Trevor, P. L.; Lan, B. Y. *J. Am. chem. Soc.* **103** (1981) 2203–2206.
2. Lammel, G.; Perner, D.; Warneck, P. *J. Phys. Chem.* **94** (1990) 6141–6144.
3. Loegager, T.; Sehested, K. *J. Phys. Chem.* **97** (1993) 10047.
4. Appelman, E. H.; Gosztola, D. J. *Inorg. Chem.* **34** (1995) 787–789.
5. Drexler, C.; Elias, H.; Fecher, B.; Wannowius, K. J. *Fresenius J. Anal. Chem.* **340** (1991) 605–615.
6. Drexler, C.; Elias, H.; Fecher, B.; Wannowius, K. J. *Ber. Bunsenges. Phys. Chem.* **96** (1992) 481–485.
7. Elias, H.; Götz, U.; Wannowius, K. J. *Atmos. Environ.* **28** (1994) 439–448.

8. Finnlayson-Pitts, B. J.; Pitts, J. N. jr. *Atmosheric Chemistry: Fundamentals and Experimental Techniques*, J. Wiley and Sons, New York.
9. Barlow, S.; Buxton, G. V.; Salmon, G. A.in: *Proc Symp '94* pp. 1005–1009.
10. Rinsland, C. P.; Zander, R.; Farmer, C. B.; Norton, R. H.; Brown, L. R.; Russel, J. M.; Park, J. H. *Geophys. Res. Lett.* **13** (1986) 761–764.
11. Goldman, A.; Murcray, F. J.; Blatherwick, R. D.; Kosters, J. J.; Murcray, F. H.; Murcray, D. G. *J. Geophys. Res.* **94** (1989) 14945–14955.

3.2 Electrolyte Effects on Aqueous Oxidation of Sulfur Dioxide by Strong Oxidants

Janine Lagrange, Philippe Lagrange, C. Pallarès and G. Wenger

URA 405 au CNRS, Laboratoire de Cinétique et Analyse, EHICS, Université de Strasbourg, 1 rue Blaise Pascal, F-67000 Strasbourg, France

Summary

The kinetic laws and the mechanisms of the oxidation of dissolved SO_2 by H_2O_2, O_3 and ClO^- are reported. The influences of ionic medium, metallic traces and temperature are studied. The competition of S(IV) oxidation by O_3 and H_2O_2 as a function of H^+ concentration and of ionic environment is discussed.

Aims of the research

Sulfur(IV) oxidation in the atmosphere is complex and can occur both in the gas phase and in liquid droplets. The mechanism leading to the formation of sulfuric acid includes both homogeneous and heterogeneous processes. SO_2 in the gas phase of the atmosphere leads by dissolution in aqueous phase to very reactive anions HSO_3^- and SO_3^{2-}. Aqueous phase oxidations of SO_2 by O_2 (with metallic traces),O_3 and H_2O_2 are generally thought to be the primary mechanism of production of the acidity and SO_4^{2-} in the continental clouds and fogs. Our studies describe, by laboratories measurements, the kinetics of the oxidation reaction of dissolved SO_2 by three strong oxidants: O_3, H_2O_2 and HClO and determine their kinetic laws. Since atmospheric water contains a great variety of inorganic and organic substances that might have kinetic effects on dissolved SO_2 oxidation, influences of trace components and of ionic medium on this oxidation have been evaluated.

Principal scientific results

Rate laws and mechanisms of hydrogen sulfite and sulfite oxidations

Hydrogen sulfite and sulfite oxidations by the studied oxidants were performed with a stopped-flow spectrophotometer (fast reactions) or with a classical UV spectrophotometer. In all kinetic studies a series of experimental measurements were carried out at a given constant ionic strength with a given supporting electrolyte. Solutions of the oxidant in a given electrolyte medium and at fixed H^+ concentration were mixed with solutions of sulfite in the same medium and at the same H^+ concentration. During each kinetic run, H^+ concentration was generally maintained constant with or without the addition of buffers. In a few cases, the H^+ concentration was not maintained constant, and a pH indicator was used to follow the H^+ concentration during the reaction (nitrate medium). The rate laws are expressed, as usual, as a function of the concentration of the reactants. In the case of the H^+ reactant, our technique allows measurement of the concentration of H^+ instead of the activity.

Aqueous oxidation of dissolved SO_2 by H_2O_2 (studied range $3 < pH < 5$)

Kinetic measurements in presence of acetate buffer ([acetate] ≤ 0.04 mol L^{-1}) lead to the following rate law [1]:

$$\text{rate} = r = k(H_2O_2)\,[HSO_3^-]\,[H_2O_2]\,[H^+] \quad (1)$$

At constant ionic environment, the effect of the acetate buffer on the kinetics of oxidation of HSO_3^- by H_2O_2 may be neglected. The proposed mechanism suggests the formation of a peroxymonosulfurous acid intermediate (fast pre-equilibrium: constant = K_{Pe}). This intermediate is subject to specific acid catalysis (rate determining step: rate constant = k_H) with protons transferred from H_3O^+. Generalized acid catalysis does not seem to be taken into account in the proposed mechanism (protons transferred from the donating proton of the buffer). The third order rate constant may be expressed as: $k(H_2O_2) = k_H\,K_{Pe}$.

Aqueous oxidation of dissolved SO_2 by O_3 (studied range $3 < pH < 5$)

Kinetic measurements of observed rate constants without buffer lead to the following rate law [2]:

$$r = k(O_3)\,[HSO_3^-]\,[O_3]\,[H^+]^{-0.5} \quad (2)$$

In the literature, two types of mechanisms are proposed. Penkett *et al.* [3] think that a free radical mechanism involving OH, HSO_3, HSO_5 may occur, whereas Maahs [4] and Hoffmann [5] favour polar mechanisms that proceed via electrophilic attacks of O_3. As our supporting electrolyte effects suggest that S(IV) oxidation proceeds via a radical chain reaction involving OH-radical, and as the Penkett *et al.* mechanism accounts for our rate law expression, this free radical chain mechanism may be chosen. In this mechanism, two molecules of ozone are consumed for each OH-radical produced, respecting the results of Hoigné *et al.* [6].

Aqueous oxidation of dissolved SO_2 by H_2O_2 (studied range $8 < pH < 13$)

In this range of H^+ concentration, the predominant species of S(IV) is SO_3^{2-} and the oxidant H_2O_2 as well as its conjugate base HO_2^- have to be taken into account in solution. The experimental results show first order kinetics with reference to total hydrogen peroxide ($[H_2O_2]_{total} = [H_2O_2] + [HO_2^-]$) and to SO_3^{2-}. The following kinetic law is checked in the studied range of H^+-concentration:

$$r = k'(H_2O_2)_{total}\,[SO_3^{2-}]\,[H_2O_2]_{total} \quad (3)$$

$k'(H_2O_2)_{total}$ is a hyperbolic function of H^+ concentration:

$k'(H_2O_2)_{total} = (a + b\,[H^+]) / (1 + c\,[H^+])$,

where a, b and c are constants. The experimental data may be interpreted on the assumption that the significant reactions are two bimolecular reactions between SO_3^{2-} and H_2O_2 and between SO_3^{2-} and HO_2^- (rate constants $k'(H_2O_2)$ and $k'(HO_2)$. Then $a = k'(HO_2)$; $b = k'(H_2O_2) / K(H_2O_2)$; $c = 1 / K(H_2O_2)$, where $K(H_2O_2)$ is the equilibrium constant of the fast equilibrium between H_2O_2 and its conjugate base HO_2^-. At $8 < pH < 9.8$, the rate law takes therefore a reduced form: $r = k'(H_2O_2)\,[SO_3^{2-}]\,[H_2O_2]$.

Aqueous oxidation of dissolved SO_2 by ClO^- (studied range $10 < pH < 13$)

Experiments were performed at constant and controlled OH^- concentration without any buffer and the following kinetic law is obtained:

$$r = k_{ClO}\,[SO_3^{2-}]\,[ClO^-]\,[OH^-]^{-1} \quad (4)$$

Several authors [7, 8] observed an extremely rapid reaction between HClO and SO_3^{2-} which is of about a five order of magnitude faster than the reaction between ClO^- and SO_3^{2-}. A fast equilibrium between HClO and ClO^- and a direct oxygen atom transfer process between HClO and SO_3^{2-} explain our experimental kinetic law.

Influence of the ionic environment and of the ionic strength

The rate constants of the four studied reactions were determined for several concentrations of various supporting electrolytes ($NaClO_4$, NaCl, $NaNO_3$, Na_2SO_4, and NH_4ClO_4). An important effect of the ionic medium was found (Fig. 1). The variations of the rate constants of the kinetic equations (1), (3), and (4) exhibit, for a given supporting electrolyte, either a maximum or minimum when plotted against the ionic strength. These rate constants appear in the mechanism as a product of pre-equilibrium constants and rate constants. In the activated complex theory, the rate constant may be expressed as a function of the equilibrium constant of the activated complex. These equilibrium constants may be expressed as a function of the activity factors of the reactants. The equation of Debye-Hückel (with the hypothesis of Guggenheim) in Na_2SO_4: $\log\{k(H_2O_2)/k(H_2O_2)_o\} = 1.65\sqrt{I}/(1+\sqrt{I}) - 0.39\,I$

in $NaNO_3$: $\log\{k(H_2O_2)/k(H_2O_2)_o\} = 5.66\sqrt{I}/(1+\sqrt{I}) - 2.23\,I$

For the reaction $SO_3^{2-} + H_2O_2$, rate law (3), at 25 °C:
rate constant at zero ionic strength: $k'(H_2O_2)_o = 0.247\ L\ mol^{-1}\ s^{-1}$ and

in $NaClO_4$: $\log\{k'(H_2O_2)/k'(H_2O_2)_o\} = -0.55\sqrt{I}/(1+\sqrt{I}) + 0.24\,I$

in NaCl: $\log\{k'(H_2O_2)/k'(H_2O_2)_o\} = -0.16\sqrt{I}/(1+\sqrt{I}) + 0.14\,I$

in Na_2SO_4: $\log\{k'(H_2O_2)/k'(H_2O_2)_o\} = -0.25\sqrt{I}/(1+\sqrt{I}) + 0.19\,I$

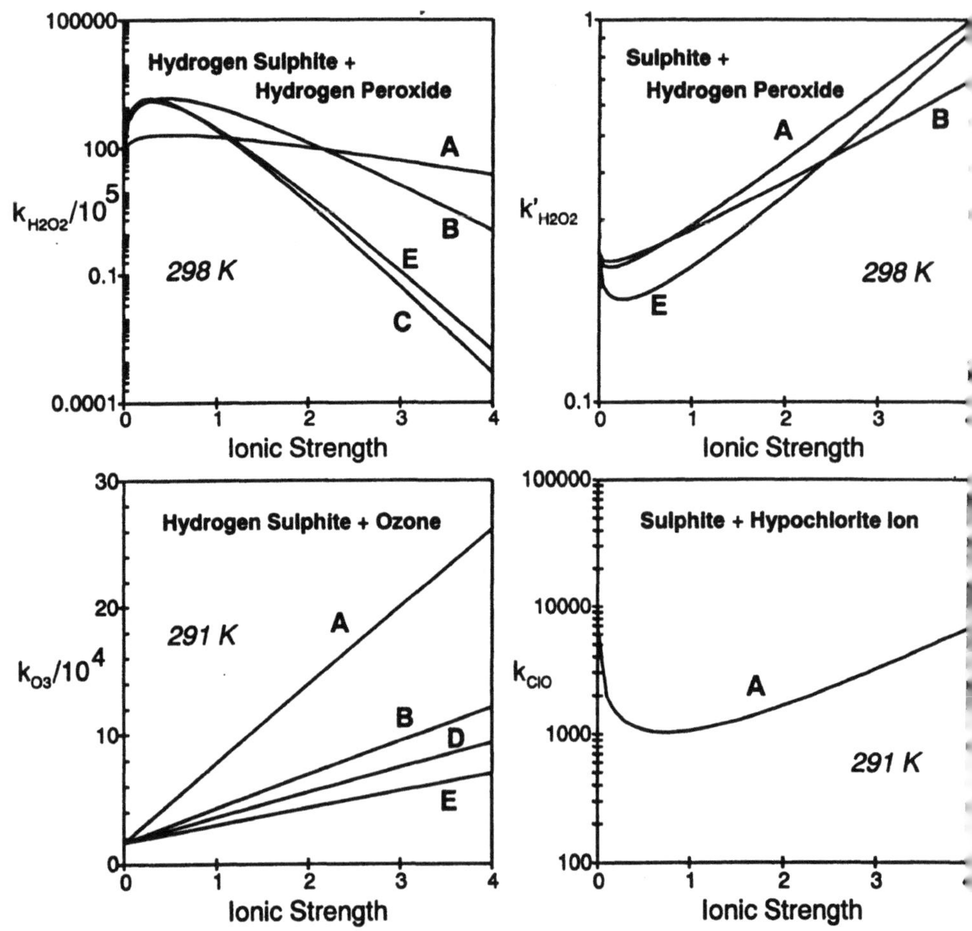

Fig. 1: Variation of the rate constants in different media: (A) Na_2SO_4; (B) NaCl; (C) $NaNO_3$; (D) NH_4ClO_4; (E) $NaClO_4$.

For the reaction $SO_3^{2-} + ClO^-$, rate law (4), at 18 °C:
rate constant at zero ionic strength: $k_{ClO,0} = 0.678 \times 10^4\ s^{-1}$ and

in Na_2SO_4: $\log (k_{ClO} / k_{ClO,0}) = -2.41 \sqrt{I} / (1 + \sqrt{I}) + 0.40\, I$

The rate constant $\{k(O_3)\}$ of the HSO_3^- oxidation by O_3 (kinetic law (2)) does not follow the same effect of the ionic strength. The proposed mechanism is also very different. A linear empirical law describes the variations of the rate constant. The rate constants can be expressed from the rate constant at zero ionic strength as:

For the reaction $HSO_3^- + O_3$, rate law (2), at 18 °C:
rate constant at zero ionic strength: $k(O_3)_0 = 1.65 \times 10^4\ L^{0.5}\ mol^{-0.5}\ s^{-1}$ and
in $NaClO_4$: $k(O_3) = k(O_3)_0 + 1.34 \times 10^4\, I$
in NaCl: $k(O_3) = k(O_3)_0 + 2.63 \times 10^4\, I$
in Na_2SO_4: $k(O_3) = k(O_3)_0 + 6.13 \times 10^4\, I$
in NH_4ClO_4: $k(O_3) = k(O_3)_0 + 1.94 \times 10^4\, I$

Influence of temperature

For each reaction studied, the activation parameters were determined as a function of the nature and of the concentration of a given supporting electrolyte. Taking into account the precision of the determination of the Arrhenius activation energy and the small effects of the supporting electrolytes on the activation energy E_a (in our studied experimental conditions:
$0 \le I \le 4$ M and 5 °C ≤ temperature ≤ 30 °C), we may propose:

for the reaction HSO_3^-	+ H_2O_2 (rate law (1)),	$E_a \approx 34\ kJ\ mol^{-1}$
for the reaction HSO_3^-	+ O_3 (rate law (2)),	$E_a \approx 25\ kJ\ mol^{-1}$
for the reaction SO_3^{2-}	+ H_2O_2 (rate law (3)),	$E_a \approx 14\ kJ\ mol^{-1}$
for the reaction SO_3^{2-}	+ ClO^- (rate law (4)),	$E_a \approx 39\ kJ\ mol^{-1}$

Influence of the metal ion traces

The effects of the metal ions: Mn(II), Cu(II), Fe(II), Fe(III), and Cr(III) on the rate constants of the four reactions were studied. Only one small catalytic effect of the Fe^{2+} ion was observed on the reaction: $HSO_3^- + H_2O_2$ (rate law (1)).

Relevance to atmospheric chemistry

From our results we can give some indications to the important influence of the medium on the aqueous oxidation of sulfur dioxide: two examples are given.

Profile of SO_2 oxidation rate by H_2O_2 as a function of $[H^+]$, supporting electrolyte and ionic strength.

Fig. 2 is a calculation of the apparent second order rate constant k:
"dissolved SO_2 oxidation rate constant" $= -d[S(IV)] / dt = k\, [H_2O_2]_{total}\, [S(IV)]_{total}$

as a function of the ionic environment and [H^+]. A salt effect is observed for $I = 0.1$ M and this effect is practically independent of the nature of the 1:1 supporting electrolyte. The rate of dissolved S(IV) oxidation by hydrogen peroxide is about ten times that at zero ionic strength ($3 < pH < 6.5$). At higher constant ionic strength (for example, $I = 1$ M), the nature of the supporting electrolyte has a measurable effect. The oxidation rate in sodium chloride is eight times that in sodium sulfate ($6 < pH < 7$, $I = 1$ M, 25 °C).

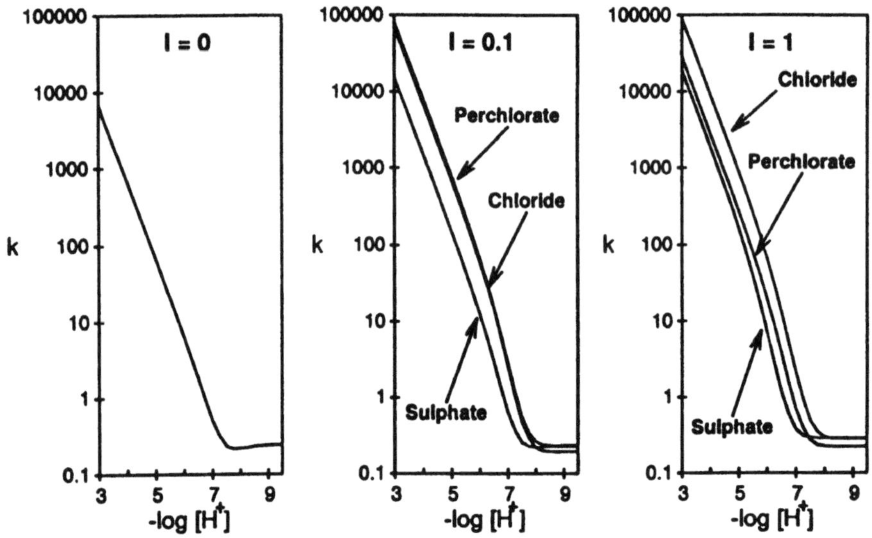

Fig. 2: Second order apparent rate constant (k) of the oxidation reaction of dissolved SO_2 by H_2O_2 as a function of H^+ concentration and ionic medium.

Comparison of the dissolved SO_2 oxidation rates by H_2O_2 and O_3 under atmospheric conditions.

The atmospheric oxidation of dissolved SO_2 by H_2O_2 and O_3 in aqueous phase is considered to contribute significantly to the overall production of sulfate. In the literature, two investigators compare the specific reaction rate, r/[S(IV)], obtained for the conversion of sulfur(IV) between both strong oxidants (H_2O_2, O_3) at given levels [H_2O_2] and [O_3]. Penkett *et al.* [3] have calculated the production of sulfur(VI) {in µg/(mL min)} due to each oxidant. He observed that the rate of production of sulfate by ozone becomes predominant for pH higher than 5.8 with the conditions in the aqueous phase of the atmosphere: [H_2O_2] = 4.31×10^{-5} mol L^{-1} and [O_3] = 1.17×10^{-9} mol L^{-1}. These concentrations correspond to 1 part per billion by volume (hydrogen peroxide) and 50 parts per billion by volume (ozone), respectively, in the gas phase. Maahs [4], with slightly different data and kinetic rate laws has found that the conversion rate by ozone was competitive with that by hydrogen peroxide at pH about 4.5 with $[H_2O_2] = 4.3 \times 10^{-6}$ mol L^{-1} and $[O_3] = 5 \times 10^{-10}$ mol L^{-1} or at pH about 5 at higher H_2O_2 concentration, $[H_2O_2] = 4.3 \times 10^{-5}$ mol L^{-1} and $[O_3] = 5 \times 10^{-10}$ mol L^{-1}. From our present results, in two media: NaCl and Na_2SO_4, and at 18 °C, a characteristic value of the concentration of H^+ ($[H^+]_e$), where the specific oxidation rates by ozone and by hydrogen peroxide are equal, is determined. Below this value the ozone reaction is faster of the two and above this value the hydrogen peroxide oxidation predominates. The reactant concentrations for ozone and hydrogen peroxide that were used for this calculation were 1.17×10^{-9} and 4.31×10^{-5} mol L^{-1}, respectively (values selected by Penkett *et al.* [3]). These values are representative of the values observed under atmospheric conditions. Both electrolytes studied, SO_4^{2-} and Cl^-, are present in the aqueous phase of clouds (SO_4^{2-} may represent an urban environment, Cl^- a marine environment). The variations of $-\log [H^+]_e$ in function of the concentration of the supporting electrolytes are given in Fig. 3.

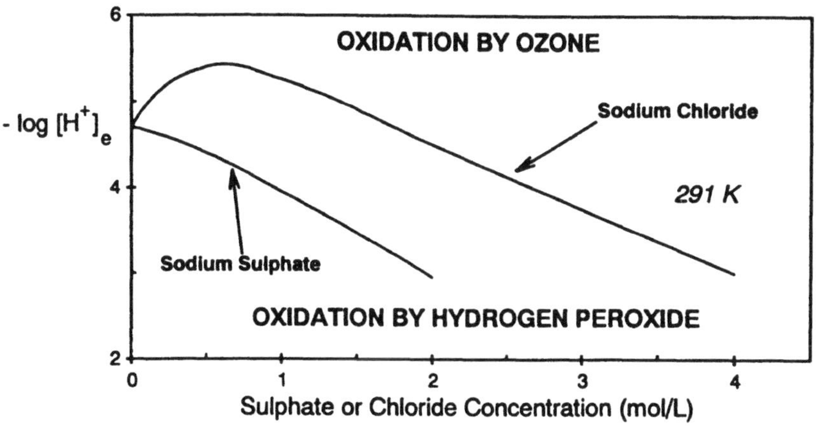

Fig. 3: Plot of the characteristic H^+ concentration, $[H^+]_e$, where the specific oxidation rates by H_2O_2 and O_3 are equal (under atmospheric conditions: $[O_3] = 1.17 \times 10^{-9}$ mol L^{-1} and $[H_2O_2] = 4.31 \times 10^{-5}$ mol L^{-1}) as a function of the concentration of NaCl or Na_2SO_4 at 18 °C.

At zero ionic strength the ozone reaction becomes faster than the hydrogen peroxide reaction above pH = 4.7. This effect appears at higher H^+ concentration when the ionic strength increases. In a chloride medium we notice that the $[H^+]_e$ concentration exhibits a minimum for a concentration of NaCl of about 0.5 mol L^{-1} ($-\log [H^+]_e = 5.4$). When [NaCl] > 0.5 mol L^{-1}, $[H^+]_e$ increases with the increase of Cl^- concentration ($-\log [H^+]_e = 3$ for $[Cl^-] = 4$ mol L^{-1}). In a sulfate medium the $[H^+]_e$ values increase with the sulfate concentration $-\log [H^+]_e = 3$ for $[SO_4^{2-}] = 2$ mol L^{-1}. For a one molar sulfate concentration, ozone becomes competitive at $-\log [H^+]_e$ above 3.8, a value much lower than in one molar chloride medium ($-\log [H^+]_e = 5.2$). With high ionic strength conditions we observed that O_3 is more competitive, in the oxidation reaction of S(IV), than

H_2O_2 even if H^+ concentration is higher than 10^{-5} mol L^{-1}. The high concentrations of electrolyte used in our study may be observed in atmospheric conditions either in the hygroscopic phase which surrounds aerosols or in the several successive phases of condensation/evaporation during the water droplet's life. These results are part of a program to model atmospheric acidity formation.

Conclusions

In the oxidation processes of SO_2 in aqueous phase, ionic strength and ionic environment are not negligible parameters. In the model simulation of the conversion of SO_2 to SO_4^{2-} in clouds, it now seems necessary to take these factors into account.

Acknowledgements

Commission of the European Communities (Step contract 0005C), Ministère de l'Environnement (Paris), Ministère de l'Education Nationale (Paris), Centre National de la Recherche Scientifique and Programme Environnement, PIREN, (France).

References

1. J. Lagrange, C. Pallarès, G. Wenger, P. Lagrange: *Atmos. Environ.* **27**A (1993) 129.
2. J. Lagrange, C. Pallarès, P. Lagrange: *J. Geophys. Res.* **99** (1994) 14595.
3. S. A. Penkett, B. M. R. Jones, K. A. Brice, A. E. J. Eggleton: *Atmos. Environ.* **13** (1979) 123.
4. H. G. Maahs: *J. Geophys. Res.* **88** (1983) 10721.
5. M. R. Hoffmann: *Atmos. Environ.* **20** (1986) 1145.
6. J. Hoigné, H. Bader, W. R. Haag, J. Staehelin: *Water Res.* **19** (1985) 993.
7. M. W. Lister, P. Rosenblum: *Can. J. Chem.* **41** (1963) 3013.
8. K. D. Fogelman, D. M. Walker, D. W. Margerum: *Inorg. Chem.* **28** (1989) 986.

Chapter 4

Transition Metal Ions in Atmospheric Waters

4.1 Metal Ion Catalysed Autoxidation of Dissolved Sulfur Dioxide

Johan Berglund and Lars I. Elding

Inorganic Chemistry 1, Chemical Center, University of Lund, P. O. Box 124, S-221 00 Lund, Sweden

Summary

The overall kinetics for manganese(III/II) catalysed autoxidation of dissolved sulfur dioxide has been studied by UV–visible spectrophotometry. A free-radical mechanism has been proposed, and the rate constant for one of the key reactions between manganese(II) and peroxomonosulfate has been determined by pulse radiolysis. Previous literature on manganese catalysed autoxidation of $SO_2(aq)$ has been thoroughly discussed and explanations of some of the earlier discrepancies have been offered.

The oxidation of $SO_2(aq)$ by complexes of Pt(IV), Au(III) and Tl(III) has also been investigated. These two electron oxidants oxidize SO_2 (aq) only in stoichiometric amounts and show no catalytic activity. The mechanistic differences observed between one and two electron oxidants have proved fruitful in order to explain the lack of catalytic activity of some atmospherically important metal ions such as Ni(II).

Aims of the research

Many atmospheric model calculations have been performed on the relative importance of different pathways for oxidation of sulfur dioxide, and the role of transition metal ion catalysis has been discussed based on the results. However, model calculations can never be expected to be more accurate than the input data in terms of rate laws and rate constants for the different pathways considered. Despite the large amount of literature on metal ion catalysis it has been difficult to come to an agreement on the mechanisms governing the autoxidation of SO_2(aq). This is particularly true for the manganese catalysis.

The aim of our contribution has been to improve the knowledge of metal ion catalysed autoxidation of SO_2(aq). In particular, the catalytic effect displayed by the manganese(III/II) redox couple has been studied in order to provide a better basis for the incorporation of the manganese catalysed pathway in atmospheric model calculations. The following summary of our results is based on work described in more detail elsewhere [1–7].

Principal scientific results

After this period, the rest of the kinetic trace can be evaluated according to first order kinetics. Addition of a small amount of manganese(III) has a major impact on the reaction. The induction period disappears gradually when the concentration of manganese(III) is increased and finally vanishes completely. Furthermore, the rate of oxidation increases with the concentration of manganese(III). At pH 2.4 the observed rate constant is linearly dependent on [Mn(II)] at low Mn(II) concentrations, [Mn(II)] < 2×10^{-5}, and reaches a saturation value when the concentration of manganese(II) is increased above approximately 1×10^{-4} M. The mechanism outlined in Scheme I is based on a free-radical chain mechanism for autoxidation of sulfur(IV) initiated by photons or hydroxo radicals [8] and the recent results obtained for re-oxidation of divalent metal ions by HSO_5^- and oxosulfur radicals [1, 9]. In the first step, manganese(II) forms a hydrogen sulfite complex, $[Mn(HSO_3)]^+$, with a stability constant β_1. This is the only way to explain the saturation of the rate of reaction observed for increasing concentrations of Mn(II). In the second reaction step $[Mn(HSO_3)]^+$ reacts with Mn(III) forming SO_3^- radicals. These radicals rapidly react with dissolved molecular oxygen producing SO_5^- radicals. Reaction between SO_5^- and Mn(II) regenerates Mn(III) and forms HSO_5^-, see below. Hydrogen peroxomonosulfate oxidizes complex bound hydrogen sulfite to S(VI) as the final product. This mechanism describes a closed catalytic cycle in which S(IV) is oxidized to S(VI) by molecular oxygen in a simple stoichiometric overall reaction. It also rationalizes the experimental observations at pH 2.4. In the absence of initially added Mn(III) it is proposed that low concentrations of iron(III), present as unavoidable impurities in water and chemicals, are responsible for the initiation. The initiation by iron(III) is due to reaction between Fe(III) and HSO_3^- in which sulfite radicals are formed according to reaction 6.

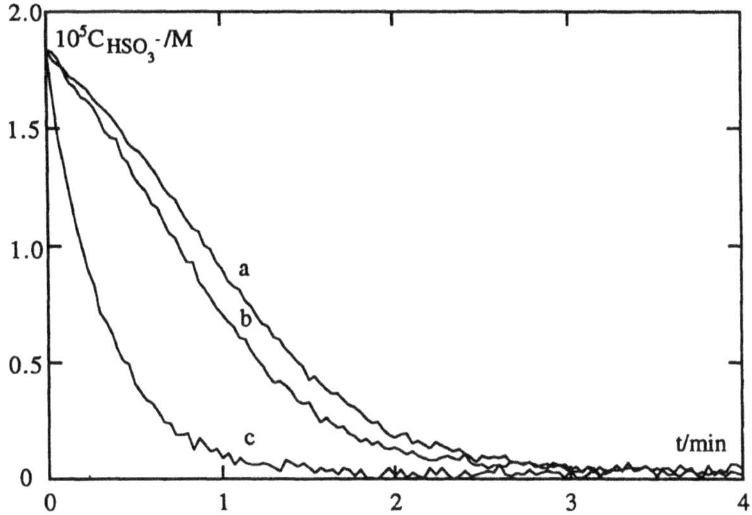

Fig. 1: Illustration of the induction period observed for the Mn(II) catalysed autoxidation and the effect of small Mn(III) additions [1]. Conditions: [Mn(II)] = 3.64×10^{-5} M, $[O_2]$ 2.5×10^{-4} M, pH = 2.4, Temp. 25 °C. $[Mn(III)]_0$ =0 (a), 0.57×10^{-8} M (b), 1.13×10^{-8} M (c). Published with permission from Inorganic Chemistry. Copyright 1993 American Chemical Society.

The manganese(III/II) catalysed autoxidation of sulfur(IV) (SO_2(aq), HSO_3^-, SO_3^{2-}) is autocatalytic with an induction period, Fig. 1.

$$k_6 \qquad Fe(III) + HSO_3^- \rightarrow Fe(II) + SO_3^- + H^+ \qquad (6)$$

These radicals enter the chain and react with molecular oxygen forming SO_5^- radicals, reaction 3, which subsequently predominantly oxidize Mn(II) to Mn(III), reaction 4. Considering the low concentration of iron and the large rate constant for reaction between Mn(II) and SO_5^-, $k \approx 10^8$–10^9 $M^{-1}s^{-1}$ (see below), reaction between Fe(II) and SO_5^- can be neglected. The induction period observed in the absence of initially added manganese(III) can be explained by the build-up of manganese(III) following the rather slow oxidation of sulfur(IV) by iron(III), k_6 = 318 $M^{-1}s^{-1}$ [10]. By employing the steady state approximation to the intermediates SO_3^-, SO_5^- and HSO_5^-, rate expression 7 can be derived [1] in agreement with the experimental rate law.

$$-dC(HSO_3^-)/dt = \frac{2k_2\,[Mn(II)]}{\beta_1^{-1}+[HSO_3^-]+[Mn(II)]}\,([Mn(III)]_0+[Fe(III)]_0)\,C(HSO_3^-) \qquad (7)$$

Here, $[Mn(III)]_0$ denotes the concentration of initially added manganese(III) and $C(HSO_3^-) = [HSO_3^-] + [MnHSO_3^+]$. Hexa-aquamanganese(III) is a strong acid and at pH 2.4 the predominant forms of manganese(III) are hydrolysed complexes such as $[Mn(H_2O)_5(OH)]^{2+}$ and $[Mn(H_2O)_4(OH)_2]^+$. Mn(III) in Scheme I and Eq. 7 represents hydrolysed manganese(III) without any further specification. $[Fe(III)]_0$ denotes the concentration of iron(III) impurities. k_2 is the rate constant for the redox reaction between Mn(III) and $[Mn(HSO_3)]^+$, Scheme I. The values of the parameters have been derived as follows: $k_2 = (1.3 \pm 0.6) \times 10^6\ M^{-1}s^{-1}$, $\beta_1 = (3 \pm 1) \times 10^4\ M^{-1}$ and $[Fe(III)]_0 = 2 \times 10^{-8}$ M. Previous analyses show that the iron(III) concentration in highly purified water is in the range 10^{-8} to 5×10^{-7} M [11, 12]. The trace concentration of iron(III) calculated from the present results is of the same order of magnitude.

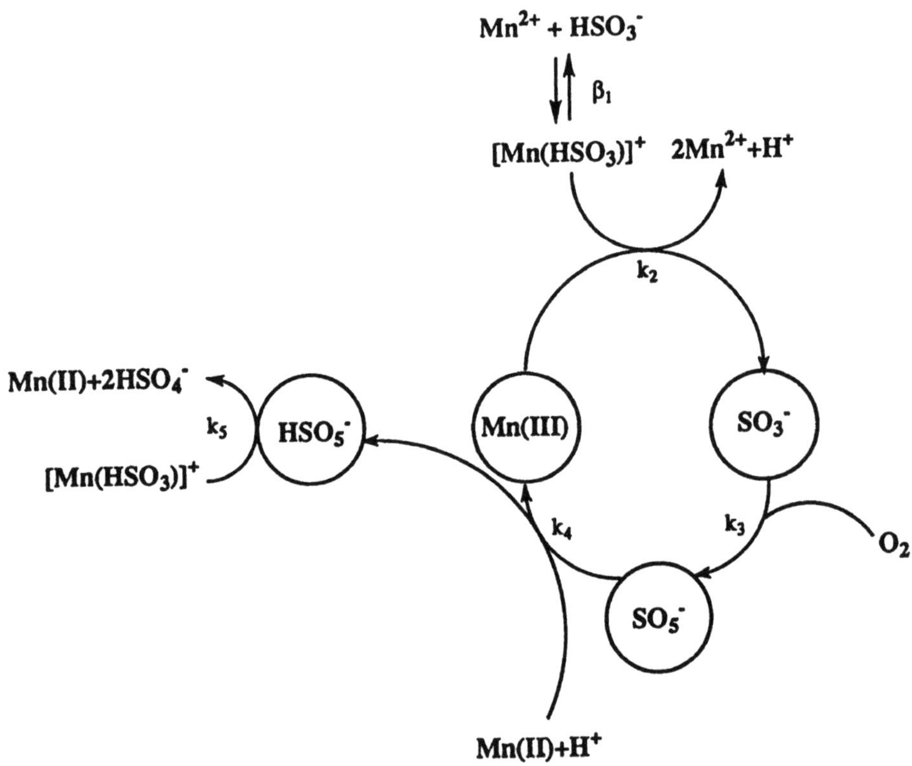

Scheme 1: Proposed mechanism for the manganese(III/II) catalysed autoxidation of hydrogen sulfite.

Three conclusions can be drawn from the rate law of Eq. 7.

(i) The observed rate constant, k_{obsd}, is proportional to [Mn(II)] in the limit [Mn(II)]→0.
(ii) In the limit [Mn(II)]→0, k_{obsd} reaches a saturation value.
(iii) k_{obsd} is a linear function of the concentration of initially added Mn(III). Computer simulations using the FACSIMILE chemical modelling package have qualitatively confirmed the postulated mechanism.

At pH 4.0, the experimental rate law of eq. 8 includes a quadratic term in [Mn(II)]. k, k' and A are constants.

$$-dC(HSO_3^-)/dt = \frac{k\,[Mn(II)] + k'\,[Mn(II)]^2}{A + [Mn(II)]} \quad C(HSO_3^-) \qquad (8)$$

The quadratic term can be rationalized by formation of a bridged sulfito complex $[Mn(SO_3)Mn]^{2+}$ according to the pH dependent reaction (9).

$$[Mn(HSO_3)]^+ + Mn(aq)^{2+} \rightleftharpoons [Mn(SO_3)Mn]^{2+} + H^+ \qquad (9)$$

This complex is assumed to react with manganese(III) generating sulfite radicals, reaction 10, in a reaction parallel to the reaction between Mn(III) and $[MnHSO_3]^+$.

$$[Mn(SO_3)Mn]^{2+} + Mn(III) \rightarrow 3\,Mn(aq)^{2+} + SO_3^- \qquad (10)$$

Those reactions most likely take place by interaction between hydrolysed manganese(III) and the sulfito complexes forming oxo- or hydroxo-bridged precursor complexes, Mn^{III}–O–Mn^{II}–HSO_3 and Mn^{III}–O–Mn^{II}–SO_3–Mn^{II}, respectively. Within these complexes, bridged electron transfer from the manganese(II) sulfito moieties to manganese(III) most likely is fast. Rapid decomposition of the successor complexes gives Mn(II) and sulfite radicals.

Reaction between peroxomonosulfate radicals and manganese(II)

Pulse radiolysis traces in the wavelength region 380 to 570 nm show an initial increase in absorbance attributed to formation of Mn(III) [2], in agreement with spectra previously reported for Mn(III) complexes [13]. The kinetic data have been interpreted qualitatively by a mechanism in which $Mn(aq)^{2+}$, $[Mn(HSO_3)]^+$ and $[Mn(SO_3)Mn]^{2+}$ are oxidized by SO_5^- in three parallel reactions. The second order rate constant is in the range 10^8 to 10^9 $M^{-1}s^{-1}$ depending on the distribution of the Mn(II) complexes. A subsequent slower change in absorbance is rationalized by disproportionation of Mn(III) and reactions of Mn(III) with $SO_2(aq)$. Generation of Mn(III) triggers the catalysed autoxidation of $SO_2(aq)$. The reactions involved in this oxidation are partly responsible for the complicated processes following the formation of Mn(III) and for disturbing the kinetic investigation.

Oxidation of dissolved sulfur dioxide by two electron metal ion oxidants

There have been no reports on autoxidation of sulfur(IV) catalysed by two electron metal ion oxidants. Mechanistic studies of complementary redox reactions between sulfur(IV) and complexes of platinum(IV) [3], thallium(III) [4] and gold(III) [5] in the presence of dissolved molecular oxygen have shown that only a stoichiometric amount of sulfur(IV) is oxidised and that the oxygen concentration is constant throughout the reaction. It has been concluded that two electron transfer takes place within the encounter complex formed between these metal complexes and hydrogen sulfite or sulfite. Thus, no sulfite radicals are released into the bulk solution and the free-radical chain is not initiated. This explains the lack of catalytic activity for these oxidants and provides additional support for the free-radical chain mechanism operating in the catalytic systems.

Previous studies have shown that metal ions without redox chemistry most probably have no or very poor catalytic activity. For example, a study of the influence of nickel(II) salts performed in our laboratory showed that the catalytic effect observed could be attributed to trace metal impurities, especially iron, and not to the nickel(II) itself. Indications that only redox active metal ions are catalytically active can be rationalized by use of the free-radical chain mechanism. Thus, metal centres with only one oxidation state fail to initiate the chain while metal centres with access to oxidation states one apart may allow generation of sulfite radicals according to a reaction similar to reaction 2.

Comparison with previous literature

Considering the complex processes governing the autoxidation of sulfur(IV) it is not surprising that so many different rate laws and inconsistent data have been obtained, since changing the experimental conditions may change the predominant pathway for the reaction. In this report it is only possible to indicate the previous discrepancies. A more thorough discussion and a full list of references is offered in [6].

(1) Induction periods have been reported in some previous studies. Neglect of this phenomenon in the evaluation of the kinetics may be fatal, especially when the method of initial rates is used to evaluate rate constants.

(2) Due to the complex formation between Mn(II) and SO_2(aq), increasing the concentration of SO_2(aq) leads to saturation. For $\beta_1 \approx 10^4$ M^{-1}, the reaction order with respect to sulfur(IV) therefore changes from one at low sulfur(IV) concentrations to zero at higher ones. It should start deviate from 1.0 at ca. 10^{-4} M and should definitely be zero above 10^{-3} M.

(3) The formation of the sulfito bridged manganese(II) complex $[Mn(SO_3)Mn]^{2+}$ may rationalize rate laws including quadratic [Mn(II)] terms.

(4) The critical role of Mn(III) should be emphasized. This has been overlooked many times in the past.

(5) It seems to be generally agreed that the reaction order with respect to oxygen is zero. This is also what should be expected if the mechanism in Scheme I is valid.

(6) According to the mechanism, Scheme I, there should be a pH dependence in acidic solutions (pH < 2.5) since $[Mn(HSO_3)]^+$ is the reactive species and SO_2(aq) most likely does not form strong complexes with Mn(II). We strongly recommend further investigations of the pH dependence under different conditions in order to definitely solve the inconsistent data reported.

Conclusions

From the atmospheric chemistry point of view, the most important result of our contribution is that the manganese catalysis can be described by the simple rate law of Eq. 11 under atmospheric conditions, in agreement with previous literature [14].

$$-d[S(IV)]/dt = k[Mn(II)][S(IV)] \qquad (11)$$

Most likely, the second-order rate constant k is not dependent on pH in the region most commonly found in atmospheric droplets, $2.5 < pH < 5$. We recommend that the value $k=1.4 \times 10^3\ M^{-1}s^{-1}$ should be used in atmospheric model calculations under conditions of low concentrations of SO_2(aq) and Mn(II)($[S(IV)] \leq 10^{-5}$ M, $[Mn(II)] \approx 2 \times 10^{-5}$ M). Previous reports of different rate laws than Eq. 11 are most probably not valid under atmospheric conditions since many of them were obtained using much higher concentrations of reactants. As emphasized above, the complicated mechanism of the manganese catalysed autoxidation give totally different rate laws under such conditions.

The synergism observed in a mixed iron(III) and manganese(II) system must be taken into serious consideration, in order to obtain a fair estimate of the contributions from manganese and iron catalysis. Further kinetic investigation of this effect seems to be called for, since a more complete mechanistic explanation is still lacking. Furthermore, other combinations of metal ions present in significant concentrations in atmospheric droplets should also be investigated in a pursuit of important synergistic rate enhancements.

Acknowledgements

This work has been performed within the EUROTRAC subproject HALIPP. Financial support from the Swedish Natural Research Council and from the Commission of the European Communities (STEP and ENVIRONMENT programs) is gratefully acknowledged. We thank Prof. Sture Fronaeus for stimulating discussions during the whole project and Mrs Bodil Eliasson, Ms Anna Ericson, Ms Pia Werndrup, Mr Robert Voigt and Mr Thor-Björn Bladh for valuable technical assistance.

References

1. J. Berglund, S. Fronaeus, L. I. Elding, *Inorg. Chem.* **32** (1993) 4527–4537.
2. J. Berglund G. V. Buxton, L. I. Elding, S. McGowan, G. A. Salmon, *J. Chem. Soc., Faraday Trans.* (1994) in press.
3. J. Berglund, R. Voigt, S. Fronaeus, L. I. Elding, *Inorg. Chem.* **33** (1994) 3346–3353.
4. J. Berglund, P. Werndrup, L. I. Elding, *J. Chem. Soc., Dalton Trans.* (1994) 1435–1439.
5. J. Berglund, L. I. Elding, *Inorg. Chem.* **33** (1994) *in press.*
6. J. Berglund, L. I. Elding, *Atmos. Environ.* (1995) *in press.*
7. J. Berglund Ph.D. Thesis, University of Lund, Sweden 1994.
8. U. Deister, P. Warneck P. *J. Phys. Chem.* **94** (1990) 2191–2198.
9. R. van Eldik, N. Coichev, K. Bal Reddy, A. Gerhard, *Ber. Bunsenges. Phys. Chem.***96** (1992) 478–481.
10. K. Bal Reddy, R. van Eldik, *Atmos. Environ.* **26A** (1992) 661–665.
11. P. Brimblecombe, D. J. Spedding *Atmos. Environ.* **8** (1974) 937–945.
12. A. Huss Jr., P. K. Lim, C. A. Eckert, *J. Am. Chem. Soc.* **100** (1978) 6252–6253.
13. G. Davies, *Coord. Chem. Rev.* **4** (1969) 199.
14. T. Ibusuki, H. M. Barnes, *Atmos. Environ.* **18** (1984) 145–151.

4.2 Catalytic Effects of Atmospheric Aerosol Particles in the Aqueous Phase of Trace Gases

Irena Grgic, Mateja Poje, Milko Novic, Mirko Bizjak and Vida Hudnik

National Institute of Chemistry, P.O.B. 30, SLO-61115 Ljubljana, Slovenia

Summary

The participation of the National Institute of Chemistry in Ljubljana started at the end of 1992 after the acceptance of our contribution by SSC and after the approval by the High Level Group of EUREKA.

The S(IV) oxidation by oxygen in the presence of Fe(II, III) at initial pH of 4.4, which is very often observed in atmospheric waters, was studied. Our results confirm the redox cycling of Fe(II)/Fe(III) during the catalysed reaction. But at this pH the mechanism of S(IV) oxidation becomes still more complicated due to the formation of an insoluble hydroxo Fe(III) species. The results also clearly demonstrate the influence of the pH on the rate of re-establishment of the catalytic cycle Fe(II)/Fe(III) and thus on the reaction rate of aqueous S(IV) oxidation by oxygen. Laboratory experiments were also conducted with atmospheric aerosol particles collected in Ljubljana. Some preliminary results on aqueous S(IV) oxidation in the presence of these particles are presented.

Aims of the research

Our previous results showed that the chemical form of metal ions leached from coal soot particles are important for aqueous S(IV) oxidation [1]. Iron is one of the most abundant elements found in atmospheric waters [2]. The reactivity of dissolved iron compounds towards different photo-oxidants and pollutants depends upon the oxidation state and speciation of iron [3]. The aim of the present research was to contribute knowledge on the S(IV) oxidation by oxygen in aqueous solution in the presence of iron species in wider pH range. The experimental conditions (pH, the concentration of S(IV) and Fe species) were similar to those found in urban atmospheres. The main goal was to elucidate the role of particular parameters in the redox cycling of Fe(II/III) during the Fe-catalysed oxidation of S(IV) in aqueous phase at pH values typically observed in atmospheric waters. These results are important for our research of catalytic effects of atmospheric aerosol particles in aqueous-phase S(IV) oxidation [4]

Principal scientific results

The experiments on aqueous S(IV) oxidation by oxygen in the presence of iron (II, III) species were performed in the laboratory reactor under controlled conditions (T, pH, air flow rate, stirring). The concentration of Fe(II) and total iron (Fe(II + III)) during the experimental course was followed on-line with continuous flow analyser with spectrophotometric detection. The concentration of S(IV) and S(VI) was followed by ion chromatography and pH by measurements with a glass electrode.

Aqueous S(IV) *oxidation in the presence of* Fe(II, III) *at initial pH of 4.4*

The oxidation of S(IV) by dissolved oxygen in aqueous solution at initial pH of 4.4 ± 0.1 and temperature of 25 ± 1 °C in the presence of iron in the form of Fe(II) or Fe(III) was investigated. A detailed kinetic study of the Fe-catalysed S(IV) oxidation was outside the scope of this work. In the absence of other affects such as UV radiation, the S(IV) oxidation by oxygen in the presence of Fe ions is highly accelerated. The reaction rate depends upon the actual concentration of Fe(III) in the solution. It was found that the reaction rate of S(IV) oxidation is not much affected by the oxidation state of iron at the beginning of the reaction. Namely, in both cases, when iron is added as Fe(II) (which always has a low concentration of Fe(III)) or as Fe(III), the ratio of Fe(II) to Fe(III) reaches a steady-state very fast, after first few minutes of reaction (Figs. 1a, b). An excess of particular iron species depends on the concentration of S(IV) (HSO_3^- at present pH) in the reaction solution. Besides the oxidation-reduction process of iron at the initial step of the S(IV) oxidation, also the formation of insoluble hydroxo Fe(III) species proceeds, which can be seen in decreasing of the total (Fe(II) + Fe(III)) concentration of iron (Fig. 1c). This process is more expressive at lower concentration of S(IV). Because of the oxidation of S(IV), the concentration of H_3O^+ ions increases and pH of the solution decreases, what is more evident at higher initial S(IV) concentration. With higher concentration of H_3O^+ ions the insoluble hydroxo Fe(III) species start to dissolve and Fe(III) ions return in to the catalytic cycle. Anyhow, the final product of the S(IV) oxidation by dissolved oxygen in aqueous solution in the presence of Fe(II) or Fe(III) is the formation of S(VI) and Fe(III) ions. The concentration of Fe(III) ions in the solution at the end of the reaction depends upon the initial concentration of S(IV).

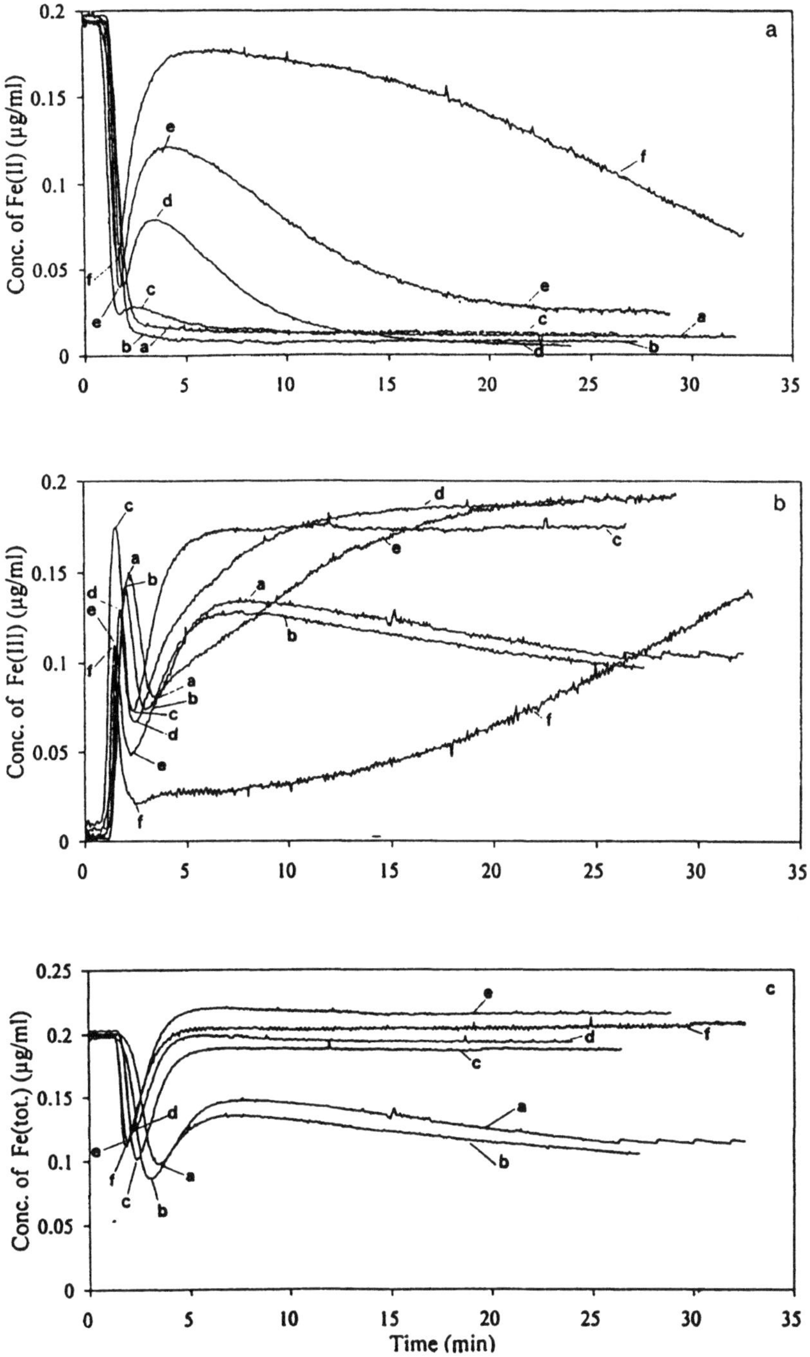

Fig. 1: Time dependence of Fe(II) (a), Fe(III) (b) and Fe(tot) (c) for catalytic S(IV) oxidation by O_2 in the presence of Fe(II) at different initial concentration of S(IV) (a: 2.7, b: 4, c: 8, d: 24, e: 40, f: 80 μg mL^{-1}), $pH_i = 4.4 \pm 0.1$.

Aqueous* S(IV) *oxidation by oxygen in the presence of* Fe(II, III) *at initial pH of 5.2 and 2.8

In order to prove the influence of the pH of the solution on the rate of re-establishment of the catalytic cycle of Fe(II)/Fe(III) and consequently on the reaction rate of aqueous S(IV) oxidation by dissolved oxygen, the experiments were also performed in aqueous S(IV) solution at initial pH of 5.2 ± 0.1 and 2.8 ± 0.1. At high initial pH values the formation of insoluble hydroxo complexes of Fe(III) was promoted, while at low pH values the formation of these complexes was excluded.

When Fe(II) is added in to the S(IV) solution at initial pH of 5.2, Fe(II) is immediately oxidised in Fe(III). Soon as Fe(III) is presented in the S(IV) solution, the formation of insoluble complexes of Fe(III) occur. Therefore, the catalytic cycle of Fe(II)/Fe(III) is very slow and the S(IV) oxidation nearly stops. The reaction proceeds slowly until the concentration of S(IV) is high enough to produce H_3O^+ ions, which decrease the pH of the solution and dissolve the inert hydroxo complexes of Fe(III). In this way, the catalytic cycle of Fe(II)/Fe(III) is again re-established.

The results of the experiments of aqueous S(IV) oxidation in the presence of Fe(II) or Fe(III) at initial pH of 2.8 were confirmed, that during the S(IV) oxidation the formation of insoluble hydroxo complexes of Fe(III) does not proceed. Namely, the concentration of the total Fe is constant during the entire experiment. On the other hand, the concentration of reactive HSO_3^- ions at this pH decreases (90 % of HSO_3^-), and therefore the S(IV) oxidation is slower in comparison with the S(IV) oxidation at initial pH of 4.4.

Aqueous* S(IV) *oxidation in the presence of atmospheric aerosol particles

Some results concerning studies of aqueous S(IV) oxidation by dissolved oxygen in the presence of aerosol particles were also obtained. To exclude the possible effect of the surface of aerosol particles (particle size range: 0.4–1.6 μm), the most experiments were performed in leachate solutions of aerosol particles in pH range 2.8–6.1. In comparison with noncatalytic oxidation of S(IV) by oxygen, the S(IV) oxidation in leachate solutions is accelerated and becomes faster with lower pH values. With regard to the concentration of soluble fraction of metals (especially [Fe]: 0.05–0.25 μg mL^{-1}) higher rates of S(IV) oxidation were expected. The observed oxidation rates were much lower than those expected for Fe-catalysed oxidation by oxygen in clean water. By use of on-line determination of Fe(II) and total dissolved Fe in suspension of aerosol particles, it was established that almost all dissolved Fe was present in the form of Fe(II). After addition of Fe(III) standard solution to the suspension, it was found out that Fe(III) was immediately reduced to Fe(II). It seems, that the conditions in the solution did not allow the formation of Fe(III) and therefore higher reaction rates. On the other hand, the inhibition of the oxidation of S(IV) is expected in the presence of organics [5, 6].

For instance, acetate and formate can act as scavengers of radical intermediates. Inhibition could also be attributed to the complexation of Fe(III) by oxalate. These organics were also determined in our suspensions of atmospheric aerosol particles. So, the aim of our further investigations are the reactions of S(IV) oxidation in the presence of aerosol particles, with emphasise on the role of organic fraction of aerosol particles.

Conclusions

An assessment of the achievements of the contribution with respect to the original aims and to those of the subproject shows that our research has proved the redox cycling of Fe(II/III) during the Fe-catalysed oxidation of S(IV) by dissolved oxygen in aqueous solution. In order to come closer to the aerosol particle situation, the experiments were conducted at higher initial pH values. Experiments confirmed that the pH of the solution is a crucial factor, which controls the concentration of Fe(III) ions and thus the catalytic cycle of Fe(II)/Fe(III) in the aqueous solution of S(IV) and consequently the reaction rate of S(IV) oxidation.
The results concerning studies of aqueous S(IV) oxidation in the presence of atmospheric aerosol particles indicate that the reaction rate is accelerated and becomes faster with lower pHs. Regarding the concentration of soluble fraction of metals (especially Fe), higher rates were expected. To contribute to the knowledge on the potential importance of aerosol particles in heterogeneous processes under specific environmental conditions much more work will be required.

Acknowledgements

This work was supported by the Ministry for Science and Technology of the Republic of Slovenia which is gratefully acknowledged.

References

1. I. Grgic, V. Hudnik, M. Bizjak, J. Levec: Aqueous S(IV) oxidation - III. Catalytic effect of soot particles, *Atmos. Environ.* **27**A (1993) 1409.
2. P. Behra, L. Sigg: Evidence for redox cycling of iron in atmospheric water droplets, *Nature* **344** (1990) 419.
3. D. L. Sedlak, J. Hoigné: The oxidation of S(IV) in atmospheric water by photooxidants and iron in the presence of copper, *Environ. Sci. Technol.* **28** (1994) 1898.
4. I. Grgi}, M. Poje, M. Novic, V. Hudnik: Catalytic effects of atmospheric aerosol particles in aqueous-phase oxidation of S(IV)-oxides, in: *Proc. Joint LACTOZ/HALIPP, EC/EUROTRAC/ GDCh Workshop "Tropospheric Oxidation Mechanisms"*, Leipzig 1994, in press.
5. L. R. Martin, M. W. Hill A. F. Tai, T. W. Good: The iron catalyzed oxidation of sulfur (IV) in aqueous solution: Differing effects of organics at high and low pH, *J. Geophys. Res.* **96** (1991) 3085.
6. A. Kotronarou, L. Sigg: SO_2 oxidation in atmospheric water: Role of Fe(II) and effect of ligands, *Environ. Sci. Technol.* **27** (1993) 2725.

4.3 Kinetics of Ozonation Processes in Cloud and Fogwater: Effect of Transition Metal Ions on Free Radical Reactions

Jürg Hoigné[1] and Rolf Bühler[2]

[1]EAWAG, CH-8600 Dübendorf,
[2]Laboratory for Physical Chemistry, ETH-Zentrum, CH- 8092 Zürich

Summary

Ubiquitous atmospheric iron and copper species can control a significant part of the oxidation processes that occur in atmospheric waters. Hydroperoxyl radicals (HO_2) are steadily transferred from the daytime gaseous atmosphere into droplets or are produced during the fast photolysis of iron complexes such as Fe(III) oxalates. A fraction of HO_2 dissociates into superoxide (O_2^-). This is a main reactant for ozone. However, copper ions are generally even more important reactants for O_2^- because they are present at a higher concentration than ozone, and O_2^- reduces Cu^{2+} and oxidizes Cu^{1+} with rate constants that are nearly diffusion controlled. Therefore, during daytime a photo-stationary fraction of the dissolved copper (II) is present as copper(I). This copper(I) not only accelerates the atmospheric transformation of HO_2/O_2^- into hydrogen peroxide but also steadily reduces iron(III) to iron(II). By this the daytime concentration of iron(III) is highly decreased. HO_2/O_2^- seems not to oxidize S(IV) species. An example situation is modelled in Fig. 1.

Aims of the research

The aim of our research project performed within the HALIPP project has been to extend our earlier kinetic measurements made on the reactions of ozone [1–3] and also to quantify the role of ubiquitous dissolved copper and iron species as reactants for atmospheric photo-oxidants. Such photo-oxidants also control the rate of oxidation of aqueous SO_2 species and therefore the acidification processes. For improving the models describing the kinetics of such processes the photolytic transformations of iron(III) and the related formation of OH and HO_2/O_2^- radicals were determined [4–8]. Additional rate constants for the reactions of O_2^- with Cu(II) and Cu(I) as well as for the reaction between Cu(I) and Fe(III) and for the reaction between HO_2/O_2^- and aqueous species of SO_2 were required [9–14]. A further aim of the project has also been to investigate how pulse radiolytic methods or continuous γ irradiation can be applied to simulate the free radical chemistry that occurs in atmospheric waters. A comprehensive set of reactions and rate constants should allow for an improved modelling of the processes of photo-oxidants that proceed in atmospheric waters and for critical testing of the reaction models by both, laboratory experiments and field studies. On the other hand, such models will help to interpret the field data measured within complementary projects performed at EAWAG and at other sites [16, 17].

Principal scientific results

All results were extensively published in peer reviewed journals, dealing with atmospheric research or chemical reaction kinetics. They were periodically discussed with colleagues from further EUROTRAC projects. Table 1 compiles new rate data that resulted from our studies. It also lists the publications where the essential experimental information can be found. Fig. 1 gives an example for the application of such data to estimate daytime redox fluxes and oxidation states of transition metals in atmospheric droplets.

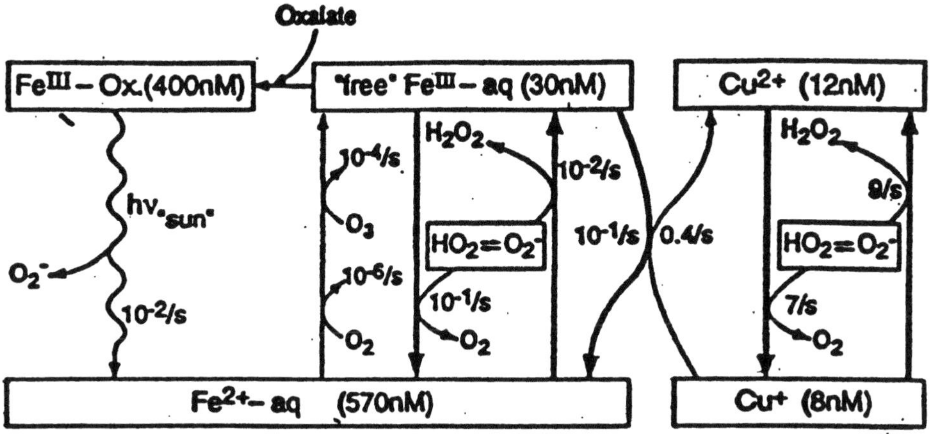

Fig. 1: Steady-state concentrations and pseudo first-order rate constants for the transformation of iron and copper in presence of O_2^-. Assumptions: pH = 4.5; $[O_2^-]$ = 0.8 nM; $[O_3]$ = 0.5 nM; [oxalate] = 400 nM. The concentrations of oxidized and reduced iron and copper calculated for the steady-state mediated by the presence of HO_2/O_2^- are indicated in the boxes. (From: Hoigné, Sedlak and Bühler, *Proc. EUROTRAC Conf.*, Leipzig, Sept. 94, in press.).

Table 1: Reaction rate data determined within our contribution to the HALIPP program-

Reactions	Rate constant*	Method	Ref.
$O_2^- + Cu^{2+} + \ldots \rightarrow O_2 + Cu^+$ $O_2^- + Cu^+ + \rightarrow H_2O_2 + Cu^{2+}$	$(8.0 \pm 0.5) \times 10^9\ M^{-1}\ s^{-1}$ $(9.4 \pm 0.8) \times 10^9\ M^{-1}\ s^{-1}$	pulse radiolysis	[9]
$HO_2 / O_2^- + HSO_3^- + \rightarrow \ldots \rightarrow \ldots.$	very small	cont. γ radiolysis	[11]
$Cu^+ + O_3 \rightarrow Cu^{2+} + \ldots$	$3 \times 10^7\ M^{-1}\ s^{-1}$ (approx)	contin. flow method	[19]
$+\ Fe^{III}(OH)^{2+} \rightarrow$	$> 10^7\ M^{-1}\ s^{-1}$ (approx)	cont. γ radiol. and modelling	[10]
$Fe^{2+} + O_3 \rightarrow FeO^{2+}$	$2 \times 10^5\ M^{-1}\ s^{-1}$ (approx)	cont. flow method $2 < pH < 5$	[7]
H_2O_2 + Fe(II)oxalalte → ... + OH	$3.1 \times 10^4\ M^{-1}\ s^{-1}$	photochem.	[18]
$NO_3^- \xrightarrow[H^+]{h\nu} OH +$	$\tau_{solar\ light} = 7 \times 10^5$ s Φ(OH, 313 nm) = 0.013 ± 0.002	photochem.	[20]
$Fe^{III}(OH)^{2+} \xrightarrow{h\nu} Fe^{2+} + OH + \ldots$	τ solar light = 1.5×10^5 s $\Phi(Fe^{2+})$ at 313 nm 0.14 ± 0.04 at 360 nm 0.017±0.003	Sept. noon sun expos; photochem.	[4]
$Fe(III)(oxalate)_{1-3} \xrightarrow{h\nu} Fe^{2+} + O_2^- + CO_2 +$	τ solar light = 270 s	Sept. noon sun expos; pH=4; [oxalate]= 5×10^{-6} M	[5]
$(CH_3)_3C\cdot + O_2 \rightarrow (CH_3)_3COO$ $2\ (CH_3)_3COO \rightarrow \ldots$	$1.4 \times 10^9\ M^{-1}\ s^{-1}$ $1.8 \times 10^8\ M^{-1}\ s^{-1}$	pulse radiolysis	[12]
$CH_2\text{–}C(CH_3)_2COO^-$ } } $+ O_2 \rightarrow OOCH_2\text{–}C(CH_3)_2COO^-/H$ $CH_2\text{–}C(CH_3)_2COOH$ }	$2\text{–}3 \times 10^9\ M^{-1}\ s^{-1}$	pulse radiolysis	[14]
$CH_2C(CH_3)_2COO^-$ } → } ⟶ $CH_2\text{–}C(CH_3)_2COOH$ } →	$7 \times 10^7\ M^{-1}\ s^{-1}$ $2.2 \times 10^8\ M^{-1}\ s^{-1}$ $2.2 \times 10^8\ M^{-1}\ s^{-1}$	pulse radiolysis	[13]
$Cu^+ + \cdot OOCH_2\text{–}C(CH_3)_2COOH$	$10^9\ M^{-1}\ s^{-1}$	pulse radiolysis	[13]

* Numerical data are given only for an overview. For citations or applications the original literature should be consulted. There, the essential informations on error estimates, temperature, ionic strength, and experimental conditions are given.

Pulse radiolysis is used to deal with the copper ion kinetics in the presence of oxygen and, eventually, with ozone. An improved set of rate data for the so-called ping-pong reaction was initially derived, with formic acid to convert radiolytically produced OH to O_2^-:

$$Cu^{2+} + O_2^- \rightarrow Cu^+ + O_2$$
$$Cu^+ + O_2^- \rightarrow Cu^{2+} + O_2^{2-} \ (\rightarrow H_2O_2)$$

As the usual phosphate buffer chelates Cu^{2+} and substantially limits the concentration of free copper ions, it is replaced by pivalic acid, which simultaneously acts as OH radical scavenger to avoid formation of Cu(III) [14]. Its application allows us to omit the addition of formic acid, which might interfere with ozone.

From experiments with copper ions and pivalic acid, the peroxy radical ($O_2CH_2(CH_3)_2COOH$) formed is found to react with Cu^+ with a rate constant of about $10^9\ M^{-1}\ s^{-1}$ [13]. This indicates that organic peroxy radicals, in general, might interfere with the copper redox mechanism.

The data compiled in Table 1 show *e.g.* that for converting atmospheric HO_2/O_2^- copper(I) and copper(II) ions become more important than dissolved iron whenever the sum of their concentrations exceeds 1 % of that of the iron species [7, 10], and they control the transformation of O_2^- long before their concentrations become comparable to that of ozone, *i.e.* about 10^{-9} M.

In presence of HO_2/O_2^- the concentration ratio of Cu(I)/Cu(II) assumes a ratio that is controlled by the inverse of the relative rate constants with which these species react with HO_2/O_2^- [9]. The thereby occurring daytime steady-state concentration of Cu(I) in atmospheric droplets may control the reduction of aqueous iron(III) to iron(II) [10].

Because, in atmospheric droplets, oxalate may assume concentrations that are comparable to those of iron(III), it chelates a significant fraction of it and forms a highly photo-labile complex. In daylight this quickly (within minutes) photolyses to produce iron(II) and HO_2/O_2^- [5, 7, 8].

The copper-catalysed accelerated disproportionation of HO_2/O_2^- yields an enhanced formation of hydrogen peroxide. This acts as a main oxidant for aqueous SO_2 [11]. Because HO_2/O_2^- seems not to oxidize aquatic species of SO_2, its depletion by copper ions does not interfere with the SO_2 oxidation [11].

Conclusions

Modelling of chemical redox reactions in cloud systems must take into account the reactions of copper and iron species. For these reactions relevant kinetic data are now available. However, many other parameters that control the in-droplet or in-cloud processes are difficult to predict and therefore models must still be critically tested and calibrated by appropriate experiments. Without experimental tests the required assumptions cannot be qualified. However, the models that can now be formulated should help to design good experiments and interpret many of the field data [15–17].

Acknowledgements

We thank the ETHZ (Swiss Federal Institute of Technology, Zürich) and EAWAG (Swiss Federal Institute for Environmental Science and Technology) for the continuous support of this project. Important additional financial support was given by the President of the Board of the Swiss Federal Institute of Technology (1989–91) and by the "Bundesamt für Bildung und Wissenschaft" (1993–95).

References

1. R. E. Bühler, J. Staehelin, J. Hoigné: Ozone decomposition in water studied by pulse radiolysis 1. HO_2/O_2- and HO_3/O_3- as intermediates. *J. Phys. Chem.* **88** (1984) 2560–2564.
2. J. Staehelin, R. E. Bühler, J. Hoigné: Ozone decomposition in water studied by pulse radiolysis 2. OH and HO_4 as chain intermediates. *J. Phys. Chem.* **88** (1984) 5999–6004.
3. J. Hoigné, H. Bader, W. R. Haag, J. Staehelin: Rate constants of reactions of ozone with organic and inorganic compounds in water-III: Inorganic compounds and radicals, *Water Res.* **19** (1985) 993–1004.
4. B. C. Faust, J. Hoigné: Photolysis of Fe(III)-hydroxy complexes as sources of OH radicals in clouds, fog and rain, *Atmos. Environ.* **24**A (1990) 79–89.
5. Y. Zuo, J. Hoigné: Formation of hydroxyl peroxide and depletion of oxalic acid in atmospheric water by photolysis of iron(III)-oxalato complexes, *Environ. Sci. Technol.* **26** (1992) 1014–1022.
6. R. G. Zepp, B. C. Faust, J. Hoigné: Hydroxyl radical formation in aqueous reaction (pH 3–8) of iron(II) with hydrogen peroxide: The Photo-Fenton Reaction, *Environ. Sci. Technol.* **26** (1992) 313–319.
7. J. Hoigné, Y. Zuo, L. Nowell: Photochemical reactions in atmospheric waters: role of dissolved iron species, in: G. R. Helz, R. G. Zepp, D. G. Crosby (eds), *Aquatic and Surface Photochemistry,* Lewis Publ., Ann Arbor 1994, pp. 75–84.
8. Y. Zuo, J. Hoigné: Photochemical decomposition of oxalic, glyoxalic and pyruvic acid catalyzed by iron in atmospheric water, *Atmos. Environ.* **28** (1994) 1231–1239.
9. M. von Piechowski, T. Nauser, J. Hoigné, R. E. Bühler: O_2-decay catalyzed by Cu^{2+} and Cu^{+} ions in aqueous solutions: a pulse radiolysis study for atmospheric chemistry, *Ber. Bunsen-Ges. Phys. Chem.* **97** (1993) 762–771.
10. D. Sedlak, J. Hoigné: The role of copper and oxalate in the redox cycling of iron in atmospheric waters, *Atmos. Environ.* **27**A (1993) 2173–2185.
11. D. Sedlak, J. Hoigné: The oxidation of S(IV) in atmospheric water by photo-oxidants and iron in the presence of copper, *Environ. Sci. Technol.* **38** (1994) 57–66.
12. M. von Piechowski, M.-A. Thelen, J. Hoigné, R. Bühler: *Tert*-butanol as an OH-scavenger in the pulse radiolysis of oxygenated aqueous systems, *Ber. Bunsen-Ges. Phys. Chem.* **96** (1992) 1448–1454.
13. T. Nauser, R. E. Bühler: to be published.
14. T. Nauser, R. E. Bühler: Pivalic acid as combined buffer and scavenger for studies of cloud water chemistry by pulse radiolysis, *J. Chem. Soc. Faraday Trans.* **90** (1994) 3651.
15. J. Matthijsen, P. J. H. Builtjes, D. L. Sedlak: Cloud model experiments of the effect of iron and copper on tropospheric ozone and marine and continental conditions, to be published.

16. D. L. Sedlak, M. M. David, J. Hoigné: The chemistry of iron and copper in cloud water collected at Great Dunn Fell, U. K., to be published.
17. Y. Zuo, J. Hoigné: Evidence for photochemical formation of H_2O_2 and oxidation of SO_2 in authentic fog water, *Science* **260** (1993) 71–73.
18. R. G. Zepp, B. C. Faust, J. Hoigné: Hydroxyl radical formation in aqueous reactions (pH 3–8) of iron(II) with hydrogen peroxide: The photo-Fenton reaction, *Environ. Sci. Technol.* **26** (1992) 313–319.
19. M. von Piechowski: Der Einfluss von Kupferionen auf die Redoxchemie atmosphärischen Wassers. Kinetische Untersuchungen, *Ph.D. Thesis*, Swiss Federal Institute of Technology, Zürich 1991.
20. R. G. Zepp, J. Hoigné, H. Bader: Nitrate-induced photo-oxidation of trace organic chemicals in water, *Environ. Sci. Technol.* **21** (1987) 443–449.

4.4 Kinetics of S(IV) Oxidation in Aqueous Solution: Impacts of Transition Metal Ion Transformations

Wanda Pasiuk-Bronikowska, K.J. Rudzinski, T. Bronikowski and J. Ziajka

Institute of Physical Chemistry, Polish Academy of Sciences, Kasprzaka 44/52, 01-224 Warsaw, Poland

Summary

In the years 1992–1995 we investigated homogeneous catalysis by Co, Cu, Mn and Fe in the reaction of aqueous SO_2 with dioxygen. The reaction was performed in various laboratory reactors supposedly designed to explain the role of the metals in the mechanism of S(IV) oxidation producing cloud acidity. We showed by nonstationary (chain development) and stationary (chain developed) modelling that the cooperative oxidation of divalent metal ions by dioxygen and by peroxomonosulfate radical anions takes place, the latter being reversible. The influence of two simultaneous metal catalysts and of metal complexes with porphin-type ligands on the S(IV) oxidation rate was also examined, which came out, respectively, as the negative synergism and the ligand destruction accompanied by metal ion liberation. The reactivity of transients was studied by scavenging Co(III) with nitroso-R-salt, SO_5^- with various aryl compounds, SO_4^- with benzene, and SO_3^- with stable radicals. The S(IV) oxidation was also carried out with $S_2O_8^{2-}$ replacing the metal(III) initiator. Based on these experiments new kinetic and mechanistic judgements were made. The chemical route leading to solubilization of water-insoluble materials and formation of airborne surfactants was pointed out and the analytical procedure useful for determination of S(IV) in field samples was elaborated. Values of some rate constants for steps engaging transition metal species were found.

Aims of the research

Existing knowledge in the area of transition metal catalysed oxidation of aqueous SO_2 to sulfuric acid (acid rain) varied greatly in understanding the mechanism of catalysis and was highly incomplete in kinetic data. Our research, which we report here, attempts to elucidate the contribution of transition metals (Me = Fe, Mn, Co, Cu) to this chain reaction in diluted systems, as related to tropospheric cloud chemistry. We intended to view the role of metals in terms of the following phenomena: direct oxidation of metal ions by dioxygen as the step balancing the termination losses, redox interaction of two different metal ions as responsible for synergetic effects, catalysis by organic complexes of metals to check the potential impact of biogenic aerosols, and differentiation between radicals as transients in metal catalysed oxidation of aqueous SO_2.

Principal scientific results

We formulated the following keystones to organise the data obtained in kinetic experiments on the studied reaction, a chain one. The reaction may occur either in the nonstationary (chain development) or stationary (chain developed) mode, depending on the lifetime of aerosol droplets. These modes were accomplished in our laboratory experiments.

(i) The autocatalytic behaviour of the reaction observed in the chain development period imposes that two parallel steps producing the chain initiator (Me^{III} species) take place, one of which must be noncatalytic. We postulated that the latter is the reaction between Me^{II} species and molecular oxygen.

(ii) The stationary mode is characterized by a long chain which means that the chain initiation and termination are both very slow in comparison to the formation of sulfate anion. The initiation step is here understood as one in which radicals are formed without engaging other radicals. In the termination step radicals are irretrievably lost.

(iii) Although very slow, the initiation and termination decisively influence the overall rate of aqueous SO_2 oxidation. The fractional orders found in the oxidation rate equation indicate that the dominating termination is quadratic.

Based on these principles we derived simplified, yet realistic kinetic models for both reaction modes. The considerably reduced set of step reactions in the long chain model is:

initiation	Me(IIi)	$+ SO_3^{2-}$	$\rightarrow$	$Me(II) + SO_3^-$	k_i
propagation	SO_3^-	$+ O_2$	$\rightarrow$	SO_5^-	k_{p1}
plus SO_4^{2-} formation	SO_5^-	$+ SO_3^{2-}$	$\rightarrow$	$SO_5^{2-} + SO_3^-$	k_{p2a}
			$\rightarrow$	$SO_4^{2-} + SO_4^-$	k_{p2b}
	SO_4^-	$+ SO_3^{2-}$	$\rightarrow$	$SO_4^{2-} + SO_3^-$	k_{p3}
	SO_5^{2-}	$+ SO_3^{2-}$	$\rightarrow$	$2SO_4^{2-}$	k_f
termination	SO_5^-	$+ SO_5^-$	$\rightarrow$	inerts	k_t

initiator formation (controlled by sulfito complex formation)

$$Me(II) + \tfrac{1}{2}O_2 + H^+ \rightarrow Me(III) + \tfrac{1}{2}H_2O_2 \qquad k_{ox}$$

$$Me(II) + SO_5^- \Longleftrightarrow Me(III) + SO_5^{2-} \qquad k_+, k_-$$

The rate of S^{IV} oxidation, r_{aut}, derived from this set is:

$$r_{aut} = k_{obs}\,[Me]^{1/2}\,[S(IV)]^{3/2}$$

where k_{obs} given in Table 1 is equal under the specified conditions to $k_{p2}(k_i/2k_t)^{1/2}$ at Me = Fe(III), $k_{p2}(k_{ox}/2k_t)^{1/2}$ at Me = Mn(II), and $(k_{p2}k_ik_fK/2k_t)^{1/2}$ at Me = Co(II) or Cu(II), where $k_{p2} = k_{p2a} + k_{p2b}$ and $K = k_+/k_-$. Rate constants estimated in this research for steps engaging metal species are collected in Table 2. Other rate constants used in our modelling are shown in Table 3.

Table 1: The composed rate constant for S(IV) autoxidation.

	k_{obs} / $dm^3\ mol^{-1}\ s^{-1}$	pH	Temp./°C	Concentration / mmol dm^{-3} Metal	S^{IV}	O_2	
Fe	6.8[a]	3.0	25	0.008	0.8–2	0.25	b
	1.6×10^2	6.5–7.0	22–23	0.017–0.700	0.56	0.25	c
Co	0.72×10^4	8.5–8.7	13–32	0.0008–0.40	0.5–70	0.25	d
	1.3×10^4	7.3	20	$(0.03–0.5) \times 10^{-3}$	0.56	0.25	e
	1.3×10^4	7–8	23	$(0.066–1.7) \times 10^{-3}$	0.56	0.25	c
	7.0×10^4	9.2	25	1.0	2.1	0.13	f
Cu	6.7×10^3	7–8	22–23	$(0.13–0.63) \times 10^{-3}$	0.56	0.25	c
Mn	7.5×10^2	7–8	22–23	0.002–0.130	0.56	0.25	c

[a] calculated from data of Beer *et al.*, *EUROTRAC Annual Report 1992, Pt 6*, pp 64–69 with respect to actual Fe^{III} concentration.
[b] monitoring pH in a stirred tank reactor.
[c] conductometry in a gas absorber
[d] flow reactor, titration of unreacted S^{IV}
[e] impinger, titration of unreacted $CaSO_3$
[f] stopped-flow with UV detection

Table 2: Rate constants (in dm^3/mol s) estimated for steps engaging metal species at 25 °C

Metal	pH	k_{ox}	k_i	k_+	$k_+/k_- = K$	Method
Fe	3.0	1.8				a
	3.0		0.03	2.6×10^3		b
	6.5 – 7.0		0.40	1.3×10^6		c
Co	8.7	0.57				d
	8.7	0.10				e
	9.0	0.27				f
	9.2	200	1.4×10^6			g
	8.7		1.4×10^6		5.3×10^{-3}	g
	7 – 8		1.4×10^6		1.7×10^{-2}	c
Mn	7 – 8	0.04				c

[a] SO_4^- scavenging with Bz, HPLC
[b] calculated from data of Beer et al., *ibid*
[c] O_2 absorption in a conductometric cell
[d] SO_5^- scavenging with 1,5 NDS, IC
[e] Co^{III} scavenging with nRs, UV–visible spectroscopy
[f] chain development traced with an O_2 probe
[g] stopped-flow with UV detection
[h] O_2 absorption in a flow reactor, iodometry

Table 3: Step rate constants (in dm^3/mol s) used in S(IV) autoxidation modelling.

Metal	pH	k_{p2}	k_{p2a}	k_{p2b}	k_f	k_t
Fe	3.0	$4.9 \times 10^{5,a}$	$4.4 \times 10^{5,a}$	$5.2 \times 10^{4,a}$		7×10^7
	6.5–7.0	$3.0 \times 10^{6,b}$				7×10^7
Co	7.0–9.2	$1.3 \times 10^{7,c}$			$5.9 \times 10^{3,b}$	5.5×10^6
Mn	7–8	$1.3 \times 10^{7,c}$				5.5×10^6

[a] with $S_2O_8^{2-}$ as the only initiator
[b] Brandt and van Eldik, *Chem. Rev.* **95** (1995) 119
[c] R.E. Huie and P. Neta, *Atmos. Environ.* **21** (1987) 1743

Fig. 1, comparing the catalytic activity of various transition metals, illustrates the validity of the model. It also gives precipitation thresholds for metal catalysts, at S(IV) concentration determined by the $CaSO_3$ solubility.

Differences in the catalytic activity of metals are also evident in the chain development period, as given for Co and Cu in Fig. 2 (points). Successful mathematical-kinetic modelling of this period is also shown (line) [1].

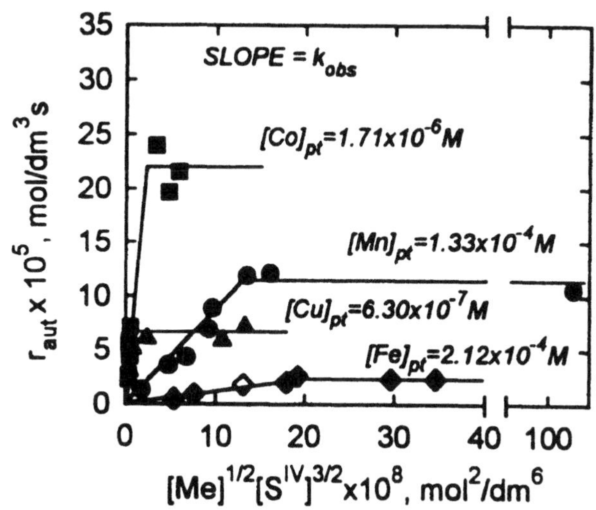

Fig. 1: Comparison of S^{IV} oxidation rates in the presence of various metal catalysts (gas-liquid(slurry) reactor, $[S^{IV}] = 5.6 \times 10^{-4}$ mol/dm^3).

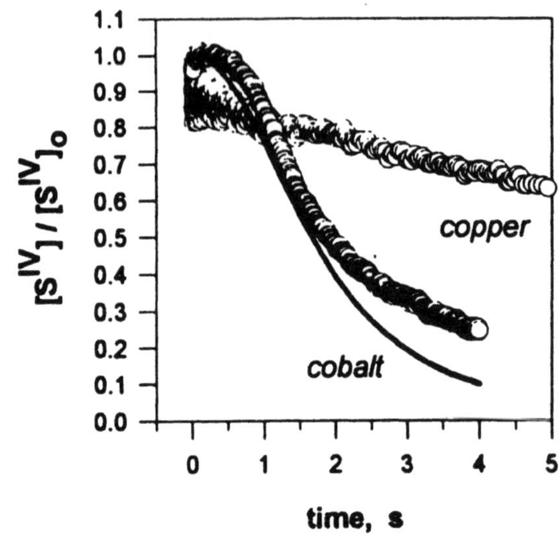

Fig. 2: Consumption of sulfite in Co- and Cu-catalysed oxidation with the chain development period revealed (; stopped-flow, -- kinetic model, $[S^{IV}]_o = 6.5 \times 10^{-5}$, $[O_2]_o = 2.5 \times 10^{-4}$, $[Me] = 1.7 \times 10^{-6}$ mol/dm^3, pH 8.1–8.5, 25 °C).

The rate of the overall reaction producing acidity may be drastically reduced in the presence of two transition metals. We observed negative synergism for Co-Fe and Co-Mn catalysts [2]. It was explained by including in the model a redox step for the initiators:

$$Me_1(III) + Me_2(II) \rightarrow Me_1(II) + Me_2(III)$$

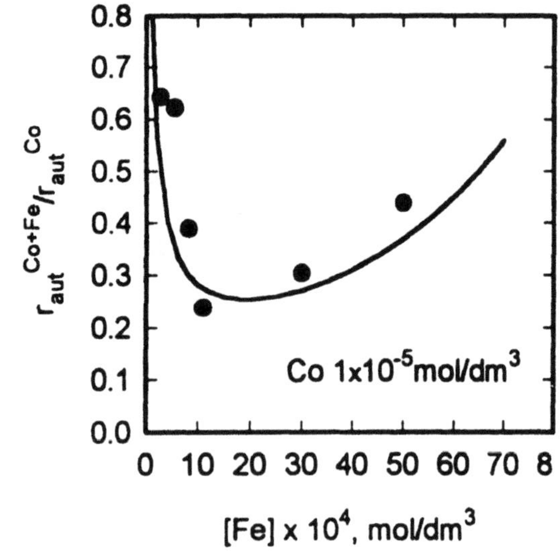

Fig. 3: Negative effect of Co-Fe synergism in S^{IV} oxidation ($[S^{IV}] = 4.0 \times 10^{-2}$ mol/dm^3, pH 9.5, 25 °C).

Experimental points and computed line obtained for Co-Fe are given in Fig. 3. We also investigated the catalytic activity of Co and Mn tetrasulfophtalocyanine complexes (CoTSP and MnTSP) as representing biogenic aerosols. We confirmed the ability of these porphin-type complexes to enhance the oxidation of S(IV), which however decreased with time. The decrease corresponded to the decay of TSP structure observed spectrophotometrically as a result of radical scavenging. The net rate of S(IV) oxidation in the presence of CoTSP is given in Fig. 4, correlated with the decay of CoTSP active dimeric form
(D = TSPCo–O–O–CoTSP) and with the rise of released Co ions.

The radical character of the Me catalysed oxidation of S(IV) implies that this reaction may be hampered by aerosol constituents, provided they can eliminate radicals. By studying side products of the partially hampered reaction at stationary conditions, it is possible to distinguish the sulfoxo radicals. Thus we proved indirectly the occurrence of the following chain carriers.

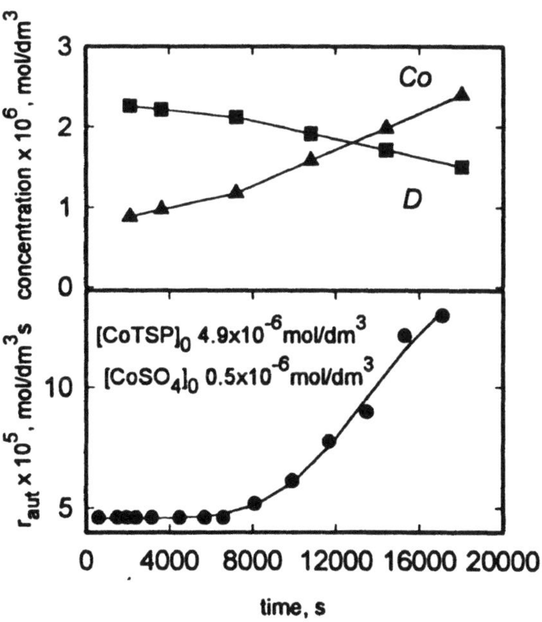

Fig. 4: Release of Co ions from CoTSP modelling a biogenic compound in radical side reaction ($[S^{IV}] = 5.6 \times 10^{-4}$, $[O_2] = 2.5 \times 10^{-4}$ mol/dm^3, pH 7–8, 25 °C).

Peroxomonosulfate radical anion $SO_5^{-}\cdot$

Rates of S(IV) oxidation were studied using aryl compounds (water soluble or insoluble) as inhibitors, [3] transformations of which were traced by UV–visible spectrophotometry and HPLC. We interpreted the results, exemplified in Fig. 5 for 1,5-naphthalenedisulfonate (1,5-NDS), postulating the sulfonation scheme:

$$-\overset{|}{\underset{|}{C}}-H + SO_5^{-\cdot} \rightarrow -\overset{|}{\underset{|}{C}}\cdot + HSO_5^{-} \qquad \text{hydrogen abstraction by } SO_5^{-\cdot}$$

$$-\overset{|}{\underset{|}{C}}\cdot + MeSO_3^{+} \rightarrow -\overset{|}{\underset{|}{C}}-SO_3^{+} + Me^{2+} \qquad \text{incorporation of } SO_3^{-\cdot}$$

This is a possible route of airborne surfactants formation, *e.g.* from natural oil-based pollutants thus solubilized, as well as in atmospheric chemistry involving coal particles and SO_2 [4]. To gain insight into the solubilization step we are investigating selenium etching with SO_5^- at the atomic scale using an atomic-force microscope (AFM). We chose selenium because of its similarity to sulfur and ubiquity in tropospheric aerosols.

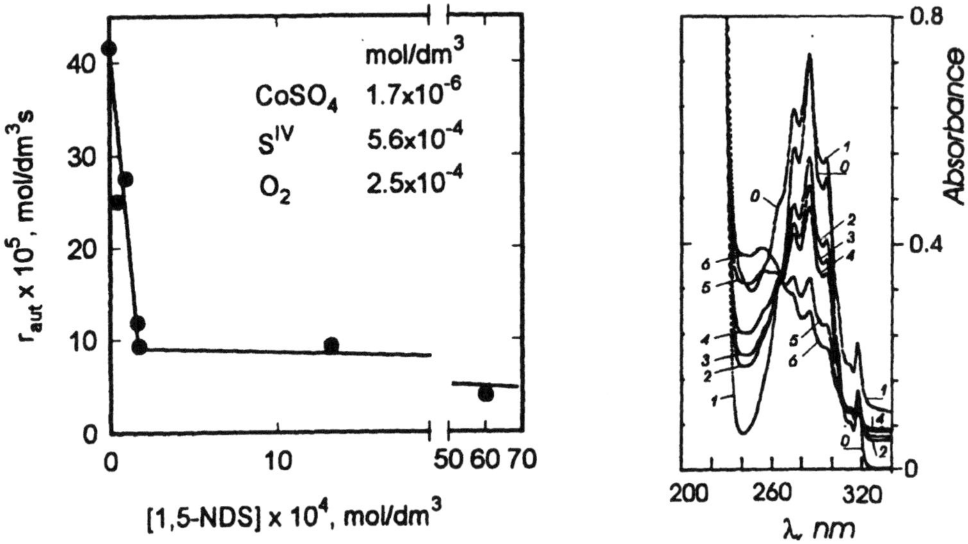

Fig. 5: Results of interaction between 1,5-NDS and S^{IV} oxidation transients (pH 7–8, 25 °C).

Sulfate radical anion $SO_4^{-}\cdot$

We adopted the technique of SO_4^- scavenging with benzene (Bz = C_6H_6) worked out by the research group in the Max-Planck Institut für Chemie (Mainz). In the cooperation with the group in Mainz we obtained experimental evidence for the involvement of $SO_4^-\cdot$ in the S(IV) oxidation chain [5].

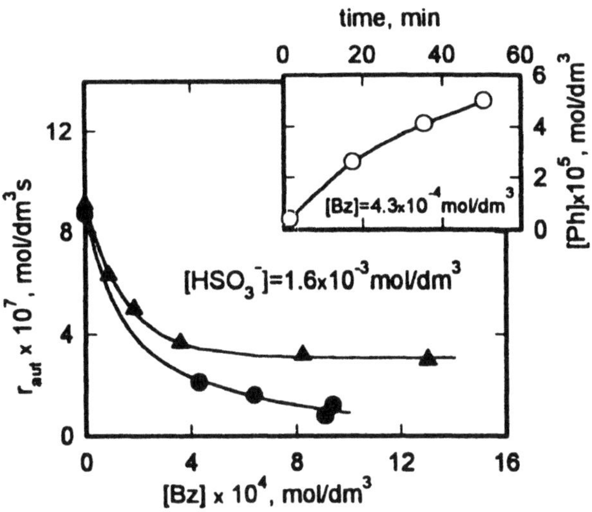

Fig. 6: S^{IV} oxidation hampered by Bz as a selective SO_4^- scavenger (Σ [$Fe(ClO_4)_3$]$_o$ = 1.0 $\times 10^{-5}$, ● and Σ [$S_2O_8^{2-}$] = 1.0 $\times 10^{-2}$ mol/dm^3, pH 3.0, 25 °C).

The reaction:

$$C_6H_6 + SO_4^- \xrightarrow{O_2,H_2O} C_6H_5O^- + SO_4^{2-} + ...$$

was concluded from the decrease in the rate of S(IV) oxidation with the rise in the concentration of benzene and from the increase in the concentration of phenol (Ph) determined by HPLC (Fig. 6). These observations were made both in the presence and absence of Fe. In the latter case, the chain was initiated by decomposition of $S_2O_8^{2-}$

Sulfite radical anion **$SO_3^-\cdot$**

To eliminate $SO_3^-\cdot$, and thus practically stop the chain propagation, we used stable radicals discriminating other sulfoxo radicals due to steric hindrance [6] In particular, we successfully applied 2,2,6,6-tetramethylpiperidine-N-oxyl (TEMPO) postulating the following reaction:

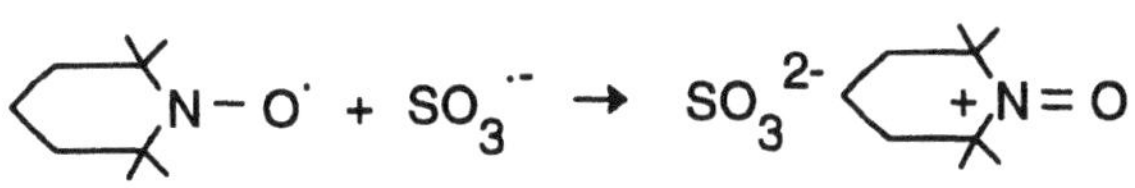

inert complex

This redox step deprives the system of chain carriers.

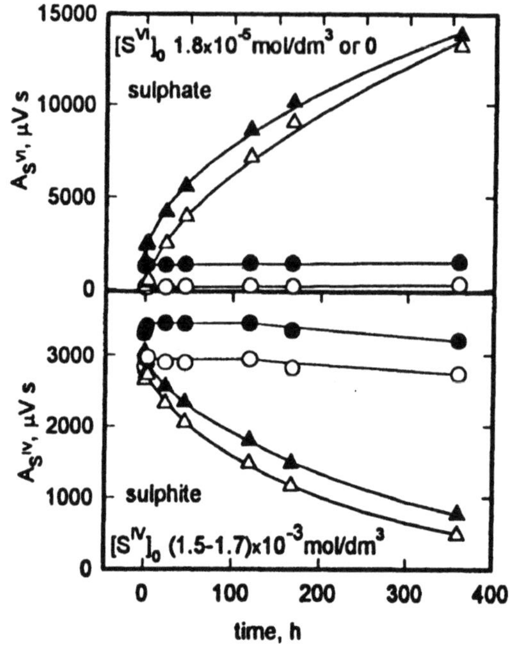

Fig. 7: Resistance of sulfite solutions to oxidation (O ● samples stabilized with TEMPO 1.0×10^{-5} mol/dm^3, Δ Σ samples without the stabilizer, black points indicate samples containing the catalyst $CoSO_4$ 1.8×10^{-5} mol/dm^3)

Fig. 7 shows changes in sulfite and sulfate contents in samples with TEMPO (circles) and without it (triangles). The IC measurements were made to work out a new analytical procedure for the determination of S^{IV} in diluted solutions (especially in samples from field experiments).

In the reported period, we improved our laboratory base necessary for extensive studies of complex reaction kinetics by acquiring the new equipment: HPLC device, ionic chromatograph and stopped-flow apparatus, and gaining access to the atomic force microscope. We also transferred the techniques of gas-absorption with chemical reaction from chemical engineering into the study of reaction kinetics with the environmental bend [7].

Conclusions

Simplified chemical-kinetic models for the transition metal catalysed oxidation of aqueous SO_2 in closed nonstationary and open stationary systems are given which may be used as subroutines in modelling of cloud processes.
Water solubilization of organics by the reaction coupled with S(IV) oxidation has been revealed as a possible route of airborne surfactants formation.

The possibility of blocking the S(IV) oxidation towards production of sulfate anions by shape selective scavenging of radical anions SO_3^- has been proved, with particular attention to analytical applications.

Acknowledgements

Our contribution to the EUROTRAC subproject HALIPP was possible owing to the support from the State Committee for Scientific Research in Poland, which we gratefully acknowledge. We thank the Max-Planck Institut für Chemie in Mainz for the assistance under the scientific cooperation agreement. We are also grateful to the Federal German Ministry of Research and Technology (BMFT), EUROTRAC International Scientific Secretariat (ISS) and Saxon Minister of Research and Art for aid towards our participation in EUROTRAC conferences.

References

1. W. Pasiuk-Bronikowska, K.J. Rudzinski: Simulation of chain development in metal-catalysed autoxidation of S(IV), in: K.H.Becker (ed), *Proc. Joint Workshop LACTOZ/HALIPP (EC/EUROTRAC/GDCh) on Tropospheric Oxidation Mechanism,* ECSC-EC-EAEC, Brussels-Luxembourg, in press.
2. W. Pasiuk-Bronikowska, A.Sokolowski: Mixed cobalt-based catalysts in autoxidation of aqueous SO_2, *Chem. Eng. Commun.*, in press.
3. W. Pasiuk-Bronikowska, T.Bronikowski, M.Ulejczyk, The role of aqueous SO_2 in formation of airborne surfactants, in: K.H. Becker (ed), *Proc. Joint Workshop LACTOZ/HALIPP (EC/EUROTRAC/GDCh) on Tropospheric Oxidation Mechanism,* ECSC-EC-EAEC, Brussels-Luxembourg, in press.
4. W. Pasiuk-Bronikowska, T.Bronikowski, M.Ulejczyk, Oxidation of coal refuse in the presence of tetravalent sulfur, *Fuel*, in press.
5. P. Warneck, J.Ziajka, Reaction mechanism of the iron(III)-catalyzed autoxidation of bisulfite in aqueous solution: steady state description for benzene as radical scavenger, *Ber. Bunsenges. Phys. Chem.* **99** (1995) 59
6. W. Pasiuk-Bronikowska, T.Bronikowski, J.Królik: Nitrosyl radicals as stabilizers of tetravalent sulfur in aqueous solutions, to be published.
7. W. Pasiuk-Bronikowska, T.Bronikowski, M.Ulejczyk, Mechanism and kinetics of autoxidation of calcium sulfite slurries, *Environ. Sci. Technol.* **26** (1992) 1976.

4.5. The Interaction of SO_x and NO_y Species in Aqueous Solution: Photo and Metal Ion Catalysis

Rudi van Eldik, M. Geißler, C. Brandt and V. Lepentsiotis

Institute for Inorganic Chemistry, University of Erlangen-Nürnberg, Egerlandstr. 1, D-91058 Erlangen, Germany

Summary

A number of instrumental developments were undertaken within the project and formed the basis for the subsequent experimental work. These included a three component stopped-flow spectrometer, a flow-through reactor, an ATR-FT-IR coupling, a rapid oxygen detection system, and an ionchromatographic method for the analysis of mixed N-S oxides. These techniques were used to study the autoxidation of S(IV) oxides, the interaction of metal complexes with NO_y species, the redox reactions of SO_x and NO_y species, and the metal ion catalysis of the SO_x/NO_y interaction. The main emphasis fell on the elucidation of the role played by metal ions and complexes in such systems. The sulfite-induced autoxidation of metal ions could be identified as an important step in the overall catalytic cycle. The formation and decomposition of mixed N-S oxides in the presence of metal ions and complexes could be clarified. A systematic control over the reactivity of the metal centre was achieved by using different polyaminocarboxylate chelates. The participation of redox cycles on the catalytic centre could be identified and are throughout very important. The project could, with exception of the photocatalysis, reach its goals, and the overall scientific progress was very good.

Aims of the research

It was the goal of our work to contribute towards a better understanding of metal ion catalysed oxidation reactions of S(IV) oxides, since such reactions have been suggested to play an important role in the autoxidation of S(IV) oxides [1–3]. A systematic spectroscopic and kinetic study was undertaken to investigate the catalytic role of Fe(III) ions and complexes during such oxidation reactions. It was a further goal to extend the work to the interaction of nitrogen oxides (NO_y) with metal ions and complexes, and to study the interaction of SO_x and NO_y species in the absence and presence of metal ions and complexes. In this way we wanted to contribute towards a better understanding of photo- and metal ion-catalysed autoxidation reactions involving sulfur and nitrogen oxides in aqueous solution.

Principal experimental findings

Instrumental developments and modifications

An important prerequisite for the successful conductance of the proposed investigations was the development and modifications of a number of instruments. This included a three component stopped-flow instrument that enables the rapid mixing of three solutions in different sequences, such that the reactions of intermediate species can be followed and pH-jump experiments are possible. This instrument was converted into a double beam mode, and a special double channel amplifier enabled the direct measurement of absorbance as a function of time. A flow-through reactor was constructed to enable the study of the catalytic systems over a longer reaction time [4]. In addition, the available FT-IR spectrophotometer was equipped with an ATR attachment that made it possible to record IR spectra of aqueous solutions. In collaboration with the company Oxytec we developed a system that allowed us to measure the oxygen content of an aqueous solution on a relatively short time basis [5]. An ion chromatographic procedure was developed for the analysis of mixed N-S oxides produced during the reaction of NO/ NO_2^-/ HONO with HSO_3^-/SO_3^{2-} (see Fig. 1) [6]. This method can identify $HON(SO_3)_2^{2-}$ (HADS), $HONH(SO_3)^-$ (HAMS), $N(SO_3)_3^{3-}$ (NTS), $HN(SO_3)_2^{2-}$ (IDS), $H_2N(SO_3)^-$ (SA) and $^-NO(NO)SO_3^-$ (NHAS) in a single analysis (see Fig. 2).

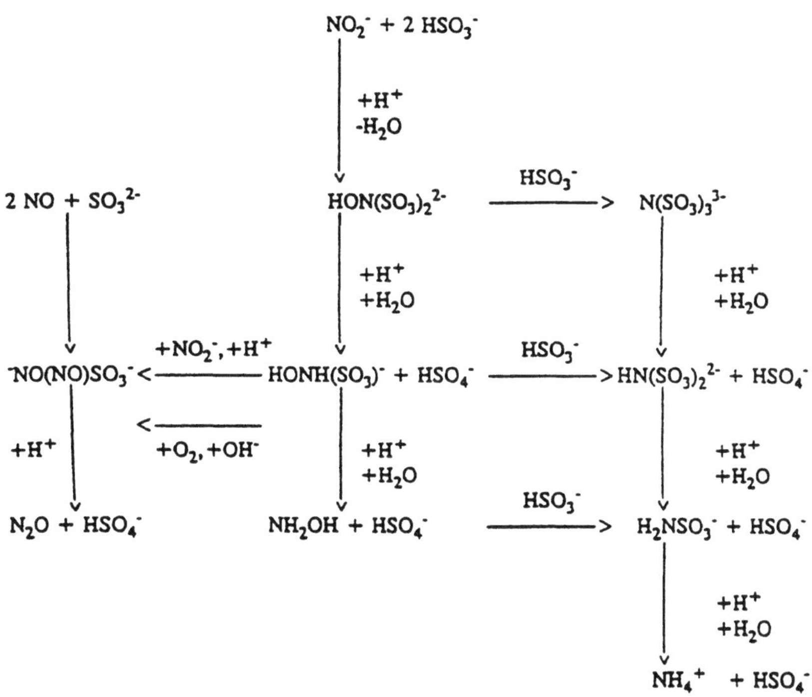

Fig. 1: Formation and decomposition reactions of N-S oxides.

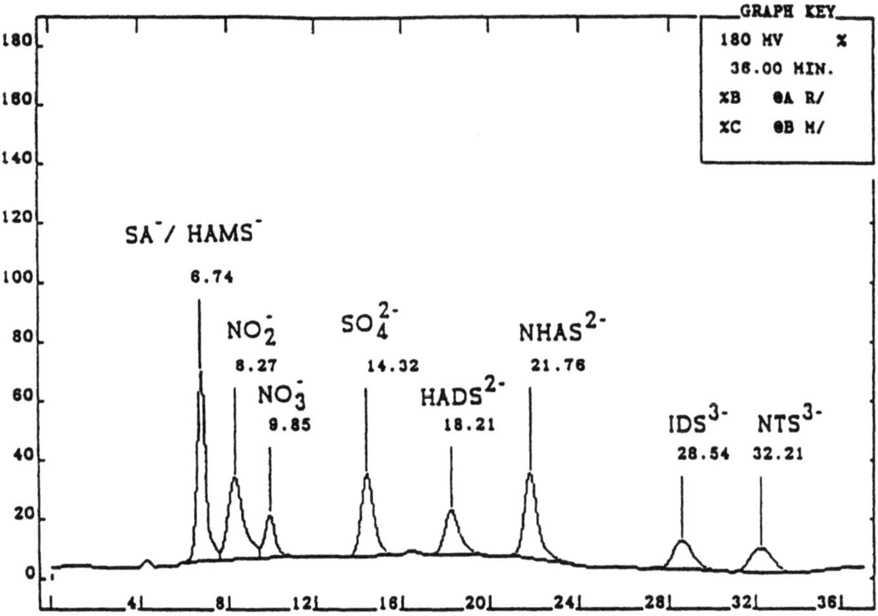

Fig. 2: Typical chromatogram for the analysis of N-S oxides [6].

Metal-catalysed autoxidation of S(IV) *oxides*

The sulfite induced redox cycling of metal ions such as Co(II/III) [7], Fe(II/III) [8] and Mn(II/III) [9] was studied in much detail. The higher oxidation state metal ions are known to catalyse the autoxidation of sulfite via a radical initiation step. In the presence of oxygen, sulfite can induce the autoxidation of Co(II), Mn(II) and Fe(II) according to the general mechanism:

$$M(III) + HSO_3^- \rightarrow M(II) + SO_3^- + H^+$$
$$SO_3^- + O_2 \rightarrow SO_5^-$$
$$SO_5^- + M(II) + H^+ \rightarrow HSO_5^- + M(III)$$
$$HSO_5^- + M(II) \rightarrow SO_4^- + OH^- + M(III)$$
$$SO_4^- + M(II) \rightarrow SO_4^{2-} + M(III)$$

The redox cycling of Fe(II/III) could also be observed directly in the Fe(III) catalysed autoxidation of S(IV) oxides [10]. Modelling of the overall reaction system successfully led to a reproduction of the observed kinetic traces, the simulated concentration profiles are reported in Fig. 3. This work was extended to the acidity range $1.5 \leq pH \leq 4.0$. In general a maximum reaction rate was observed at pH 3.0, with a significant decrease in catalytic activity due to the formation of less reactive hydroxo species of Fe(III) under such conditions.

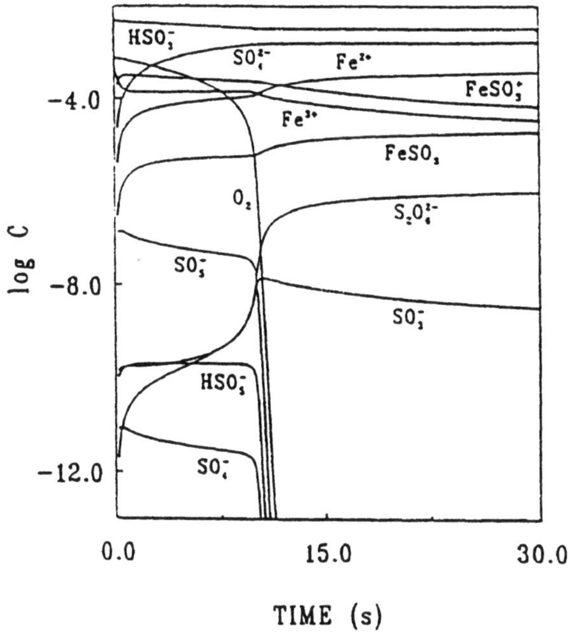

Fig. 3: Simulated concentration-time profiles. Conditions: [Fe(III)] = 5.0×10^{-4} M; [S(IV)] = 5.0×10^{-3} M; $[O_2] = 7.5 \times 10^{-4}$ M; pH = 2.5; Temp. = 25 °C

Interaction of metal complexes with NO_y *species*

Complexes of the type Fe^{II}(L), L = polyaminocarboxylate, react with NO, HONO and NO_2^- to produce Fe^{III}(L)NO. A detailed study of the formation and stability of such complexes and their reactions with bisulfite was undertaken [4, 11]. Such complexes of Fe(II) are very effective scavengers for NO and catalyse the interaction of HSO_3^- and coordinated NO. The reactions result in the formation of HADS, N_2O and Fe^{III}(L) as indicated in the following scheme.

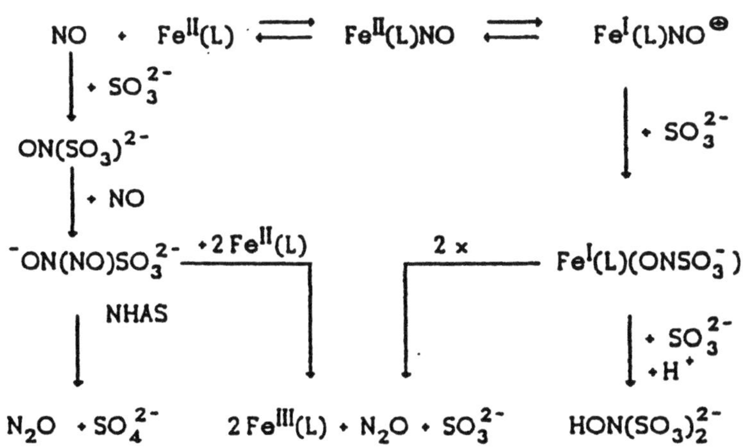

These findings made it essential for us to investigate the interaction of SO_x and NO_y species in the absence and presence of metal ions and complexes in more detail.

Redox reactions of SO_x and NO_y species

The formation and decomposition reactions of the N-S oxides summarized in Fig. 1 were studied as a function of pH using the developed ion chromatographic technique outlined above (see Fig. 2). The decomposition reactions exhibit a clear acid-catalysed path, such that most of the N-S oxides are stable at pH > 5.

In the presence of Mn(III) and Fe(III) ions and complexes the hydrolysis reactions of the N-S oxides are significantly accelerated. In general HADS rapidly hydrolyses to HAMS in the presence of these metal ions, during which they are reduced from the +3 to the +2 oxidation state. N_2O was detected as oxidation product of the N-S oxides. The rate of the metal ion catalysed hydrolysis strongly depends on pH and the buffer selected, since the buffer components will tend to form stable complexes with the metal ions and reduce their catalytic activity. In the case of NHAS, a redox cycling of Mn(III/II) could be observed in the presence of oxygen. Detailed kinetic studies enabled us to suggest reaction mechanisms for these reactions [12–14].

The role of these reactions in atmospheric oxidation processes is presently unknown. In principle, any species which affect the oxidation state of the metal ions responsible for the metal-catalysed autoxidation of sulfur and nitrogen oxides, will indirectly have an effect on the catalytic process. Rain and snow water samples from various parts of Europe were analysed for the presence of N-S oxides. In all cases the concentration levels were below the detection limit of the ion chromatographic technique [15]. This does not rule out the participation of such species as reactive intermediates. Our investigations clearly showed the formation of such species in the presence of SO_x and NO_y, and further studies are needed to identify their role in atmospheric oxidation processes [16].

Achievements

The major goals of this investigation, *i.e.* to clarify the catalytic role of metal ions and complexes in autoxidation reactions of SO_x and NO_y species, have been achieved. The extrapolation of these data to atmospheric conditions was not done, mainly due to the formation of precipitates at pH >3 which complicates the prediction of homogeneous rate constants. More information from heterogeneous processes under such conditions is urgently needed in order to complete the picture. Another aspect that could not be covered during our studies deals with the effect of photochemical irradiation on such processes. Some of this work is presently being performed in our laboratories such that more information on photo-induced redox reactions should become available in the near future. All in all, good progress was made with the project, especially considering the closely related parallel studies undertaken by various participating groups in HALIPP. Throughout this time we could benefit from a lively exchange of ideas and information during stimulating conferences and workshops.

Acknowledgements

The author gratefully acknowledges the financial support from the Deutsche Forschungsgemeinschaft and the Bundesministerium für Forschung und Technologie. He also kindly acknowledges the strong support from his collaborators: A. Gerhard, C. Brandt, M. Dellert-Ritter, F. F. Prinsloo, M. Geißler, J. J. Pienaar, K. Bal Reddy, N. Coichev, V. Zang and I. Fabian.

References

1. M. R. Hoffman, J. G. Calvert, *Chemical transformation modules for Eulerian acid deposition models*. Vol. II, The aqueous-phase chemistry, Interagency agreement DW 930237, 1985.
2. T. E. Graedel, M. L. Mandlich, C. J. Weschler, *J. Geophpys. Res.* **91** (1986) 5205.
3. I. Ibusuki, K. Takeuchi, *Atmos. Environ.* **21** (1987) 1555.
4. V. Zang, R. van Eldik, *J. Chem. Soc. Dalton Trans.* (1993) 111.
5. C. Brandt, N. Coichev, E. Hostert, R. van Eldik, *GIT Fachz. Lab.* **37** (1993) 277.
6. M. Geißler, R. van Eldik, *Anal. Chem.*, **64** (1992) 3004.
7. N. Coichev, R. van Eldik, Inorg. Chem., **30** (1991) 2375.
8. K. Bal Reddy, N. Coichev, R. van Eldik, *J. Chem. Soc. Chem. Commun.* (1991) 481.
9. N. Coichev, R. van Eldik, *Inorg. Chim. Acta* **185** (1991) 69.
10. C. Brandt, I. Fabian, R. van Eldik, *Inorg. Chem.* **33** (1994) 687.
11. V. Zang and R. van Eldik, *Inorg. Chem.* **29** (1990) 4462.
12. F. F. Prinsloo, J. J. Pienaar, R. van Eldik, H. Gutberlet, *J. Chem. Soc. Dalton Trans.* (1994) 2373.
13. F. F. Prinsloo, J. J. Pienaar, R. van Eldik, *J. Chem. Soc. Dalton Trans.* (1995) 463.
14. M. Geißler, R. van Eldik, *Polyhedron* **13** (1994) 2983.
15. C. Brandt, V. Lepentsiotis, A. Paul, H. Hohmann, R. van Eldik, *GIT Fachz. Lab.* **38** (1994) 770.
16. N. Coichev, R. van Eldik, *New. J. Chem.* **18** (1994) 123.

Chapter 5

Photochemical Sources and Reactions of Free Radicals in the Aqueous Phase

5.1 Kinetics and Mechanisms of Acid Generation in Clouds and Precipitation

G. V. Buxton, G. A. Salmon, S. Barlow, T. N. Malone, S. McGowan, S. A. Murray, J. E. Williams and N. D. Wood.

Cookridge Radiation Research Centre, The University of Leeds, Cookridge Hospital, Leeds, LS16 6QB, UK.

Summary

The spectra and molar absorbancies of $SO_3^{-}\cdot$, $SO_4^{-}\cdot$ and $SO_5^{-}\cdot$ have been determined by the technique of pulse radiolysis. The rate constants of fifteen reactions that are important in the free-radical and transition-metal-catalysed oxidation of S(IV) have been determined, including the mutual reaction of $SO_5^{-}\cdot$ radicals and the propagation steps of the chain oxidation. Some reactions relevant to the free-radical induced oxidation of hydroxymethane sulfonate (HMS) have been studied and absorption spectra of $\cdot CH(OH)SO_3^{-}$ and the peroxy-radical derived from it have been recorded. The reactivities of $SO_3^{-}\cdot$, $SO_4^{-}\cdot$ and $SO_5^{-}\cdot$ with Fe(II) have been studied and also the reaction of $SO_5^{-}\cdot$ with Mn(II).

Aims of the research

The aims of the research were to characterise the species involved in the free-radical oxidation of S(IV) in the aqueous phase and to measure the rate constants of the important steps in the mechanism and the closely related transition-metal-catalysed oxidation. In addition, the reactions of organic atmospheric pollutants, notably formaldehyde and HMS, with the free radicals involved were to be studied. The reactivity of the radicals $SO_3^-\cdot$, $SO_4^-\cdot$ and $SO_5^-\cdot$ with the lower valency states of Fe, Mn and Cu were to be investigated.

Principal scientific results

Most of the studies have involved the radicals $\cdot OH$, $SO_3^-\cdot$, $SO_4^-\cdot$ and $SO_5^-\cdot$ which have been generated by pulse radiolysis or steady-state radiolysis. Where oxygenated solutions of S(IV) were used, they were prepared immediately (~ 1s) before pulse radiolysis by mixing equal volumes of an argon or nitrous oxide solution of S(IV) with an oxygen-saturated solution of the other reagent(s).

Absorption spectra of $SO_3^-\cdot$, $SO_4^-\cdot$ *and* $SO_5^-\cdot$

The spectra determined in this study are shown in Fig. 1 and their λ-max values and molar absorbances listed in Table 1. In general, the spectra have similar shapes to those given in previous reports, but, in determining molar absorbancies, particular attention was paid to using radiolytic yields appropriate to the systems studied.

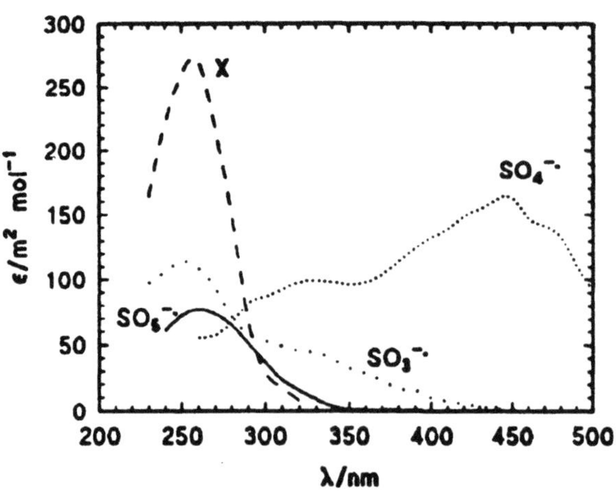

Fig. 1: Absorption spectra of $SO_3^-\cdot$, $SO_4^-\cdot$, $SO_5^-\cdot$ and species X (see text).

Table 1: Spectral Parameters for $SO_3^-\cdot$, $SO_4^-\cdot$ and $SO_5^-\cdot$.

	λ_{max}/nm	$\varepsilon(\lambda_{max})$/m^2 mol^{-1} [a]	Previous values of $\varepsilon(\lambda_{max})$/m^2 mol^{-1}
$SO_3^-\cdot$	260	114 ± 2	138 [1]
			130 [2]
$SO_4^-\cdot$	450	163 ± 5	160 [3]
			110 [1]
$SO_5^-\cdot$	260	77 ± 2	103 [1]

[a] This work.

Kinetics

The main reactions involved in the free-radical chain oxidation and their rate constants determined in this study are listed in Table 2 together with literature values.

Table 2: Rate constants for the free-radical oxidation of S(IV).

No	Reaction			k^a/dm^3 mol^{-1} s^{-1}	Literature value	Ref
1	$\cdot OH + HSO_3^-$	$\rightarrow$	$H_2O + SO_3^-\cdot$	$(2.7 \pm 0.2) \times 10^9$	4.5×10^9	4
2	$\cdot OH + HSO_3^{2-}$	$\rightarrow$	$OH^- + SO_3^-\cdot$	$(4.6 \pm 0.2) \times 10^9$	5.5×10^9	4
3	$SO_3^-\cdot + O_2$	$\rightarrow$	$SO_5^-\cdot$	$(2.5 \pm 0.1) \times 10^9$	1.1×10^9	1
4	$SO_5^-\cdot + HSO_3^-$	$\rightarrow$	$HSO_5^-\cdot + SO_3^-\cdot$	$(8.6 \pm 0.6) \times 10^3$ [c]	$< 3 \times 10^5$	4
5	$SO_5^-\cdot + SO_3^{2-}$	$\xrightarrow{H^+}$	$HSO_5^-\cdot + SO_3^-\cdot$	$(2.15 \pm 0.07) \times 10^5$ [d]	1.3×10^7	4
6	$SO_5^-\cdot + HSO_3^-$	$\rightarrow$	$SO_4^-\cdot + H^+ + SO_4^{2-}$	$(3.6 \pm 0.25) \times 10^2$ [c]		
7	$SO_5^-\cdot + SO_3^{2-}$	$\rightarrow$	$SO_4^-\cdot + SO_4^{2-}$	$(5.5 \pm 0.2) \times 10^5$ [d]		
8	$SO_4^-\cdot + HSO_3^-$	$\rightarrow$	$SO_4^{2-} + H^+ + SO_3^-\cdot$	$(6.8 \pm 0.1) \times 10^8$	7.5×10^8	7
9	$SO_4^-\cdot + SO_3^{2-}$	$\rightarrow$	$SO_4^{2-} + SO_3^-\cdot$	$(3.1 \pm 0.1) \times 10^8$	5.5×10^8	6
10	$SO_5^-\cdot + SO_5^-$	$\rightarrow$	$S_2O_8^{2-} + O_2$	$(1.00 \pm 0.06) \times 10^8$ [b]	$(1.86 \pm 0.02) \times 10^8$ [b]	8
					$(2.6 \pm 0.6) \times 10^8$ [b]	9
11	$SO_5^-\cdot + SO_5^-\cdot$	$\rightarrow$	$2SO_4^-\cdot + O_2$	$\sim 3 \times 10^8$ [b]	$(7 \pm 1) \times k_{10}$	10
12	$SO_3^-\cdot + SO_3^-\cdot$	$\rightarrow$	$S_2O_6^{2-}$	$(4.2 \pm 0.15) \times 10^8$ [b]	4.8×10^8	1
13	$SO_4^-\cdot + SO_4^-\cdot$	$\rightarrow$	$S_2O_8^{2-}$		8.8×10^8 [b]	3
14	$SO_5^-\cdot + HO_2\cdot$	$\rightarrow$	X	$(1.7 \pm 0.1) \times 10^9$	—	
15	$SO_5^-\cdot + O_2^-$	$\xrightarrow{H^+}$	$HSO_5^- + O_2$	$(2.7 \pm 0.2) \times 10^8$	—	

[a] This work; [b] $2k$; [c] Uses the branching ratio in ref. [5]; [d] Uses branching ratio in ref. [6].

Our steady-state measurements on the chain oxidation yield $(2k_4 + k_6)/(2k_{10})^{½} = 1.76 \pm 0.07$ and $(2k_5 + k_7)/(2k_{10})^{½} = 97.5 \pm 6.0$. It is noteworthy that the resulting values for the rate constants of the propagation reactions at pH 9 (k_5 and k_7) are much less than a previous estimate, [4] as is also the case at pH 4 (k_4 and k_6) where only an upper limit was given.

The reaction between $SO_5^- \cdot$ and $HO_2 \cdot$, reaction (14), was shown to lead to the formation of an intermediate, X, which we presume to be a tetroxide species and whose spectrum is shown in Fig. 1. X has been shown to react with S(IV) and a more detailed investigation of this reaction is in progress.

During our studies of the decay of $SO_5^- \cdot$ by the self-reaction, we also observed the formation of $SO_4^- \cdot$. The low values of k_6 and k_7 observed in this study are not compatible with these reactions being the source of $SO_4^- \cdot$ but suggest that reaction (11) is responsible and that $k_{11} \sim 3.0 \times k_{10}$. We consider that the evaluation of k_{10} in ref. [8] was based on an invalid mechanism. Also, it is not clear whether the value of k_{10} reported in ref. [9] might be $k_{10} + k_{11}$ since $SO_4^- \cdot$ was not detected.

Oxidation of HMS

The rate constant for the reaction of ·OH with HMS at pH 4, k_{16}, has been determined by competition with SCN^- and $Fe(CN)_6^{4-}$ and found to be $(3.0 \pm 0.3) \times 10^8\ dm^3\ mol^{-1}\ s^{-1}$.

$$\cdot OH + CH_2(OH)SO_3^- \rightarrow H_2O + \cdot CH(OH)SO_3^- \qquad (16)$$

The spectrum of the product radical is pH dependent (see Fig. 2) due to the occurrence of equilibrium (17) with $pK_{17} = 5.9$, $k_{17} = (2.4 \pm 0.5) \times 10^5\ s^{-1}$ and $k_{-17} = 2.0 \times 10^{11}\ dm^3\ mol^{-1}\ s^{-1}$

$$\cdot CH(OH)SO_3^- \quad <=> \quad \cdot CH(O^-)SO_3^- + H^+ \qquad (17)$$

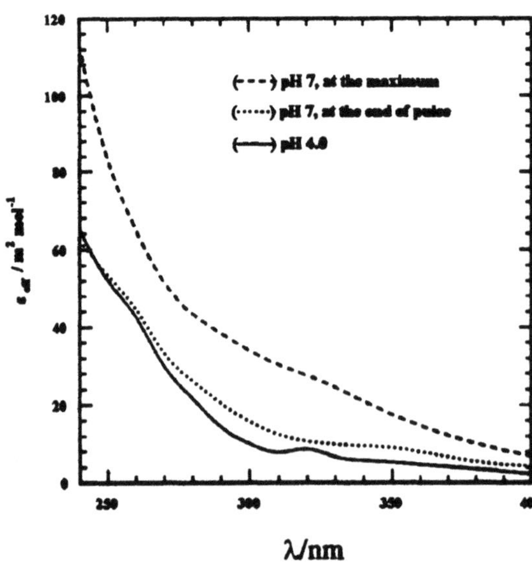

Fig. 2: Absorption spectra of the free radical obtained by attack of ·OH on HMS.

At pH 4 the radical $\cdot CH(OH)SO_3^-$ reacts with oxygen with $k_{18} = (2.6 \pm 0.3) \times 10^9\ dm^3\ mol^{-1}\ s^{-1}$ to form a peroxy radical (see Fig. 3). The peroxy radical decays with $k_{19} = (3.5 \pm 0.2) \times 10^4\ s^{-1}$ by a split path, one path yielding $HO_2\cdot$ and the other, though less certain, probably $SO_4^-\cdot$ and HCOOH. At pH 7 the $\cdot CH(O^-)SO_3^-$ radical reacts with O_2 to form $O_2^-\cdot$ (see Fig. 3) with $k_{20} = (1.64 \pm 0.08) \times 10^9\ dm^3\ mol^{-1}\ s^{-1}$.

$$\cdot CH(OH^-)SO_3^- + O_2 \rightarrow \cdot OOCH(OH)SO_3^- \quad (18)$$
$$\cdot OOCH(OH)SO_3^- \rightarrow HO_2\cdot + CHO\cdot SO_3^- \quad (19a)$$
$$\cdot OOCH(OH)SO_3^- \rightarrow SO_4^-\cdot + HCOOH \quad (19b)$$
$$\cdot CH(O^-)SO_3^- + O_2 \rightarrow CHO\cdot SO_3^- + O_2^- \quad (20)$$

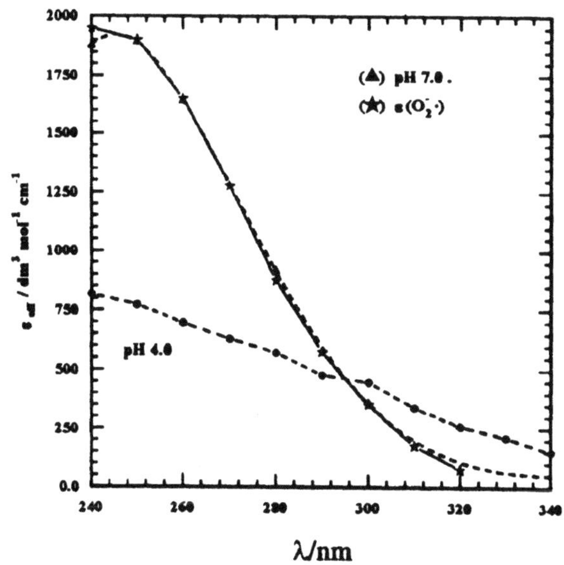

Fig. 3: Comparison of the spectrum of $O_2^-\cdot$ with the experimental spectrum at pH 7.0 and the spectrum of $OOCH(OH)SO_3^-\cdot$ at pH 4.

Reactions of $SO_3^-\cdot$, $SO_4^-\cdot$ ***and*** $SO_5^-\cdot$ ***with TMI***

$SO_3^-\cdot + Fe_2^+$

The reaction results in the formation of a product which absorbs over the spectral region 250–500 nm. The formation of the product monitored at 450 nm obeyed a first-order rate law, but the observed rate constant showed a non-linear dependence on $[Fe^{2+}]$, and also depended on [S(IV)]. From these dependencies it was concluded (i) that Fe^{2+} forms a complex with HSO_3^-, equilibrium (21), with $K_{21} = 770\ dm^3\ mol^{-1}\ s^{-1}$ at pH 4 and (ii) that $SO_3^-\cdot$ reacts with the uncomplexed and complexed Fe(II) with rate constants k_{22} and k_{23} of 1.4×10^7 and $3.9 \times 10^6\ dm^3\ mol^{-1}\ s^{-1}$, respectively.

$$Fe^{2+}(aq) + HSO_3^- \Leftrightarrow [FeHSO_3]^+ \quad (21)$$
$$Fe^{2+}(aq) + SO_3^-\cdot \rightarrow Fe(III) + SO_3^{2-} \quad (22)$$
$$[FeHSO_3]^+ + SO_3^-\cdot \rightarrow Fe(III) + SO_3^{2-} \quad (23)$$

$SO_4^-\cdot + Fe^{2+}$

The decay of $SO_4^-\cdot$ monitored at 450 nm obeyed first-order kinetics with a rate constant proportional to [Fe(II)], the slope of which gave $k_{24} = (4.8 \pm 0.2) \times 10^9$ dm^3 mol^{-1} s^{-1}, independent of ionic strength. This value is very different from that of 3.8×10^8 dm^3 mol^{-1} s^{-1} measured by McElroy and Waygood [3]. The formation of Fe(III) monitored at 300 nm was found to be irreproducible from pulse to pulse, possibly due to hydrolysis.

$SO_5^-\cdot + Fe^{2+}$

The formation of product monitored at 460 nm was observed to proceed in two stages, a fast step due to reaction (24) and a slower stage, reactions (25) and (26).

$$Fe(II) + SO_5^-\cdot \rightarrow Fe(III) + HSO_5^- \quad (24)$$
$$Fe(II) + HSO_5^- \rightarrow Fe(III) + SO_4^-\cdot + OH^- \quad (25)$$
$$Fe(II) + HSO_5^- \rightarrow Fe(III) + \cdot OH + SO_4^{2-} \quad (26)$$

For $[HSO_3^-] = 5$ mmol dm^{-3}, we determined $k_{24} = 3.2 \times 10^6$ dm^3 mol^{-1} s^{-1} and $(k_{25} + k_{26}) = (3.6 \pm 0.2) \times 10^4$ dm^3 mol^{-1} s^{-1}.

$SO_5^-\cdot + Mn(II)$

The reaction was observed at pH 3 in solutions with $[HSO_3^-]$ in the range 2×10^{-4} to 5×10^{-3} mol dm^{-3} by monitoring the formation of product at 440 nm. The observed rate constant decreased with increasing [Mn(II)], which was interpreted as being due to the Mn(II) being distributed amongst the complexes $Mn^{2+}(aq)$, $[Mn(HSO_3)]^+$ and $[Mn(SO_3)Mn]^{2+}$, each of which can react with $SO_5^-\cdot$. Values of the second-order rate constant in the range 2×10^8–2×10^{10} were determined, depending on the distribution of the Mn(II) amongst the complexes.

Conclusions

Our studies of the reactivities of $SO_3^-\cdot$, $SO_4^-\cdot$ and $SO_5^-\cdot$ have resulted in the main steps in the free-radical chain oxidation of S(IV) being elucidated, along with important reactions in the TMI catalysed reaction. The rate constants evaluated will contribute significantly to the effective computer modelling of the oxidation of S(IV) and other processes in cloud droplets.

Acknowledgements

We wish to acknowledge support from the Commission of the European Communities under contracts EV4V-0080-C(AM), STEP-0005-C(MB) and EV5V-CT93-0317 and from the SERC/NERC Atmospheric Chemistry Initiative.

References

1. R. E. Huie, C. L. Clifton, N. Alstein, *Radiat. Phys. Chem.* **33** (1989) 293.
2. Z. P. Zagorski, K. Sehested, S. O. Nielson, *J. Phys. Chem.* **75** (1971) 3501.
3. W. J. McElroy, S. J. Waygood, *J. Chem. Soc. Faraday Trans.* **96** (1990) 2557.
4. R. E. Huie, P. Neta, *Atmos. Environ.* **21** (1987) 1743.
5. P. Warneck, J. Ziajka, *Ber. Bunsenges. Phys. Chem.*
6. U. Deister, P. Warneck, *J. Phys. Chem.* **94** (1990) 2191.
7. P. H. Wine, Y. Tang, R. P. Thorn, J. R. Wells, D. D. Davis, *J. Geophys. Res.* **94**D (1989) 1085.
8. S. J. Waygood (1992) in: R. A. Cox (ed), *Air Pollution Research Report* **42**, pp. 23–26.
9. R. Zellner, H. Herrmann, G. Raabe, A. Reese, in: *EUROTRAC Annual Report 1992, Part 6,* EUROTRAC ISS, Garmisch-Partenkirchen 1993, p. 70.
10. A. N. Yermakov, B. V. Zhitomirsky, G. A. Poskrebyshev, D. M. Sozurakov, *J. Phys. Chem.* **97** (1993) 10712.

5.2 Chain Oxidation Mechanisms of S(IV) in Aqueous Solution: Application of Radical Scavenger Techniques

Peter Warneck[1], H.-J. Benkelberg[1], U. Deister[1], M. Fischer[1], A. Schäfer[1] and J. Ziajka[2]

[1]Max-Planck-Institut für Chemie, Mainz, Germany ,
[2]Institute of Physical Chemistry, Polish Academy of Sciences, Warsaw

Summary

The radical scavenging technique has been employed to determine quantum yields for radical formation in the photodecomposition in aqueous solution of $FeOH_2^+$, $FeSO_4^+$, NO_2^-, HNO_2, and H_2O_2 as a function of wavelength, and of SO_3^{2-} and HSO_3^- at 254 and 214 nm, respectively. Photodissociation of $FeOH_2^+$ appears to be the most effective process for the generation of radicals in clouds ($j_{OH} = 2.3 \times 10^{-3}$ s^{-1}, in mid-latitude summer). Rate coefficients and branching ratios for several reactions involved in the photochemically induced chain oxidation of S(IV) in aqueous solution have been determined. The iron-catalysed oxidation of S(IV) has been demonstrated to involve SO_4^- radicals, and a chain mechanism capable of quantitatively explaining the experimental data has been developed.

Aims of the research

A sound knowledge of cloud chemical mechanisms is necessary to achieve an understanding of the formation of acidity in clouds, and it is a prerequisite for the development of models capable of simulating cloud chemical processes. The general aim of this study was to help establishing cloud chemical mechanisms in the aqueous phase by appropriate laboratory experiments. The focus of the study originally was to determine the role of sulfate radicals in the chain oxidation of sulfur (IV), and to quantify the wavelength-dependent photochemical yield of OH radicals from iron(III)-hydroxo complexes by using the radical scavenger technique combined with product analysis. After these goals had been accomplished, the technique was applied to determine quantum yields for several other photodecomposition processes in aqueous solution. The radical scavenger technique was also applied to the iron-catalysed oxidation of sulfur (IV) in order to determine whether the mechanism involved radicals.

Principal scientific results

Photochemical processes

Table 1 summarises the photochemical processes studied, the wavelength ranges explored, the quantum yields obtained, the radical scavengers applied and the products used to determine quantum yields.

Table 1: Summary of photochemical processes and quantum yields.

Process	Wavelength range / nm	Radical quantum yield	Scavenger / Product[(a)]
$FeOH^{2+} + h\nu \rightarrow Fe^{2+} + OH$	280–370	0.06–0.31 [(b)]	2-propanol/acetone
$FeSO_4^+ + h\nu \rightarrow Fe^{2+} + SO_4^-$	280–350	$(1.5–7.5) \times 10^{-3}$ [(b)]	benzene/phenol
$SO_3^{2-} + h\nu \rightarrow SO_3^- + e_{aq}^-$	253.7	0.38 ± 0.03	sulfate/dithionate [(d)]
$HSO_3^- + h\nu \rightarrow SO_3^- + H$	213.9	0.19 ± 0.04	sulfate/dithionate [(d)]
$NO_2^- + h\nu \rightarrow NO + O^-$	280–390	0.02–0.07 [(b)]	benzene/phenol
$HNO_2 + h\nu \rightarrow NO + OH$	280–385	0.30 ± 0.02 [(c)]	benzene/phenol
$H_2O_2 + h\nu \rightarrow OH + OH$	250–310	1.43 ± 0.15 [(c)]	2-propanol/acetone

[(a)] Compound used as radical scavenger and product measured for determination of quantum yields;
[(b)] Quantum yield is rising with decreasing wavelength;
[(c)] Independent of wavelength;
[(d)] No scavenger applied, products observed in oxygen-free solution.

Photodecomposition of iron(III)-hydroxo complexes

Aqueous solutions of Fe(III) at pH ≤ 4 contain at least four ionic species in equilibrium with each other: Fe^{3+}, $FeOH^{2+}$, $Fe(OH)_2^+$ and the $Fe_2(OH)_2^{4+}$ dimer. The concentration of the dimer was minimised by choosing total Fe(III) concentrations $\leq 5 \times 10^{-4}$ mol/dm^3. The equilibrium constant and its ionic strength dependence for the ion pair $Fe^{3+}/FeOH^{2+}$ was well known at the beginning of this study, whereas that for $FeOH^{2+}/Fe(OH)_2^+$ was not. Spectrophotometric measurements were carried out to ascertain that at pH <= 3 the influence of $Fe(OH)_2^+$ on the system was negligible. The major effect of varying the pH thus was a change in the equilibrium between Fe^{3+} and $FeOH^{2+}$. The determination of overall OH quantum yields made use of a xenon arc and monochromator for wavelength selection in 10 nm steps; 2-propanol was employed as a scavenger for OH radicals, and the yield of product acetone was measured. The specific quantum yield for OH production from $FeOH^{2+}$ was calculated from the data obtained at pH 2 and 3. The photodissociation coefficient for mid-latitude summer conditions was calculated to be 2.3×10^{-3} s^{-1} [1].

Photodecomposition of the Fe(III)-sulfato complex

The $FeSO_4^+$ complex was formed by addition of sulfate to the Fe(III) system. Benzene was used as a radical scavenger, and phenol was determined as product. The addition of 2-methyl-2-propanol, which reacts rapidly with OH yet slowly with SO_4^- radicals, allowed to discriminate against OH radicals formed from

$FeOH^{2+}$ present in the system. The wavelength dependence of SO_4^- quantum yields from the photodecomposition of $FeSO_4^+$ showed a behaviour similar to that of OH from $FeOH^{2+}$, but the absolute values were much smaller. The photodissociation coefficient was calculated to be 4×10^{-5} s^{-1} under mid-latitude summer conditions. This makes the process fairly unimportant in the atmosphere.

Photodecomposition of sulfite and hydrogensulfite anions

The first of these processes has been studied by irradiating aqueous solutions of sodium sulfite with light of 254 nm derived from a mercury lamp. It was necessary to eliminate the 185 nm line by means of speciality quartz. The generation of SO_3^- radicals is indicated by the simultaneous production of sulfate and dithionate, which are addition and disproportionation products, respectively, resulting from the self-reaction of SO_3^-. The coproduction with SO_3^- of a hydrated electron was demonstrated by saturating the solution with N_2O, which reacts rapidly with hydrated electrons. This raised the quantum yield of sulfate and dithionate to twice the value observed in argon-saturated solution.

In acidic solutions sulfite is converted to the hydrogensulfite anion. Absorption of light by the HSO_3^- ion requires wavelengths shorter than 230 nm; a zinc resonance lamp with emission at 214 nm was employed to photodecompose HSO_3^- at pH ~ 4. The formation of dithionate and sulfate was again used as an indicator for the formation of SO_3^- radicals. The intensity of the zinc lamp was determined by means of hydrogen azide actinometry in acidic aqueous solution with quantification of both nitrogen and hydroxylamine as products. Neither SO_3^{2-} nor HSO_3^- would be photochemically active in sunlight under atmospheric conditions. These processes are convenient sources of SO_3^- radicals for laboratory studies.

Photodecomposition of nitrite anion and nitrous acid

The monochromator set-up was employed to study the wavelength dependence of OH quantum yields, benzene was used as scavenger and the yield of phenol was determined. Whereas the OH yield from the photodecomposition of nitrite anion rose with decreasing wavelength in harmony with results from other investigators, the yield of OH from undissociated nitrous acid was essentially independent of wavelength in the 280–385 nm range. Nitrate also evolved in fairly large yield; the latter observation is not yet fully understood.

Photodecomposition of hydrogen peroxide H_2O_2

In this case, 2-propanol was used as scavenger for OH radicals and product acetone was monitored. The wavelength dependence was studied with the monochromator arrangement. Additionally the 254 nm mercury resonance line was used as photochemical light source, and the decrease of H_2O_2 concentration with time was followed to determine the quantum yield in the absence of scavenger. Potential chain reaction processes were minimised by selecting an appropriate concentration regime for the measurements. The mechanism of 2-propanol oxidation also was studied. No wavelength dependence of the OH quantum yield was observed in the 250–310 nm range. The photochemical production of oxygen atoms was ascertained to be negligible.

Role of sulfate radicals in the chain oxidation of sulfur (IV)

In aerated aqueous solution the photochemical production of SO_3^- from sulfite generates SO_5^- radicals, which enter into a chain oxidation reaction with sulfite. Benzene in the presence and absence of 2-methyl-2-propanol was employed as scavenger to search for OH radicals in the system. There was no indication for the formation of OH. Ethanol and 2-propanol were found suitable as scavengers for SO_4^- radicals to determine from the rates of formation of acetaldehyde and acetone, respectively, relative to that of sulfate, the rate constant for the reaction of SO_4^- with sulfite and the branching ratio of the chain propagating reaction between SO_5^- and sulfite anion

$$\text{(a)}\ SO_5^- + SO_3^{2-} \rightarrow SO_5^{2-} + SO_3^- \qquad k_a/(k_a + k_b) = 0.28$$
$$\text{(b)}\ SO_5^- + SO_3^{2-} \rightarrow SO_4^{2-} + SO_4^- \qquad k_b/(k_a + k_b) = 0.72$$
$$SO_4^- + SO_3^{2-} \rightarrow SO_4^{2-} + SO_3^- \qquad k = (3.8 \pm 1.5) \times 10^8\ \text{dm}^3\ \text{mol}^{-1}\ \text{s}^{-1}$$

where the latter value was corrected and refers to zero ionic strength. The photo-oxidation of HSO_3^- was explored in mildly acidic solutions (pH ~ 4) irradiated with light at 214 nm wavelength. The chain length in the absence of scavenger was fairly short (about 1.4 *versus* 300 in alkaline medium). In addition to sulfate, peroxyodisulfate was formed, which was assumed to be the main product resulting from the self-reaction of SO_5^- radicals. Based on the known rate coefficient for this process, $7 \times 10^7\ \text{dm}^3\ \text{mol}^{-1}\ \text{s}^{-1}$ [2], the data provided approximate values for rate coefficients of the propagation reactions and the major termination reaction

$$\text{(a)}\ SO_5^- + HSO_3^- \rightarrow HSO_5^- + SO_3^- \qquad k_a + k_b \approx 1.25 \times 10^4\ \text{dm}^3\ \text{mol}^{-1}\ \text{s}^{-1}$$
$$\text{(b)}\ SO_5^- + HSO_3^- \rightarrow SO_4^{2-} + SO_4^- + H^+$$
$$HO_2 + SO_5^- \rightarrow SO_5^- + O_2 \qquad k = 2.8 \times 10^9\ \text{dm}^3\ \text{mol}^{-1}\ \text{s}^{-1}$$

It was not possible to determine the branching ratio for reaction (3). One detriment was the short chain length, another problem was that the scavengers tested suffered photo-decomposition giving products identical to those arising from the reactions of SO_4^- radicals. However, information about the branching ratio was obtained from other experiments described below.

Iron-catalysed oxidation of hydrogensulfite

The oxidation of millimolar concentrations of HSO_3^- in aerated solution at pH ≈ 3, catalysed by either ferric or ferrous ions in the 10^{-6}–10^{-5} M concentration range, was studied by following the rise of hydrogen ion activity, which was shown to be equivalent to the production of sulfate. Benzene, with or without the further addition of 2-methyl-2-propanol, was used as radical scavenger, because it does not form complexes with either Fe(III) or Fe(II). It was demonstrated that SO_4^-

was involved, whereas the presence of OH radicals was not evident. The addition of benzene to the system reduced the rate of hydrogen ion formation and gave rise to the production of phenol. An important feature of the reaction that was evident most clearly in the absence of scavenger is the autocatalytic rise in the oxidation rate when Fe(II) was the catalyst. Fe(III) as catalyst was found to be rapidly converted toward Fe(II) until a steady state was established between both species. Table 2 shows a mechanism used to explain the observations. Computer calculations based on the rate coefficients shown demonstrated that the mechanism is capable of simulating the experimental data [3]. A steady state treatment was then developed to derive from changes in the formation rates of hydrogen ion and phenol, upon changing initial concentrations, ratios of rate coefficients by competition kinetics [4]. Table 2 includes rate coefficients derived in this manner by reference to absolute rate data from independent studies. The branching ratio for the formation of SO_4^- in reaction (10) is $k_{10a}/k_{10} \leq 0.04$. This value is much smaller than that determined for the propagation reaction of SO_5^- with the sulfite anion.

Table 2: Mechanism[(a)] for the iron-catalysed oxidation of HSO_3^- HSO_3^-.

No.	Reaction	Rate coefficient[(b)] Computer simulation	Steady state
0	$FeOH^{2+} + HSO_3^- \Leftrightarrow [FeOHSO_3H]^+$	$K_0 \gg 600$ dm³/mol	
1	$[FeOHSO_3H]^+ \rightarrow Fe^{2+} + SO_3^- + H_2O$	$0.065\ s^{-1}$	
2	$SO_3^- + O_2 \rightarrow SO_5^-$	2.5×10^9	
3	$Fe^{2+} + SO_5^-\ (+H_2O) \rightarrow FeOH^{2+} + HSO_5^-$	8.0×10^5	6.5×10^5 (c)
4	$Fe^{2+} + HSO_5^- \rightarrow FeOH^{2+} + SO_4^-$	5.0×10^4	3.1×10^4 (d)
5	$FeOH^{2+} + HO_2 \rightarrow Fe^{2+} + O_2 + H_2O$	1.3×10^5	
6	$Fe^{2+} + HO_2\ (+H_2O) \rightarrow FeOH^{2+} + H_2O_2$	1.2×10^6	
7	$Fe^{2+} + H_2O_2 \rightarrow FeOH^{2+} + OH$	75	
8	$Fe^{2+} + SO_4^-\ (+H_2O) \rightarrow FeOH^{2+} + SO_4^{2-} + H^+$	3.5×10^7	
9	$Fe^{2+} + S_2O_8^{2-}\ (+H_2O) \rightarrow FeOH^{2+}\ SO_4^{2-} + SO_4^- + H^+$	12	
10(a)	$SO_5^- + HSO_3^- \rightarrow HSO_5^- + SO_3^-$	1.1×10^4	1.2×10^4
10(b)	$SO_5^- + HSO_3^- \rightarrow SO_4^{2-} + SO_4^- + H^+$	1.0×10^3	4.8×10^2 (c)
11	$SO_4^- + HSO_3^- \rightarrow SO_4^{2-} + SO_3^- + H^+$	6.8×10^8	7.5×10^8 (e)
12(f)	$SO_4^- + Bz\ (+H_2O, O_2) \rightarrow Ph + SO_4^{2-} + HO_2 + H^+$	8.0×10^8	2.4×10^9 (g)
13	$SO_5^- + SO_5^- \rightarrow S_2O_8^{2-} + O_2$	7.0×10^7	
14	$HO_2 + SO_5^- \rightarrow HSO_5^- + O_2$	4.3×10^9	
15	$HO_2 + HO_2 \rightarrow H_2O_2 + O_2$	8.3×10^5	
16	$HSO_5^- + HSO_3^- + H^+ \rightarrow 2\ SO_4^{2-} + 3\ H^+$	1.0×10^4	1.4×10^4 (h)
17	$H_2O_2 + HSO_3^- + H^+ \rightarrow SO_4^{2-} + H_2O + 2\ H^+$	3.0×10^4	

(a) The mechanism is partly based on a reaction scheme proposed by Bal Reddy and van Eldik [5];
(b) units: dm³ mol⁻¹ s⁻¹ unless otherwise noted; for sources of rate constants used in computer simulations see Ziajka *et al.* [3];
(c) relative to k_{10} as given by Buxton *et al.* [2];
(d) relative to k_{16} taken from Elias *et al.* [6]; (e) corrected for the prevailing ionic strength;
(f) Bz = Benzol, Ph = Phenol; (g) relative to k_{11}; (h) corrected to pH 2.89.

Conclusions

The aims of the projects have been achieved. It remains to test the results in cloud models.

Acknowledgements

This research has received support from the Deutsche Forschungsgemeinschaft within Sonderforschungsbereich 233 "Dynamics and Chemistry of Hydrometeors", and in part from the Commission of the European Communities. We also thank the Max-Planck-Gesellschaft for a grant in support of cooperation with the Institute of Physical Chemistry, Polish Academy of Sciences, in Warsaw.

References

1. H.-J. Benkelberg, P. Warneck, *J. Phys. Chem.* **99** (1995) in press.
2. G. V. Buxton, S. McGowan, G. A. Salmon, in: P. M. Borrell, P. Borrell, T. Cvitaš, W. Seiler (eds), *Proc. EUROTRAC Symp. '92*, SPB Academic Publishing bv, The Hague 1993, pp. 599–604.
3. J. Ziajka, F. Beer, P. Warneck, *Atmos. Environ.* **28** (1994) 2549–2552.
4. P. Warneck, J. Ziajka, *Ber. Bunsenges. Phys. Chem.* **99** (1995) 59–65.
5. K. Bal Reddy, R. van Eldik, *Atmos. Environ.* **26**A (1991) 661–665.
6. H. Elias, U. Götz, K. J. Wannowius, *Atmos. Environ.* **28** (1994) 439–448.
7. U. Deister, P. Warneck, *J. Phys. Chem.* **94** (1990) 2191–2198.

5.3 Formation and Reactions of Oxidants in the Aqueous Phase

R. Zellner, H. Herrmann, M. Exner, H.-W. Jacobi, G. Raabe and A. Reese

Institut für Physikalische und Theoretische Chemie, Universität GH Essen, Universtätsstr. 5, D-45117 Essen, Germany

Summary

This report summarizes the principal results from the contributions to the HALIPP research activity from our laboratory. The main results for the study of the kinetics and spectroscopy of the radicals NO_3, SO_x^- (x = 3, 4, 5), OH, CO_3^- and Cl_2^- are presented.

Aims of the research

To quantify the contribution of free radicals to the oxidizing capacity within the tropospheric aqueous phase, *i.e.* in the droplets of clouds, fog and rain.

Principal scientific results

NO_3 *kinetics*

The main kinetic results for reactions of the nitrate (NO_3) radical are summarized in the Table 1.

For some reactions of NO_3 the influence of ionic strength on reaction rate coefficients was studied. For the reaction with chloride
$k_1(I \rightarrow 0) = (9.1 \pm 0.9)\ 10^6$ L mol^{-1} s^{-1},
$k_1\ (I \rightarrow \infty) = (9.1 \pm 1.0)\ 10^7$ L mol^{-1} s^{-1} and
$\beta = (0.64 \pm 0.15)$ L mol^{-1} (β_7: salting coefficient) and for the reaction with formate
$k_{11}(I \rightarrow 0) = (3.6 \pm 1.0)\ 10^7$ L mol^{-1} s^{-1},
$k_{11}(I \rightarrow \infty) = (2.3 \pm 0.3)\ 10^8$ L mol^{-1} s^{-1} and
$\beta = (0.68 \pm 0.2)$ L mol^{-1} were obtained.

For the H abstraction reactions the measured reaction rate coefficients were correlated with C–H bond dissociation energies by the expression

$$\log (k_{NO_3}^{H}) = (44 \pm 8) - (0.10 \pm 0.02)\ (D_{X\text{-}H} / \text{kJ mol}^{-1})$$

Table 1a: Main results for NO_3 radical kinetics, Part 1.

	Reactant	k(298 K)/ L mol^{-1} s^{-1}	A / L mol^{-1} s^{-1}	E_A / kJ mol^{-1}
(1)	Cl^-	$(1.0 \pm 0.2) \times 10^7$	$(1.9 \pm 0.2) \times 10^{13}$	(36 ± 4)
(2)	OH^-	$(8.2 \pm 0.9) \times 10^7$	$(8.1 \pm 0.4) \times 10^{11}$	(23 ± 3)
(3)	Br^-	3.8×10^9	—	—
(4)	HSO_3^-	$(1.4 \pm 0.1) \times 10^9$	$(1.1 \pm 0.2) \times 10^{12}$	(16 ± 3)
(5)	$SO_{2(aq)}$	$(2.6 \pm 0.5) \times 10^8$	$(5.6 \pm 0.7) \times 10^{13}$	(30 ± 10)
(6)	SO_3^{2-} *	$(3.0 \pm 0.4) \times 10^8$	—	—
(7)	H_2O_2	$(7.0 \pm 1.2) \times 10^6$	$(4.0 \pm 0.5) \times 10^9$	(16 ± 7)
(8)	$(CH_3)_2CO$	$(4.4 \pm 0.6) \times 10^3$	$(8.2 \pm 0.7) \times 10^9$	(36 ± 5)
(9)	CH_3CH_2OH	$(2.4 \pm 0.5) \times 10^6$	$(1.4 \pm 0.1) \times 10^{11}$	(27 ± 4)
(10)	$H_2C(OH)_2$	$(7.8 \pm 1.2) \times 10^5$	$(3.6 \pm 0.3) \times 10^{12}$	(38 ± 6)
(11)	$HCOO^-$	$(7.0 \pm 0.4) \times 10^7$	$(8.2 \pm 0.8) \times 10^{10}$	(18 ± 6)
(12)	HCOOH	$(3.3 \pm 1.0) \times 10^5$	$(3.4 \pm 0.3) \times 10^{10}$	(28 ± 5)
(13)	CH_3COO^-	$(2.3 \pm 0.4) \times 10^6$	$(1.0 \pm 0.2) \times 10^{12}$	(32 ± 10)
(14)	CH_3COOH	$(1.3 \pm 0.3) \times 10^4$	$(4.9 \pm 0.5) \times 10^9$	(31 ± 6)

* $T = 278$ K

Table 1b: Main results for NO_3 radical kinetics, Part 2.

Reactant	k_{2nd} / L mol^{-1} s^{-1} (298 K, pH 4)	Arrhenius-parameters (A/L mol^{-1} s^{-1}; E_A/kJ mol^{-1}) Salting Coefficient (β / L mol^{-1})
(15) NO_2^-	$(1.5 \pm 0.2) \times 10^9$	
(16) HMS ($CHOSO_3H$)	$(4.2 \pm 1.2) \times 10^6$ (293 K)	$E_A = (21 \pm 14)$; $A = (2.4 \pm 0.6) \times 10^{10}$
(17) Glyoxal	$(1.1 \pm 0.1) \times 10^6$	$E_A = (28 \pm 6)$; $A = (1.0 \pm 0.1) \times 10^{11}$
(18) Oxalic Acid	$(1.7 \pm 0.5) \times 10^8$	E_A(pH 9) $= (21 \pm 14)$, $A = (3.2 \pm 0.5) \times 10^{11}$ E_A(pH 1) $= (17 \pm 6)$, $A = (2.7 \pm 0.3) \times 10^9$ $\beta = (0.10 \pm 0.03)$ $k(I \rightarrow 0) = (1.4 \pm 0.4) \times 10^8$
(19) CH_3CHO	$(1.9 \pm 0.6) \times 10^6$	$\beta = (0.19 \pm 0.11)$ $k(I \rightarrow 0) = (1.8 \pm 0.6) \times 10^6$

For electron transfer reactions rate coefficients are correlated with the redox potential by

$$\log(k_{NO_3}) = (10 \pm 1) - (1.5 \pm 0.6)(E° / V)$$

Rate coefficients for H abstraction reactions by NO_3 in both aqueous solution and the gas phase are correlated with each other by

$$\log\{k(NO_3(g)+X)/L\ mol^{-1}\ s^{-1}\} = (0.58 \pm 0.15) + (0.79 \pm 0.20)\log\{k(NO_3(aq)+X)/L\ mol^{-1}\ s^{-1}\}$$

Investigations of SO_x^- (x = 3, 4, 5) radical anions

Radicals were generated by excimer laser photolysis via the processes (20)–(22) and monitored by time-resolved UV/visible broadband diode array absorption spectroscopy.

$$S_2O_8^{2-} + h\nu(248\ \text{nm}) \rightarrow 2\ SO_4^- \qquad \text{(R-20)}$$
$$S_2O_6^{2-} + h\nu(193\ \text{nm}) \rightarrow 2\ SO_3^- \qquad \text{(R-21)}$$
$$SO_3^- + O_2 \rightarrow SO_5^- \qquad \text{(R-22)}$$

Details of the method used for the investigations of the sulfur radical anions SO_x^- can be found elsewhere [1].

Absolute decadic absorption coefficients as a function of wavelength for the SO_x^- radical anions

The absorption spectra of the SO_x^- radical anions (x = 3, 4, 5) were presented at Joint LACTOZ/HALIPP Workshop "Tropospheric Oxidation Mechanisms" in Leipzig (September 1994). Detailed data of the absorption coefficients as a function of the wavelength will be published [2]. For the radical anions SO_3^-, SO_4^- and SO_5^- maximum absorption coefficients of 1090, 1650 and 740 L mol^{-1} cm^{-1} were obtained at wavelengths of 255, 445 and 265 nm, respectively.

Results from the Study of SO_4^- kinetics

In this part of the study the dependence of the first-order rate coefficient on the persulfate concentration for the reactions (23) and (24) has been investigated

$$SO_4^- + S_2O_8^{2-} \rightarrow SO_4^{2-} + S_2O_8^- \qquad \text{(R-23)}$$
$$SO_4^- + H_2O \rightarrow HSO_4^- + OH \qquad \text{(R-24)}$$

At 293 K and pH = 5 rate constants of $k_{23} = (6.3 \pm 1.5) \times 10^5$ L mol^{-1} s^{-1} (for $I \rightarrow 0$) and $k_{24} = (6.6 \pm 0.4) \times 10^2$ s^{-1} has been determined. From T-dependent measurements the expression

$$k_{24}(T) = (4.7 \pm 0.2) \times 10^2 \exp[-(1110 \pm 60)\ \text{K} / T]\ \text{L mol}^{-1}\ \text{s}^{-1}$$

with $E_A = (9.2 \pm 0.5)$ kJ/mol was derived. For the recombination reaction (R-25)

$$SO_4^- + SO_4^- \rightarrow S_2O_8^{2-} \qquad \text{(R-25)}$$

a rate constant of $k_{25} = (1.6 \pm 0.4) \times 10^8$ L $mol^{-1}s^{-1}$ has been obtained. In another part of our study the possible formation of OH-radicals in atmospheric droplets via the process

$$SO_4^- + OH^- \rightarrow SO_4^{2-} + OH \qquad \text{(R-26)}$$

has been investigated. By variation of the OH^- concentration in a range from 3×10^{-6} to 1×10^{-4} mol l^{-1} the dependence of the first-order rate coefficient on the OH^-- concentration for reaction (26) was investigated. At 298 K a rate constant of $k_{26} = (1.4 \pm 0.2)\ 10^7$ L mol^{-1} s^{-1} (for $I \rightarrow 0$) has been found.

Results from the study of SO_5^- kinetics

For the self-reaction of the peroxymonosulfate radical anion (R-27a,b)

$$SO_5^- + SO_5^- \rightarrow 2\ SO_4^- + O_2 \qquad \text{(R-27a)}$$

$$SO_5^- + SO_5^- \rightarrow S_2O_8^{2-} + O_2 \qquad \text{(R-27b)}$$

a rate constant of $k_{27} = (1.3 \pm 0.3) \times 10^8$ L mol^{-1} s^{-1} (for $I \rightarrow 0$) was found at 298 K and pH 4. The second order rate coefficient for reaction (27) has also been investigated in the temperature range from 288 to 320 K. The temperature dependence is best described by

$$k_{27}(T) = (1.1 \pm 0.1) \times 10^{12} \exp[-(2600 \pm 1000)\ \text{K} / T]\ \text{L mol}^{-1}\ \text{s}^{-1}$$

corresponding to an activation energy of E_A = (22 ± 9) kJ/mol. The branching ratio for the two channels (R-27a and R-27b) has been estimated to $k_{27a}/(k_{27a} + k_{8b}) \leq 0.04$.

For the branching reaction (28), *i.e.*

$$SO_5^- + HSO_3^- \rightarrow HSO_5^- + SO_3^- \qquad \text{(R-28a)}$$

$$\rightarrow HSO_4^- + SO_4^- \qquad \text{(R-28b)}$$

apparently a key process in the conversion of S(IV) whenever sulfur-oxy radicals are involved, a rate coefficient for 298 K and pH 4 of
$k_{(28a+28b)} = (2.8 \pm 0.3)\ 10^6$ L mol^{-1} s^{-1} was found.

Another important reaction was found in the HALIPP mechanism [3] for the iron-catalysed oxidation of bisulfite is the reaction of SO_5^- with Fe(II):

$$SO_5^- + Fe^{2+} \rightarrow \text{Products} \qquad \text{(R-29)}$$

By variation of the iron(II) concentration the dependence of the first-order rate coefficient for the decay of SO_5^- on the iron(II) concentration was investigated. At 298 K a rate constant of $k_{29} = (4 \pm 2) \times 10^7$ L mol^{-1} s^{-1} was derived. For the reaction of SO_5^- with Mn(II)

$$SO_5^- + Mn^{2+} \rightarrow \text{Products} \qquad \text{(R-30)}$$

we obtained a rate constant of k_{30} ($I \rightarrow 0$) = $(4.6 \pm 1.0) \times 10^6$ L mol^{-1} s^{-1} at 298 K.

Spectroscopic and kinetic investigations of the carbonate radical anion (CO_3^-)

With the time-resolved UV/visible broadband diode array absorption spectroscopy technique described in [1], the absorption spectra of CO_3^-/HCO were investigated. From pH dependent measurements of the absorption spectra of the carbonate radical anion we derived a $pK_S(HCO_3) = (7.6 \pm 0.2)$ [4].
The kinetics of reactions of the carbonate radical anion relevant to the tropospheric aqueous phase have been investigated with the technique described elsewhere [1]. A compilation of the results obtained is shown in Table 2.

Table 2: Overview of kinetic results for reactions of CO_3^-.

Reaction	pH	$\frac{I}{\text{mol L}^{-1}}$	$\frac{k(298\text{ K})}{\text{L mol}^{-1}\text{ s}^{-1}}$
(31) $CO_3^- + NH_4^- \rightarrow$ products	8	0.46	$(3.3 \pm 0.4) \times 10^2$
(32) $CO_3^- + CH_3OH \rightarrow HCO_3^- + CH_2OH$	7.9	0	$(2.5 \pm 0.3) \times 10^3$
(33) $CO_3^- + CH_3COCH_3 \rightarrow HCO_3^- + CH_3COCH_2$		0.39	$(2.3 \pm 0.4) \times 10^2$
(34) $CO_3^- + CH_3COO^- \rightarrow$ products	7	0.39–0.52	$(5.8 \pm 0.8) \times 10^2$
(35) $CO_3^- + H_2C(OH)_2 \rightarrow HCO_3^- + HC(OH)_2$	7.9	0.39	$(1.3 \pm 0.3) \times 10^4$
(36) $CO_3^- + HCOO^- \rightarrow$ products	7.5	0.41	$(1.6 \pm 0.2) \times 10^5$

Spectroscopic investigations of the hydroxyl radical (OH)

In this part of the study several systems to generate OH radicals photochemically have been investigated.

Photolysis ($\lambda = 193$ nm) of hydrogenperoxide (H_2O_2)

The direct detection of OH radicals was not possible, because OH radicals react fast with the precursor H_2O_2 to $HO_2 \Leftrightarrow O_2^- + H^+$. The resulting absorption spectra show maxima between 225–250 nm.

Photolysis ($\lambda = 193$ nm) of H_2O at pH = 10–12

The variation of the pH between 10 and 12 lead to reproducibly detected absorption signals. These spectra show their maximum absorption at $\lambda = 230$–235 nm.

OH Radicals as a Product of the Reaction of SO_4^- with OH^- at pH 10–12

In the photolysis of $S_2O_8^{2-}$ at $\lambda = 193$ nm SO_4^- radical anions were generated, which react fast with OH^- to form OH. The resulting absorption spectra show a maximum at 235 nm in good agreement with the results above.

Reactions of the* Cl_2^- *radical

The reaction kinetics of the Cl_2^- radical has been investigated with a laser photolysis-long path laser absorption apparatus. Cl_2^- is produced by excimer laser photolysis of peroxodisulfate at 248 nm in the presence of excess chloride. The built-up of the radical via the reaction (R-37) $SO_4^- + Cl^- \rightarrow SO_4^{2-} + Cl$ and (R-38) $Cl^- + Cl \rightarrow Cl_2^-$ with $k_{37}(I \rightarrow 0) = 2.5 \times 10^8$ L mol^{-1} s^{-1} [5] and k_{38} (I = 0.15 mol L^{-1}) = 8 × 10^9 L mol^{-1} s^{-1} [6] is very fast. Cl_2^- is detected by time-resolved long-path laser absorption using the 325.0 nm line of HeCd laser.

The reactions of the Cl_2^- radical with methanol, ethanol, formaldehyde and formic acid have been investigated. The influence of temperature and ionic strength on the reaction rate constants of the reactions with methanol and formaldehyde has been studied. The salting coefficient b corresponds to the equation $\log k = \log k(I \rightarrow 0) + b\, I$. The rate constants of the self-reaction of Cl_2^- and the reaction with water are calculated by results obtained by variation of the excess chloride concentration [7]. The results are shown in Table 3.

Table 3: Reaction rate constants, activation energies and salting coefficients for the reactions of Cl_2^- Cl_2^- Cl_2^-.

Reaction	I / mol L^{-1}	pH	k(298 K) / L mol^{-1} s^{-1}	E_a / kJ mol^{-1}	β / L mol^{-1}
9) $Cl_2^- + CH_3OH \rightarrow H^+ + 2\,Cl^- + CH_2OH$	0	4	(5.5 ± 0.2) × 10^4	46 ± 13	− 0.24 ± 0.06
0) $Cl_2^- + CH_3CH_2OH \rightarrow H^+ + 2\,Cl^- + CH_3CHOH$	0.1	4	(1.2 ± 0.2) × 10^5	—	—
1) $Cl_2^- + CH_2(OH)_2 \rightarrow H^+ + 2\,Cl^- + CH(OH)_2$	0	4	(3.6 ± 0.1) × 10^4	36 ± 5	− 0.09 ± 0.04
2) $Cl_2^- + HCOOH \rightarrow H^+ + 2\,Cl^- + HCO_2$	0.1	0.4	(6.0 ± 0.4) × 10^3	—	—
3) $Cl_2^- + Cl_2^- \rightarrow Cl_3^- + Cl^-$	0	4	8.7 × 10^8	—	—
4) $Cl_2^- + H_2O \rightarrow HClOH + Cl^-$	0.1	4	(9 ± 2) × 10^2	—	—

Conclusions

The reactivities of the investigated radicals show the following trend: OH > NO_3 ≈ SO_4^- > Cl_2^- > CO_3^-. The obtained correlations of the rate constants of the NO_3 radical with H bond dissociation energies or redox potentials make estimations possible for reactions, which are experimentally not easy accessible. This can be very helpful in modelling studies.

Acknowledgements

The authors gratefully acknowledge the support of the research contributions to the HALIPP activity within EUROTRAC by the Bundesministerium für Bildung und Wissenschaft, Forschung und Technologie (BMBF) under contracts 07EU7801 and 07EU780A1. H.H. wishes to thank 'Stiftung Stipendien Fonds des VCI' and NATO/DAAD for additional support.

References

1. H. Herrmann, A. Reese, R. Zellner, in: J. Peeters (ed), *Air Pollution Research Report* **45**, CEC/EUROTRAC, Brussels 1993, p. 253
2. H. Herrmann, A. Reese, R. Zellner, to be published.
3. P. Warneck, *EUROTRAC Ann. Rep. Part 6*, EUROTRAC ISS Garmisch-Partenkirchen 1993, p. 10.
4. M. Exner, H. Herrmann, A. Reese, R. Zellner, in preparation.
5. R. E. Huie, C. L. Clifton, *J. Phys. Chem.* **94** (1990) 8561.
6. V. Nagarajan, R. W. Fessenden, *J. Phys. Chem.* **89** (1985) 2330.
7. A. Reese, *Master Thesis*, University of Essen, 1992.

Chapter 6

Gas - Liquid Interactions

6.1 Photocatalysis of Tropospheric Chemistry by Sea Spray

Wolfgang Behnke, Manfred Elend, Christian George, Heinz-Ulrich Krüger, Volker Scheer and Cornelius Zetzsch

Fraunhofer-Institut für Toxikologie und Aerosolforschung, Nikolai-Fuchs-Straße 1,
D-30625 Hannover, Germany

Summary

The uptake of NO_2 and N_2O_5 by deliquescent NaCl aerosol is investigated in an aerosol smog chamber. While the slow uptake of NO_2 leads to HONO in the gas phase (instead of NOCl), the fast uptake of N_2O_5 leads to $ClNO_2$ at high yields of almost unity with a high uptake coefficient of $(3.2 \pm 0.3) \times 10^{-2}$. The yield of $ClNO_2$ in the gas phase and its subsequent uptake are studied in wetted-wall flow tube experiments on various solutions containing chloride, nitrate and sulphate. From these experiments, a reaction mechanism is proposed where dissociation of N_2O_5 (in the aqueous phase) into NO_2^+ and nitrate is involved.

The formation of $ClNO_2$ from sea-spray aerosol may modify the photo-oxidising capacity of the troposphere by two effects: the lifetime of NO_x is prolonged, and photolytically formed atomic Cl may initiate smog formation soon after sunrise.

Aims of the Research

Most of the tropospheric aerosol is produced by sea spray over the oceans. It has been postulated that sea-spray causes heterogeneous reactions producing atomic Cl (or its precursors). In fact, several groups found evidence recently for the presence of atomic Cl from the degradation pattern of hydrocarbons over the ocean during field experiments [1, 2] or for the production of the photolabile Cl_2 [3], that may be produced on metal-ion containing aerosol surfaces in the presence of O_3 and Cl^- [4–6].

This activation of atomic Cl is believed to be mainly due to heterogeneous reactions of nitrogen oxides especially NO_2 and N_2O_5 (but probably also NO_3) produced during night-time by the reactions between NO_2 and NO_3. Earlier laboratory experiments showed that NO_2 and N_2O_5 react heterogeneously with dry bulk NaCl salt to form respectively nitrosyl chloride [7, 8] (NOCl) and nitryl chloride [9] ($ClNO_2$). These chloride containing gases are rapidly photolysed after sunrise producing Cl and NO_x. This means that these heterogeneous reactions of NO_2 and N_2O_5 are central to the oxidising capacity of the troposphere because they prolong the lifetime of NO_x (*e.g.* N_2O_5 may undergo other reactions than simply hydrolysis [10, 11]) and modify the radicals "speciation". In contrast to these experimental conditions, the NaCl-containing sea-spray aerosol remains liquid in the troposphere during its drying, even below the deliquescence point down to 45 % relative humidity [12]. Therefore, under typical conditions of the oceanic boundary layer, the main components of freshly generated sea-spray aerosol are water and NaCl.

In this contribution we investigated the reactions of NO_2 and N_2O_5 with dry and deliquescent NaCl aerosol, synthetic and genuine sea spray as well as the reactions of N_2O_5 and $ClNO_2$ in or on aqueous solutions in order to assess the impact of these heterogeneous processes on the photo-oxidant formation in the troposphere.
The aerosol reactions are investigated in Teflon bags (FEP 200A, Dupont) in a smog chamber facility described elsewhere [6, 13–16]. In addition, a wetted-wall flow tube facility was developed and employed during the investigation period of the project [17].

Principal scientific results

Uptake of NO_2 in aerosol chamber experiments

The uptake of NO_2 by deliquescent NaCl aerosol was determined to be slow ($\gamma < 10^{-4}$), leading to the formation of HONO in the gas phase (that was detected from the photolytic formation of OH) [18] instead of NOCl (as observed on dry NaCl) [7, 8].

Uptake of N_2O_5 in aerosol chamber experiments

The observations made in presence of O_3, N_2O_5 and NaCl aerosol in the Teflon bag are consistent with the reaction mechanism:

$$N_2O_5 \rightleftharpoons NO_2 + NO_3$$
$$NO_2 + O_3 \rightarrow NO_3 + O_2$$
$$NO_3 \xrightarrow{\text{wall}} NO_2$$
$$N_2O_5 \xrightarrow{\text{wall}} 2\ HNO_3$$
$$N_2O_5 \xrightarrow{\text{aerosol}} N_2O_5(ad)$$

completed for illumination conditions by:

$$NO_3 + h\nu \rightarrow NO_2 + O(^3P)$$
$$\rightarrow NO + O_2$$
$$NO_2 + h\nu \rightarrow NO + O(^3P)$$
$$NO + O_3 \rightarrow NO_2 + O_2$$
$$O(^3P) + O_2 \rightarrow O_3$$

These reactions are integrated numerically using the software package LARKIN [19] in order to fit the rate constant for the uptake of N_2O_5 by the aerosol to the measurements. The kinetics of the adsorption process is interpreted by the equation of Fuchs and Sutugin (1971) considering the aerosol size distribution as described elsewhere [15, 16].

The results of the smog chamber experiments are listed in Table 1. They differ slightly from previously reported data [15, 16], since wall losses of N_2O_5 and NO_3 determined in separate blank experiments were considered quantitatively and the experimental conditions were improved considerably [13, 17]. The relative humidity was varied between 77.3 % and 93.9 %. The initial NaCl concentration of the aerosol in equilibrium with these relative humidities lies in the range from 5.7 to 1.8 mol/L. A mean value of $\gamma \pm \sigma = 0.032 \pm 0.003$ for the uptake coefficient of N_2O_5 was obtained from these experiments. No change of the uptake coefficient is observed after adding Na_2SO_4 to the NaCl ($\gamma \pm \sigma = 0.032 \pm 0.002$). An additional systematic error is expected especially at high relative humidities, where the wall losses of N_2O_5 and NO_3 (see Table 3) are in the range or larger than the portion of N_2O_5 absorbed by the aerosol.

Table 1: Relative humidities, r.h., uptake coefficients, γ, N_2O_5 and O_3, initial mass density of the NaCl aerosol and final composition.

r.h. / %	γ	$[N_2O_5]_0$ / ppb	$[O_3]$ / ppb	$[NaCl]_0$ / $\mu g\ m^{-3}$	$\frac{[NaNO_3]}{[NaCl]+[NaNO_3]}$
77.3	0.0353	16.6	62	46.5	0.39
82.7	0.0299	22.3	44	137.0	0.32
87.3	0.0301	14.8	39	66.8	0.38
92.2	0.0348	15.5	50	63.5	0.43
93.9	0.0301	21.3	49	22.5	0.63
73.6	0.0295	23.3	52	99.3*	0.36
84.1	0.0328	24.0	48	106.6*	0.34
86.7	0.0353	7.8	21	89.9*	0.14
92.0	0.0307	24.7	44	103.2*	0.33
92.3	0.0318	22.0	46	98.2*	0.37

*Na_2SO_4 was added at $[Na_2SO_4]/[NaCl] = 0.2$

The last column of Table 1 lists the portion of the NaCl aerosol transformed to $NaNO_3$. $NaNO_3$ is less hygroscopic than NaCl. Therefore, the aerosol size decreases during the transformation of NaCl to $NaNO_3$. This decrease was taken into consideration for the determination of the uptake coefficient.

The NO_3 concentration formed from the thermal decomposition of N_2O_5 before illuminating the Teflon bag remains below 15 % of the N_2O_5 concentration in all experiments (as shown by model calculations). A possible uptake of NO_3 by the aerosol remained unconsidered in the evaluation (see below). Taking an uptake of NO_3 in a similar range as that of N_2O_5, the uptake would be 10–20 % smaller than evaluated here.

Yield of $ClNO_2$ *in aerosol smog chamber experiments*

Fig. 1 shows an example of the degradation rate of the hydrocarbons by atomic Cl on a logarithmic scale against the time (third experiment of Table 1).

The data are fitted by the software package MINUIT [20], leading to a photolytic rate constant of $(4.48 \pm 0.15) \times 10^{-4}\ s^{-1}$ (solid line), a value expected for the photolysis of $ClNO_2$ (by folding the UV absorption with the spectrum of our solar simulator). If the uptake of NO_3 by the NaCl aerosol would produce a photolytic precursor of atomic Cl, we would expect Cl_2, which would be photolysed very rapidly. All experiments show a small increase of the first value against the weighted exponential fit, but by fitting the degradation rate of the hydrocarbons by two exponentials, see dashed line), there is no significant proof of a second precursor of atomic Cl (see Table 2).

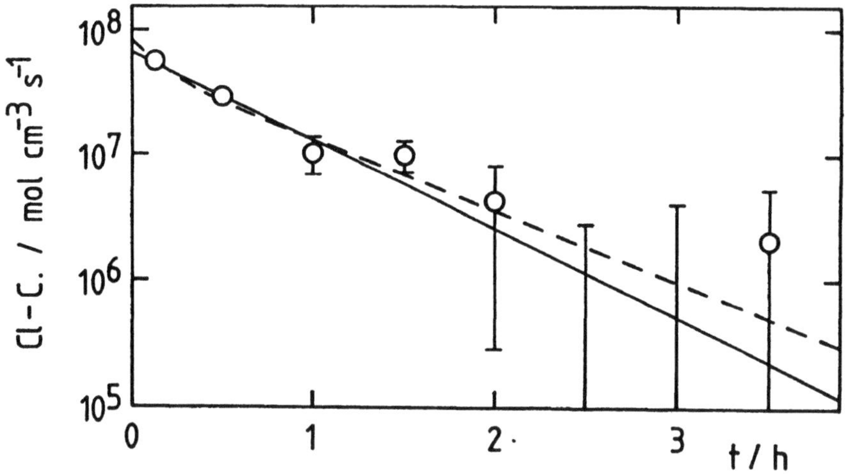

Fig. 1: Degradation rate of the hydrocarbons by atomic Cl (third experiment of table 1) and fits of the data by a single exponential and by two exponential functions.

Table 2: Fitting parameters in Fig. 1, errors are 1 *s*.

	C_1/ppb	$k_1 / 10^{-4}$ s^{-1}	C_2/ppb	$k_2 / 10^{-4}$ s^{-1}
solid line	6.0 ± 0.6	4.48 ± 0.15	—	—
dashed line	5.4 ± 0.9	3.6 ± 0.5	0.8 ± 0.9	18 ± 27

Table 3 lists the balance of nitrogen, the Cl^- content of the aerosol at the beginning of the experiments and the yield of $ClNO_2$ in the experiments of Table 1. The yield is found to be independent of the Cl^- content of the aerosol.

Table 3: Cl^- concentrations of the deliquescent aerosol, N_2O_5 injected, remaining NO_2, losses of N_2O_5 and NO_3 by wall reaction (in ppb N), and yields of $ClNO_2$ (determined from hydrocarbon-degradation by atomic Cl).

$[Cl^-]$ / mol L^{-1}	$[N_2O_5]_0$ / ppb	$[NO_2]_r$ / ppb	losses / N / ppb	$[ClNO_2]$ / ppb	yield of $ClNO_2$
5.7	16.6	8.2	4.5	6.5	0.64
4.5	22.3	5.9	8.9	10.1	0.68
3.5	14.8	4.05	9.8	6.0	0.76
2.2	15.5	3.5	7.3	5.66	0.56
1.8	21.3	9.8	17.2	5.0	0.64

Although the conditions of the experiments in the Teflon bag are a fairly realistic simulation of environmental conditions, the concentrations in the aerosol change drastically during the uptake of N_2O_5. Furthermore, there will be a change of the water and Cl^- content and of the pH value of the aerosol. This can almost completely be avoided in wetted-wall flow tube experiments.

Yield and uptake of $ClNO_2$ *in wetted-wall flow tube experiments*

Uptake coefficients of N_2O_5 cannot be evaluated from experiments in the wetted-wall flow tube since wall reactions are diffusion limited for uptake coefficients greater than approximately 10^{-4}. More than 99 % of the N_2O_5 react with the solution at the wall during passage through the first 10 cm of the wetted-wall flow tube. Fig. 2 compares the yield of $ClNO_2$ from the N_2O_5 consumed by the liquid film (open circles) or by the liquid aerosol in Teflon bag experiments (filled circles). Although the concentrations of $NaNO_3$ and NaCl change drastically during the experiments in Teflon bags (as mentioned above), the two techniques confirm the same tendency: a nearly constant ratio of $[ClNO_2]/[N_2O_5]_0$ above 1.8 mol/L NaCl.

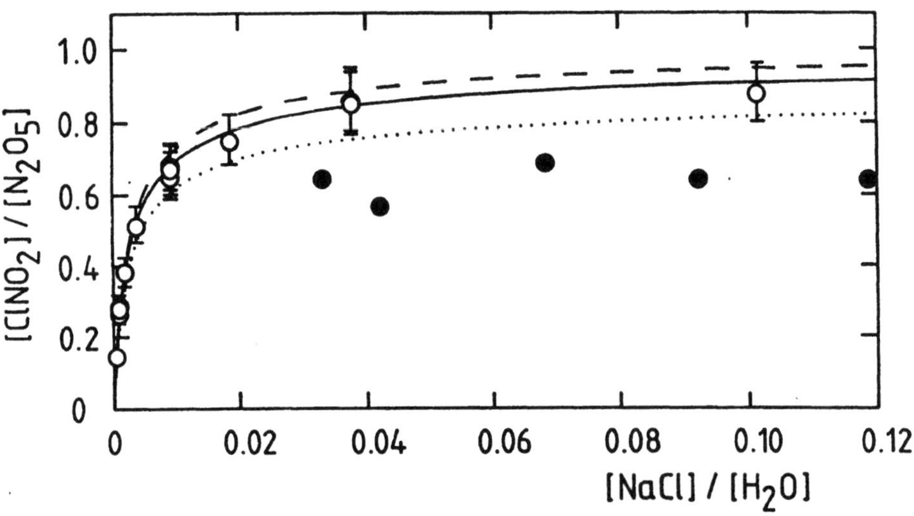

Fig. 2: Yields of $ClNO_2$ from N_2O_5 as a function of the NaCl concentration in the moist aerosol/solution obtained from experiments in Teflon bags (filled circles)/ in a wetted-wall flowtube (open symbols) at 291.2K.

The uptake of $ClNO_2$ was measured on pure deionized water and on different NaCl solutions up to 4.6 mol/l showing a strong influence of the chloride concentration on the uptake rate. In a joint effort between the University of Strasbourg and our group, we investigated the chemistry of $ClNO_2$ over solutions containing iodide. We observed that $ClNO_2$ reacts very rapidly with I^- producing ICl and HONO [21]. We expect therefore that the uptake of nitryl chloride is affected in real aerosols by the presence of bromide, *i.e.* these reactions may release BrCl.

Both, the yield of $ClNO_2$ and its uptake, can be explained by the reaction mechanism presented in Fig. 3. The yields of $ClNO_2$ from the wetted-wall flow tube and the uptake measurements of $ClNO_2$ are fitted according to this reaction scheme using again the program package MINUIT [20]. The results are listed in table 4 (middle curve of Fig. 2) with error bars of 1 σ.

At NaCl concentrations higher than 0.08 mol/l the formation of $ClNO_2$ (k_4) dominates over the hydrolysis of NO_2^+ (or of any different intermediate formed from N_2O_5). For the rate of formation of the intermediate (k_1) we obtain a lower limit of 10^4 s^{-1}. The decrease of the solubility of $ClNO_2$ in dependence of the concentration of salt ($a = 4.4 \pm 1.0$, see Table 4) is in the range of similar data for phosgene [22]. Considering the statistical errors for the measured values of the yield alone, the ratio between the direct hydrolysis of N_2O_5 (k_3) and k_1 appears to be highly significant ($k_3[H_2O]/k_1$=0.05 ± 0.01). On the other hand, the possibility of systematic errors (see Fig. 2) of the yield, due to the calibration of the FTIR response of $ClNO_2$ and N_2O_5, can not be excluded. Taking the upper limit of the yield equal to 1 (upper curve in Fig. 2, see also Table 4) for large concentrations of NaCl (*i.e.* $k_3 = 0$), we obtain a correction of the ratios $[ClNO_2]/[N_2O_5]_o$ of 1.047 ± 0.008, which is less than the possible systematic error of the calibration of the spectra for $ClNO_2$ and N_2O_5. Taking the lower limit of the possible systematic error of the yield (lower curve in Fig. 2, see also Table 1, third column), we obtain a ratio k_3 $[H_2O]/k_1$ of 0.19 ± 0.01. Overall, the direct hydrolysis of N_2O_5 (l) contributes less than 20 % of its dissociation into NO_2^+ and NO_3^-.

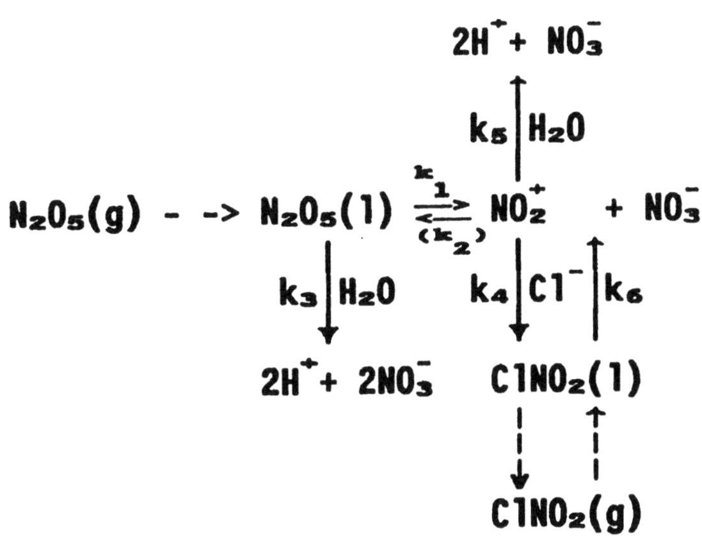

Fig. 3: Proposed reaction scheme.

Relative rate constants of NO_2^+ reactions with $OH^-/Cl^-/H_2O$ of $4.6 \times 10^4/1/1.8 \times 10^{-6}$ from the literature [23] do not agree with our relative value of Cl^-/H_2O, (k_4/k_5, obtained by experiments in acidic solutions down to pH 3): $1/1.4 \times 10^{-3}$.

Since sea water is slightly alkaline and fresh sea spray aerosol will be alkaline as well, one of our next aims will be the investigation of the influence of OH^- in basic solutions.

Table 4: Fitting parameters of the yield and uptake coefficient of $ClNO_2$ (curves see Fig. 2). The meaning of the parameters: Rate constants (see Fig. 3), Henry's law constant in the presence of NaCl $\{H = H_o / (1 + a[NaCl] / [H_2O])\}$ and yield = $F\,[ClNO_2] / [N_2O_5]$.

	upper curve	middle curve	lower curve**
F	1.047 ± 0.008	1.0*	1.0*
k_4 / k_5	691 ± 15	697 ± 16	710 ± 18
$H_o \cdot k_6$ / mol L^{-1} s$^{-½}$	0.43 ± 0.01	0.43 ± 0.01	0.43 ± 0.01
k_1 / s^{-1}	> 10^{4}*	> 10^{4}*	> 10^{4}*
$k_3\,[H_2O] / k_1$	0*	0.05 ± 0.01	0.19 ± 0.01
a	4.5 ± 1.2	4.4 ± 1.0	4.2 ± 1.2

*fixed values **lower error limits

Conclusions

In the presence of NaCl aerosol and NO_2, no NOCl is observed at relative humidities above 70 %. The production of OH radicals is larger than in the absence of NaCl aerosol indicating that instead of NOCl its hydrolysis product, HONO, has been formed.

When reacting with NaCl above 0.08 mol/L, N_2O_5 is mainly transformed to $ClNO_2$. The uptake coefficient of N_2O_5 on moist NaCl aerosol was measured to be $\gamma \pm \sigma = 0.032 \pm 0.003$. The systematic error might be several times higher than the standard deviation. The lifetime of N_2O_5 (l) after the uptake by the aerosol was estimated to be smaller than 10^{-4} s by measurements in the wetted-wall flow tube. Measurements with a droplet train technique, which were obtained in a joint campaign with the group of Mirabel [24], gave a lower limit of 10^{-5} s. The direct hydrolysis of N_2O_5 is less than 20 % of its decomposition. From the yield measurements of $ClNO_2$ and its subsequent uptake, we deduced a reaction mechanism where NO_2^+ plays a central role.

If $ClNO_2$ is released under moist tropospheric conditions, there are two important changes in the understanding of the tropospheric chemistry:

1. The NO_x cycle is not finished after the absorption of N_2O_5 on moist aerosol surfaces containing Cl^-.

2. A further pathway for the degradation of hydrocarbons exists by photolysing $ClNO_2$ to atomic Cl. This source of atomic Cl may be important for the chemistry of hydrocarbons over cold regions of the Northern Atlantic.

Indications for further sources of atomic Cl from sea-spray in the absence of NO_2 but in presence of O_3 exist [25] . These sources and other reactions involving bromide [26] may also be of importance and therefore need further investigation.

Acknowledgements

This work was supported by the Bundesminister für Forschung und Bildung (grant 07EU6761) in the framework of the EUROTRAC subproject HALIPP.

References

1. B. T. Jobson, H. Niki, Y. Yokouchi, J. Bottenheim, F. Hopper, R. Leaitch: Measurements of C_2–C_6 hydrocarbons during the Polar Sunrise 92 experiment: Evidence for Cl-atom and Br-atom chemistry, *J. Geophys. Res.***99** D (1994) 25355–25368.
2. H. B. Singh, G. L. Gregory, B. Anderson, E. Browell, G. W. Sachse, D. D. Davis, J. Crawford, J. D. Bradshaw, R. Talbot, D. R. Blake, D. Thornton, R. Newell, J. Merrill: Low ozone in the marine boundary layer of the tropical Pacific Ocean, *J. Geophys. Res. Atmos.* (1995) in press.
3. A. A. P. Pszenny, W. C. Keene, D. J. Jacob, S. Fan, J. R. Maben, M. P. Zetwo, M. Springer-Young, J. N. Galloway: Evidence of inorganic chlorine gases other than hydrogen chloride in marine surface air, *Geophys. Res. Lett.* **20** (1993) 699–702.
4. W. Behnke, F. Nolting, C. Zetzsch: Rate constants for the reactions of chlorine atoms with hydrocarbons determined by an aerosol smog chamber technique, *Abstr. XVIth Informal Conf. on Photochem.* paper T-21, Cambridge, Mass. 1984.
5. C. Zetzsch, G. Pfahler, W. Behnke: Heterogeneous formation of chlorine atoms from nacl aerosol in a photosmog system, *J. Aerosol Sci.* **19** (1988) 1203–1206.
6. C. Zetzsch, W. Behnke: Heterogeneous reaction of chlorine compounds, in: H. Niki, K. H. Becker (eds), NATO ASI Series, **17**, *Global Environmental Change*, Springer-Verlag Heidelberg 1993, pp. 291–305.
7. W. H. Schroeder, P. Urone, Formation of nitrosyl chloride from salt particles in air, *Environ. Sci. Technol.* **8** (1974) 756–758.
8. B. J. Finlayson-Pitts, Reaction of NO_2 with NaCl and atmospheric implication of NOCl formation, *Nature,* **306** (1983) 676–677.
9. B. J. Finlayson-Pitts, M. J. Ezell, J. N. Pitts, Jr., Formation of Chemically Active Chlorine Compounds by Reactions of Atmospheric NaCl Particles with Gaseous N_2O_5 and $ClONO_2$, *Nature,* **337** (1989) 241–24.
10. U. Platt, D. Perner, J. Schröder, C. Kessler, A. Toennissen, The diurnal variation of NO_3, *J. Geophys. Res.* **86C** (1981) 11965–11970.
11. A. G. Russell, G. J. McRae, G. R. Cass, The dynamics of nitric acid production and the fate of nitrogen oxides, *Atmos. Environ.* **19** (1985) 893.
12. I. N. Tang, H. R. Munkelwitz, J. G. Davis: Aerosol growth studies II. Preparation and growth measurements of monodisperse salt aerosols, *J. Aerosol Sci.* **8** (1977) 149–159.
13. W. Behnke, H.-U. Krüger, V. Scheer, C. Zetzsch: Production of $ClNO_2$ from the Reaction of N_2O_5 with NaCl- and Sea-Spray Aerosol, in: J. Peeters (ed), *Chemical Mechanisms Describing Tropospheric Processes, Air Pollution Research Report* **45**, CEC, Brussels 1992, pp. 249–252.

14. W. Behnke, C. Zetzsch: The simultaneous determination of two active species in smog chambers: evaluation of rate constants for Cl atoms in the presence of OH, in: K. H. Becker (ed), *Proc. 3rd French German Workshop on Tropospheric Chemistry by Laboratory Studies,* Rep. No. **15** of FB 8, Physikalische Chemie, Universität Gesamthochschule, Wuppertal 1987, pp. 59–60.
15. W. Behnke, H.-U. Krüger, V. Scheer, C. Zetzsch, Formation of atomic Cl from sea spray via photolysis of nitryl chloride: Determination of the sticking coefficient of N_2O_5 on NaCl aerosol, *J. Aerosol Sci.* **22** (1991) S609–S612.
16. C. Zetzsch, W. Behnke, Heterogeneous photochemical sources of atomic Cl in the troposphere, *Ber. Bunsenges. Phys. Chem.* **96** (1992) 488–493.
17. Behnke, W., H.-U. Krüger, V. Scheer, C. Zetzsch: Formation of $ClNO_2$ and HONO in the presence of NO_2, O_3 and Wet NaCl-aerosol, *J. Aerosol Sci.* **23** (1992) S933–S936.
18. W. Behnke, H.-U. Krüger, V. Scheer, C. Zetzsch, in: P. M. Borrell, P. Borrell, T. Cvitaš, W. Seiler (eds), *Proc. EUROTRAC Symp. '92*, SPB Acad. Publ. bv, The Hague 1993, pp. 565–570.
19. P. Deuflhard, U. Nowak, Efficient Numerical simulation and identification of large chemical reaction systems, *Ber. Bunsenges. Phys. Chem.* **90** (1986) 940–946.
20. CERN MINUIT Program Library entry D506, Geneva (1989).
21. Ch. George, W. Behnke, V. Scheer, C. Zetzsch, L. Magi, J. L. Ponche, Ph. Mirabel: Fate of $ClNO_2$ over aqueous solutions containing iodide, *Geophys. Res Lett.* in press.
22. W. H. Manogue, L. Pigford: The kinetics of the absorption of phosgene into water and aqueous solution, *A.I.Ch.E. Journal* **6** (1960) 494–500.
23. D. W. Johnson, D. W. Margerum, Non-metal redox kinetics: A re-examination of the mechanism of the reaction between hypochlorite and nitrite ions, *Inorg. Chem.* **30** (1991) 4845–4851.
24. Ch. George, J. L. Ponche, Ph. Mirabel, W. Behnke, V. Scheer, C. Zetzsch: Study of the uptake of N_2O_5 by water and NaCl solution, *J. Phys. Chem.* **98** (1994) 8780–8784.
25. W. Behnke, V. Scheer, C. Zetzsch: Production of a photochemical precursor of atomic Cl from aerosols and Cl^- in the presence of O_3, in: *Proc. Int. Conf. on Naturally-Produced Organo-halogens*, Delft 1993, in press.
26. W. Behnke, V. Scheer, C. Zetzsch: Production of $BrNO_2$, Br_2 and $ClNO_2$ from the reaction between sea spray aerosol and N_2O_5, *J. Aerosol Sci.* **25** (1994) S277–S278.

6.2 Phase Partitioning and Redox Processes in Fog and Cloud Systems

W. Jaeschke, J. Dierssen, S. Volkwein and J. Wohlgemuth

Zentrum für Umweltforschung, Johann Wolfgang Goethe-Universität, Frankfurt am Main, Georg-Voigt Str. 14, D-60325 Frankfurt, Germany

Summary

Our investigations were focused on phase partitioning and redox processes in fog and cloud systems. They can be divided in two parts. In the first part experiments were conducted on the phase partitioning SO_2 of trace substances HNO_3 (HNO_3, SO_2) between gas and liquid phase in a simulated multiphase system. For this purpose a flow reactor was constructed and a new nebulizer for inert droplet generation was developed. Additionally, two different methods of $HNO_3(g)$ sampling were invented. The second part of our investigations dealt with kinetic measurements in this system. The first reaction investigated was the iron-catalysed oxidation of SO_2. The experiments were carried out at different pH values and iron(III) concentrations. Concerning the liquid phase the rate constants obtained were of the same order of magnitude as values determined in homogeneous phase experiments.

Aims of the scientific research during the project

Our research was concentrated on laboratory experiments in which conditions and processes in the real atmosphere were simulated. In the first phase experiments dealt with phase partitioning of gases between droplets and surrounding air. The aim of these experiments was to show possible differences to the standard system of air above a plane water surface, which could explain deviations from theory observed in field measurements. The second phase was dedicated to investigations on reactions inside the droplets in the multiphase system. The aim was to determine decomposition rates directly in a multiphase system to prove the common procedure of calculating these rates from data obtained in homogeneous liquid phase. To achieve our objectives we had to develop special devices for droplet generation, droplet sampling and determination of gas phase concentrations of HNO_3 and SO_2.

Principal scientific results

Experimental set-up

Fig. 1 shows the most recent version of the flow reactor that was constructed for the measurements on airborne droplets. The reactor is mostly made of glass. In a typical run droplets with radii between 5 and 20 µm were generated in part A with a new developed nebulizer. By means of an air flow from an inlet at the prechamber (9) the droplets were carried to the head of the reactor (B) where they are mixed with air containing the trace gas under investigation. After the droplets had passed the thermostated part (C), they were collected by means of a round jet impactor.

A major problem that we faced was the chemical inert generation of droplets with a size distribution similar to that observed in atmospheric clouds and fog. First experiments on the phase partitioning of SO_2 were carried out with ultrasonic nebulizers. They generate droplets with the required size distribution. However the droplet generation is accompanied by the production of H_2O_2 and NO_3^-. Thus, especially for the experiments on the iron catalysed oxidation of SO_2 we had to develop a new nebulizer which is shown in Fig. 2 schematically. Air and water jet meet each other in a right angle. In contrast to venturi jets the water is pumped into the air jet by means of a peristaltic pump. In this way it was possible to generate a size distribution similar to atmospheric conditions.

Sampling devices

Up to now the phase partitioning of the nitrogen species HNO_3/NO_3^- in the atmospheric multiphase system is not completely understood. In order to study this phenomenon in more detail, sensitive sample and accumulation methods have been developed, which enable a reliable separation of the nitrogen species from the gaseous, particle and droplet phase.

For the separation of the gaseous from the particle phase a vertical wet denuder (Fig. 3) and an improved filterpack-system (Fig. 4) were invented. The vertical denuder system consists of two glass tubes. The absorption solution is continuously pumped into and out of the denuder tubes by two peristaltic pumps. Air is sampled at a flow rate of 5 L/min.

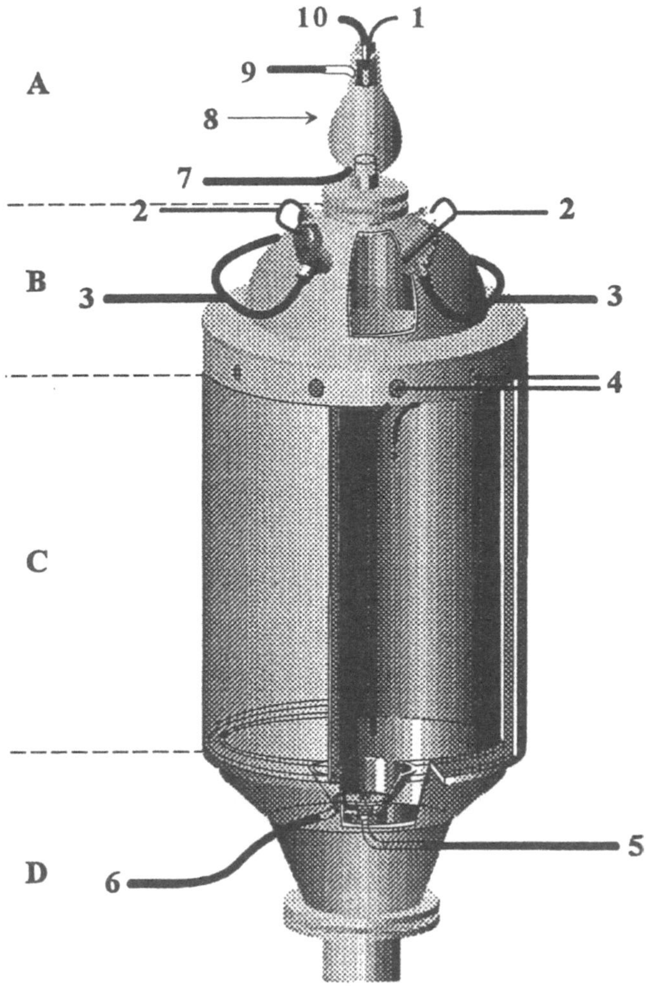

1. Solution supply for the nebulizer
2. SO_2-inlet
3. Inlet of humid air (16 L/min)
4. Connection for gas phase measurements
5. to the vakuum pump of the impactor
6. Outlet impactor
7. Outlet of wall water
8. Prechamber
9. Inlet of humid air
10. Inlet of nebulizer air (9 L/min)

Fig. 1: Flow-reactor.

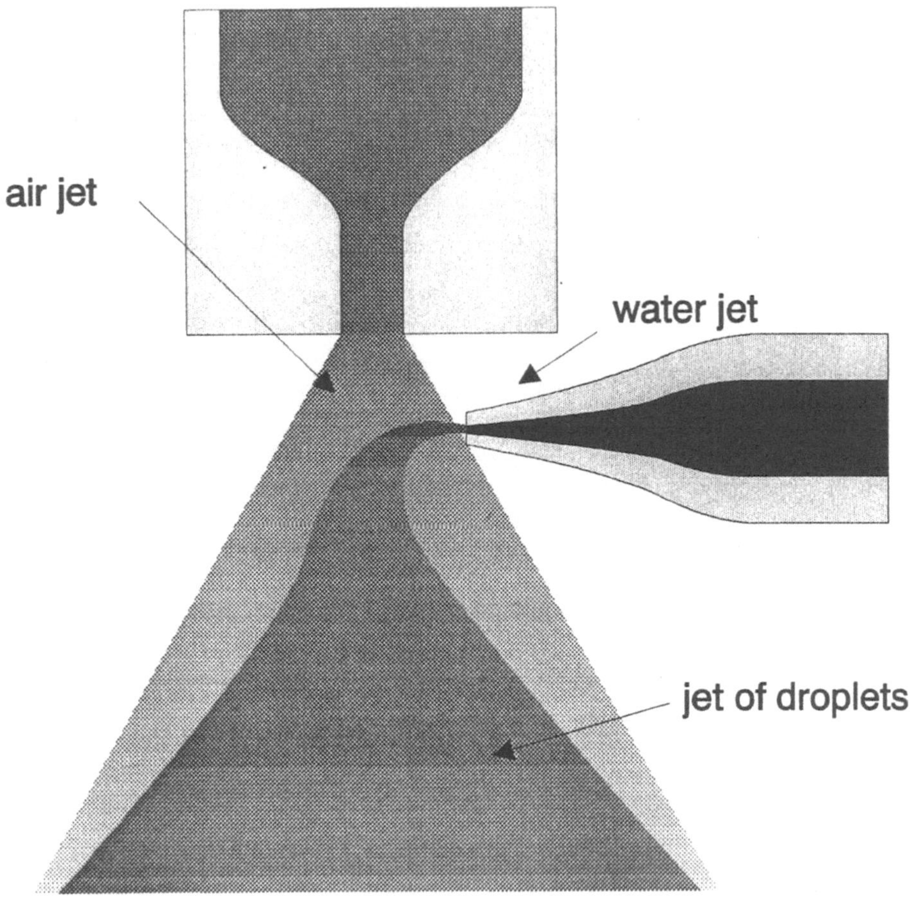

Fig. 2: Schematic representation of the new nebulizer.

The filter sampling technique consists of a three stage filterpack, loaded with filters (47 mm in diameter). Atmospheric NO_3^- in the particulate phase was retained on a Teflon prefilter while HNO_3 in the gas phase was subsequently trapped on an impregnated cellulose filter. The separation of the particles from droplet with diameter > 5 μm is done by impaction technique.

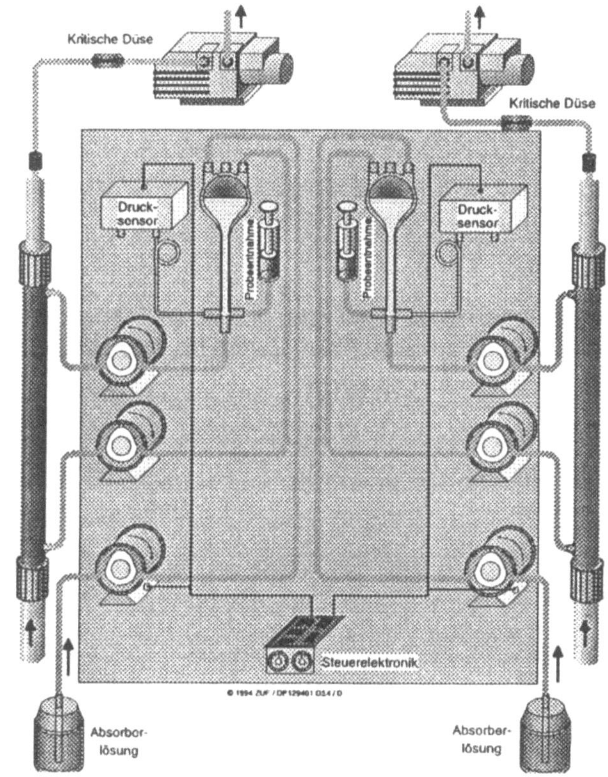

Fig. 3: Vertical denuder for HNO_3(g) sampling.

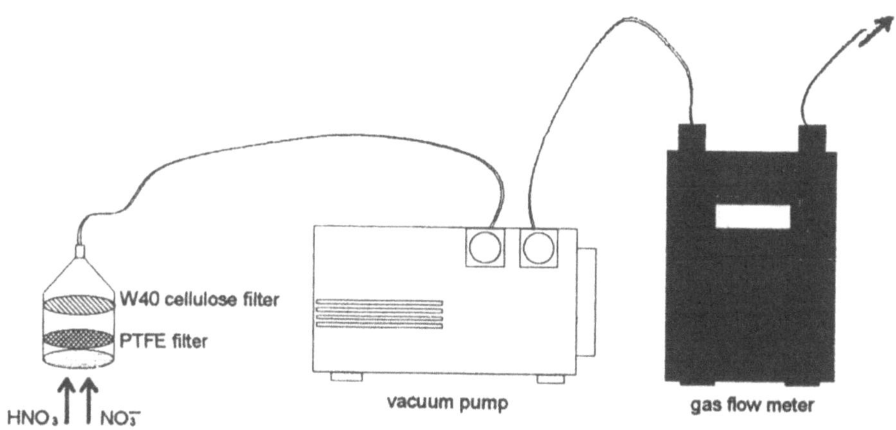

Fig. 4: Filterpack for HNO_3(g) sampling.

Results of measurements

Phase partitioning

According to Henry's law the partitioning of a component "A" between gas (g) and liquid phase (l) is given by the following equation:

$$["A"(l)] = H\, p_{"A"} \qquad (1)$$

where H is the Henry's law constant. In the case of SO_2 and HNO_3 the dissociation equilibria had to be taken into account. This leads to the formulation of a pseudo Henry's law constant which is a function of pH:

$$["A"_{tot}(l)] = H^*\, p_{"A"} \qquad (2)$$

Where the index 'tot' refers to the sum of the concentration of the dissociation products of "A".

Phase partitioning of SO_2

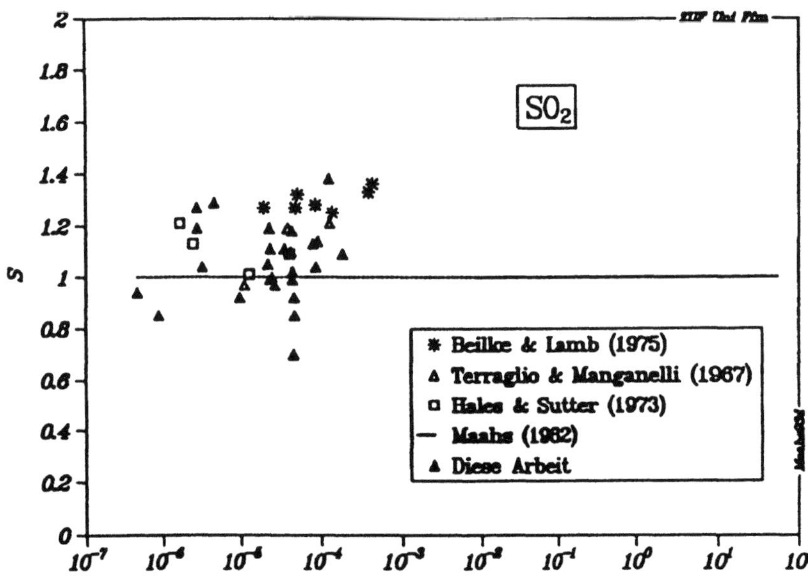

Fig. 5: Comparison of the measurements on the phase partitioning of SO_2 between droplets and air with literature data.

For the investigation on the phase partitioning of SO_2 we used ultrasonic nebulizers for droplet generation. The reactor was additionally stirred by a fan positioned in the middle of the reactor. The concentration of S(IV) in the liquid phase was monitored with a Continuous Flow/Flow Injection ChemiLuminescence analyser (CFCL) which is described elsewhere [1]. SO_2 gas phase concentrations were monitored by means of a commercial UV fluorescence analyser. The concentrations of SO_2 ranged between 10 ppb and 4.5 ppm (0.5–200 μmol m^{-3}). Fig. 5 shows a comparison of our results with literature data [2]. The saturation value S is defined as:

$$S = [S(IV)(l)]/[SO_2(g)]\ H^* \tag{3}$$

where H^* is the literature value of Maahs [2]. As can be seen Henry's law is also fulfilled for the droplet/air system. Additionally we could prove the validity of the Henrys law down to concentrations in the μmol m^{-3} region (literature: mmol m^{-3}).

Phase partitioning of HNO_3

The experiments on the phase equilibrium of HNO_3 where carried out with the new developed nebulizer without additional stirring. The concentration of HNO_3 in the gas phase ranged between 1 ppb and 6 ppb. The generated droplets consisted of deionized water with different concentrations of H_2SO_4. Analysis was performed by ion chromatography.

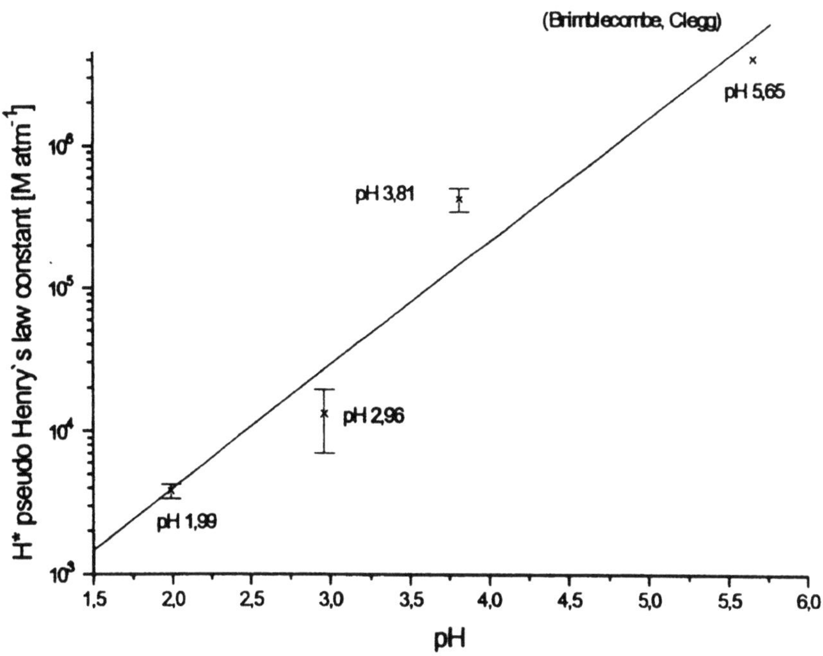

Fig. 6: Comparison of the measurements on the phase partitioning of HNO_3 between droplets and air with literature data.

In Fig. 6 the calculated values of H^* are plotted versus the pH-value. The determined values agree with literature data [3] which had been determined for individual pH values. Additionally the sampling set-up was applied during field measurements in Italy 1994 (Fig. 7). Again a linear dependency of H^* with pH was obtained.

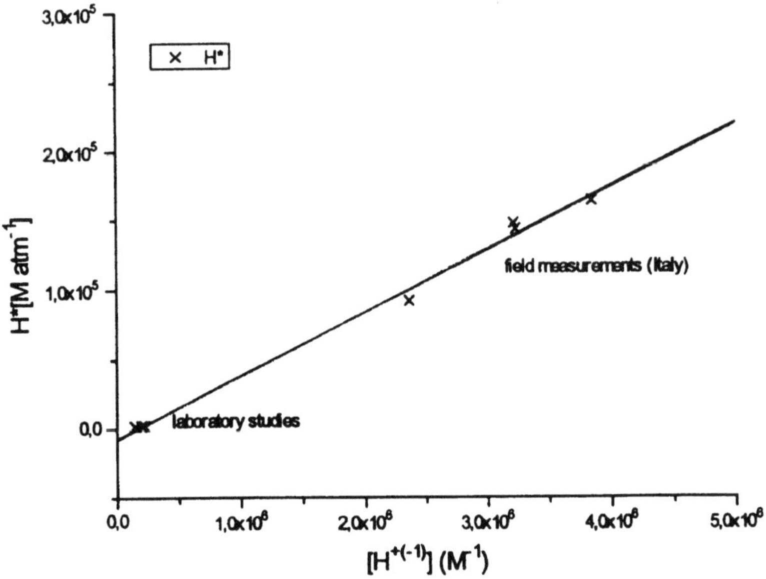

Fig. 7: Comparison of the measurements on the phase partitioning of HNO_3 between droplets and air with field measurements.

Kinetic measurements on the Fe(III)*-catalysed oxidation of* SO_2

For the decomposition of SO_2 dissolved in cloud and fog droplets the Fe(III) catalysed oxidation of SO_3^{2-}, respectively HSO_3^- may play a remarkable role [4, 5]. All investigations on the chemical kinetic of this reaction dealing with the determination of the overall or effective rate constant at atmospheric concentrations were so far carried out in homogeneous liquid phase [6–9]. These studies come to the conclusion that up to a pH value of 4 the reaction is of first order with respect to the concentration of Fe(III) and S(IV).

$$-\frac{d[S(IV)]}{dt} = k^{(2)}\,[Fe(III)]\,[S(IV)] \tag{4}$$

The second order rate constant $k^{(2)}$ is a function of the H^+ concentration. To calculate the decomposition of SO_2 in the atmosphere, for example in transport models, the pseudo-first order rate constant referring to the volume of air is usually calculated from the $k_{tot}^{(1)}$ values determined in homogeneous liquid phase by means of equation 5 [10].

$$k_{tot}^{(1)} = \frac{k^{(1)}[S(IV)(1)] \cdot LWC}{[S(IV)_{tot}]}; \qquad k^{(1)} = f([H^+], [Fe(III)]) \qquad (5)$$

Where $[S(IV)_{tot}]$ represents the total concentration of S(IV) in a given volume of air and LWC is the liquid water content in L m^{-3}. The objective of our study was to determine $k_{tot}^{(1)}$ directly from the multiphase experiment.

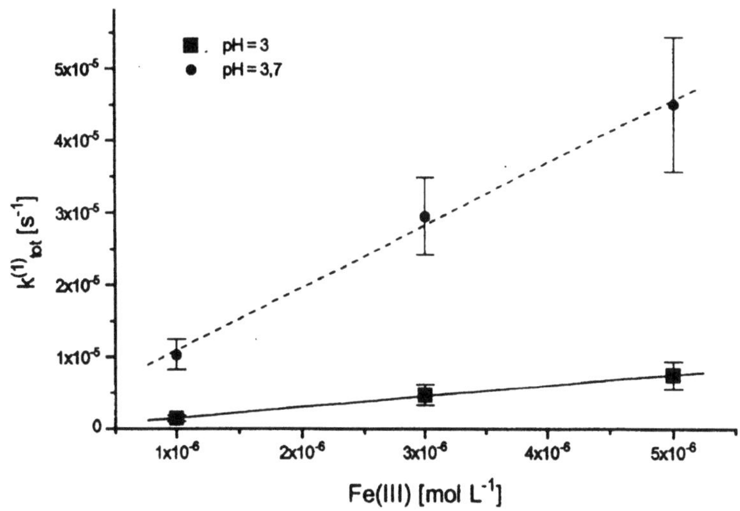

Fig. 8: Determined values of $k_{tot}^{(1)}$ at pH = 3 and 3.7 at different Fe(III) concentrations

Table 1: Mean values of $k_{tot}^{(1)}$ and $k^{(1)}$ with standard deviations in parentheses.

[Fe(III)] / µmol L^{-1}	pH = 3		pH = 3.7	
	$k_{tot}^{(1)}$ / s^{-1}	$k^{(1)}$ / s^{-1}	$k_{tot}^{(1)}$ / s^{-1}	$k^{(1)}$ / s^{-1}
1	1.44×10^{-6} ($\pm 4.4 \times 10^{-6}$)	5.70×10^{-4} ($\pm 5.6 \times 10^{-5}$)	1.04×10^{-5} ($\pm 2.1 \times 10^{-6}$)	8.13×10^{-4} ($\pm 6.1 \times 10^{-5}$)
3	4.70×10^{-6} ($\pm 1.5 \times 10^{-6}$)	1.86×10^{-3} ($\pm 1.6 \times 10^{-4}$)	2.95×10^{-5} ($\pm 5.3 \times 10^{-6}$)	2.31×10^{-3} ($\pm 8.4 \times 10^{-5}$)
5	7.50×10^{-6} ($\pm 1.9 \times 10^{-6}$)	2.99×10^{-3} ($\pm 9.0 \times 10^{-5}$)	4.51×10^{-5} ($\pm 9.4 \times 10^{-6}$)	3.51×10^{-3} ($\pm 1.8 \times 10^{-5}$)

The experiments were carried out in the flow reactor shown in Fig. 1. The mean residence time (= reaction time) was 179 s at an air flow of 30 L min^{-1}. The SO_2 concentration was 500 ppb and LWC ranged between 3–5 × 10^{-3} L m^{-3}. The droplets consist of deionized water with different concentrations of $Fe(ClO_4)_3$. The pH was adjusted with perchloric acid. To inhibit the reaction after sampling of the droplets, the excess S(IV)(l) was complexed with tetrachloromercurate. The determined values of $k_{tot}^{(l)}$ and $k^{(l)}$ (calculated from Eq. 5) are summarised in Table 1. In agreement with the literature studies the reaction rate is linear to the increase of Fe(III) concentrations (Fig. 8) and accelerates with increasing pH values.

To compare the determined rate constants quantitatively with data of studies in homogeneous liquid phase, the question of mass transport limitations has to be taken into account. If there are any limitations, the values of the determined rate constants of the multiphase experiments are incomparable with data of literature studies. According to the equations given by Schwartz [10] and Schwartz and Freiberg [11] limitations due to mass transport do not occur under the conditions of our investigations.

Fig. 9 shows a comparison of the calculated values of $k^{(l)}$ with literature values of Martin [8] and Fuzzi [7] at pH = 3. The determined values of the multiphase experiment are 2–5 times higher compared to the literature data. To estimate the effect of these higher values of $k^{(l)}$ on the decomposition of SO_2 in the atmosphere we calculated multiphase processes (Fig. 10) with the data from Fig. 9 at a Fe(III)-concentration of 5 mol L^{-1} (for additional parameters refer to the figure caption). As expected the decomposition calculated from our data (solid line) is faster than that of the reference investigations. However, after 10 hours the concentrations of SO_2 differ only up to a maximum of 5 %.

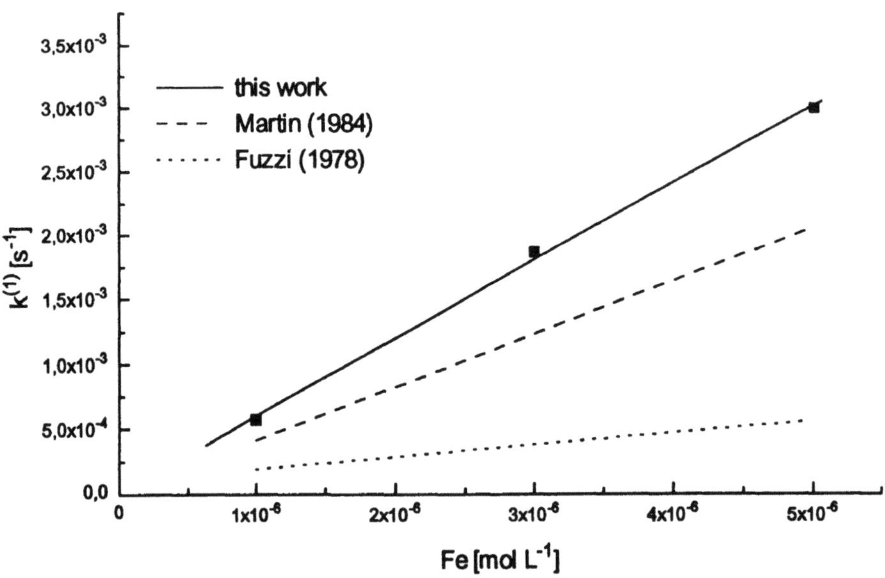

Fig. 9: Comparison of the determined values of $k^{(l)}$, the rate constant referring to the liquid phase, with literature data at pH = 3.

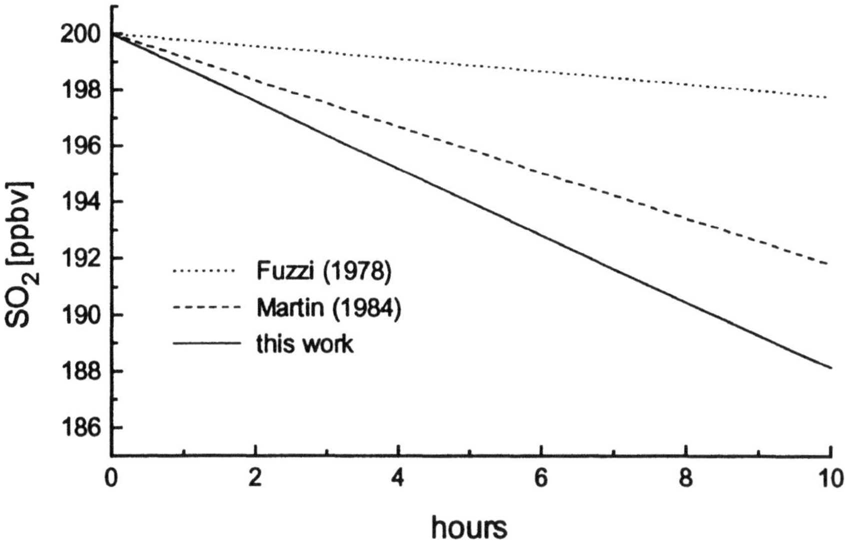

Fig. 10: Comparison of the SO_2-decomposition at pH = 3 ($LWC = 1 \times 10^{-3}$ L m^{-3}, [Fe(III)] = 5×10^{-6} mol L^{-1}, [SO_2]-initial = 200 ppb, temperature = 20 °C).

Conclusions

The results of our work in a simulated fog system showed that Henry's law is fulfilled for the phase partitioning of SO_2 and HNO_3 between droplets of deionized water and air. Thus the observed supersaturation of droplets in the real atmosphere referring to S(IV) should not be due to peculiarities of the multiphase system, like for example the curvature of the droplets surface. On the side of the kinetic measurements we could prove the validity of the theoretical basis of the calculation of the decomposition of SO_2 for the atmosphere from data of homogeneous liquid phase experiments (Eq. 5).

From the present state of our research it is possible to investigate various processes occurring in multiphase systems. Experiments are in progress on the photochemical production of peroxides in the simulated fog system. Furthermore the phase partitioning of NH_3 will be investigated.

Acknowledgements

This project was funded by the Deutsche Forschungsgemeinschaft (DFG), subproject SFB 233.

References

1. R. Maser, H. Franke, M. Preiss, W. Jaeschke: Methods provided and applied in a research aircraft for the study of cloud physics and chemistry, *Beitr. Phys. Atmos.* (1994) 321–334.

2. H. Maahs: Sulfur dioxide/water equilibria between 0 and 50 °C. An examination of data at low concentrations, in: D. Schryer (ed), *Heterogeneous Atmospheric Chemistry*, American Geophysical Union, Washington 1982, pp. 187–195.
3. P. Brimblecombe, S.L. Clegg: The solubility and behaviour of acid gases in the marine aerosol, *J. Atmos. Chem.* **7** (1988) 1–18.
4. T. E. Graedel, C. J. Weschler, M. L. Mandich: Influence of transition metal complexes on atmospheric droplet acidity, *Nature* **317** (1985) 240–242.
5. R. van Eldik, J. Kraft, *Inorg. Chem.* **28** (1989) 2297–2305.
6. P. Brimblecombe, D. J. Spedding: The catalytic oxidation of micromolecular aqueous sulphur dioxide I, *Atmos. Environ.* **8** (1974) 937–945.
7. S. Fuzzi: Study of iron(III) catalysed (sic) sulphur dioxide oxidation in aqueous solution over a wide range of pH, *Atmos. Environ.* **12** (1978) 1439–1442.
8. L. R. Martin: Kinetic studies of sulfite oxidation in aqueous solution, in: J. G. Calvert (ed), SO_2, NO *and* NO_2 *Oxidation Mechanisms: Atmospheric Considerations*, Butterworth, Boston 1984, pp. 63ff.
9. I. Grgic, V. Hudnik, M. Bizjak, J. Levec: Aqueous S(IV) oxidation I. Catalytic effects of some metal ions, *Atmos. Environ.* **25**A (1991) 1591–1597.
10. S. E. Schwartz: Gas-aqueous reactions of sulfur and nitrogen oxides in liquid-water clouds, in: J. G. Calvert (ed), SO_2, NO *and* NO_2 *Oxidation Mechanisms: Atmospheric Considerations*, Butterworth, Boston 1984, pp. 173ff.
11. S. E. Schwartz, J. E. Freiberg: *Atmos. Environ.* **15** (1981) 1129–1144.

6.3 Gas - Liquid Interactions

Ph. Mirabel, Ch. George, L. Magi and J.L. Ponche

Laboratoire de physico-chimie de l'atmosphère – Centre de Géochimie de la Surface, 28, rue Goethe, 67083 Strasbourg Cedex, France

Summary

A technique devoted to the study of uptake rates based on the principle of the droplet train has been developed and applied to 11 gases. The measured kinetics are expressed in terms of uptake coefficient γ_{obs} that represent the fraction of gas/liquid collisions that has effectively led to the uptake and/or chemical transformation. From these measurements we were able to extract information for the process of accommodation at the interface (described by the mass accommodation coefficient α) but also for bulk properties such as Henry's law constant or hydrolysis rate constants. Our results show that for most gases the accommodation does not limit the uptake rate under atmospheric conditions. We also identified a new pathway for the uptake of gases by clouds *i.e.* a surface reaction.

Aims of the research

It has been recently recognised that aerosols and hydrometeors may have a very large influence on the atmospheric chemistry and may alter the composition of the lower atmosphere. Especially, it has been identified that the main source of acid deposition is the "heterogeneous" formation of H_2SO_4 resulting from the uptake and oxidation of SO_2 in cloud droplets by O_3 and H_2O_2. The uptake of trace gases by hydrometeors in the troposphere and their subsequent chemistry may also alter the stratospheric chemistry since these heterogeneous processes may shorten their lifetimes and, in turn, reduce the amount of gases delivered to the stratosphere. More recently, it has been postulated that heterogeneous reactions of nitrogen oxides on sea-salt aerosols may activate radicals in the gas phase (especially chloride) and thus modify the oxidising capacity of the lower atmosphere (effects of such processes has been recently observed during Arctic springs). These examples show the importance of heterogeneous chemistry for the composition of the troposphere. Therefore, our goal within the HALIPP program was to determine experimentally uptake rates for various trace gases of atmospheric interest.

Principal scientific results

The rate of uptake of a gas by a liquid and/or a solid phase is a multi-step process which can be related to fundamental properties of the gas, of the interface or of the condensed phase such as mass accommodation coefficient (α), solubility and reactivity. To describe the efficiency of heterogeneous removal or transformation of trace gases by the "atmospheric condensed phase", it is necessary to have a thorough understanding of the mechanism involved in the mass transfer and chemical reactions at the air/condensed phase interface. Fig. 1 clearly shows the complexity of the reactive uptake of a trace gas by a condensed phase. It involves the combination of processes such as mass transport (*e.g.*, diffusion) in both phases to and away from the interface, the interfacial transfer (characterised by the mass accommodation coefficient α) and chemical transformations at the interface or in the bulk.

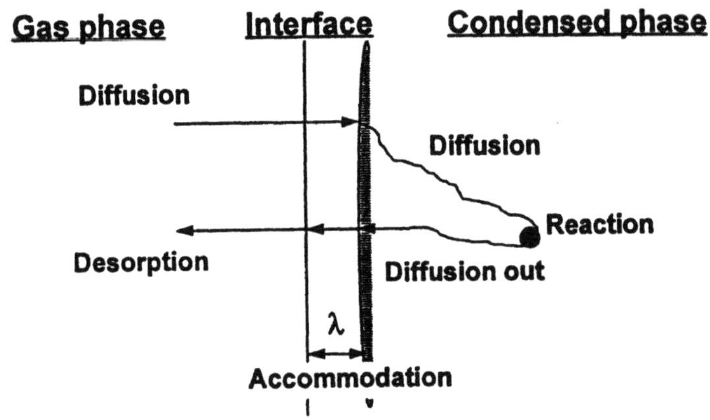

Fig. 1: Schematic representation of the reactive uptake of a trace gas by a condensed phase (λ represents the mean free path in the gas phase).

In most cases, the uptake process cannot be fully deconvoluted into its elementary steps and the overall kinetics are expressed in terms of uptake coefficients (γ_{obs}) that represent the fraction of gas/liquid collisions that has effectively led to the uptake and/or chemical transformation. In order to extract, from the uptake coefficient, further information on the elementary steps, it is necessary to experimentally investigate the effect of mass transport (controlled by the pressure and the size of the condensed phase) or the effect of chemical reactions (controlled by the composition of the exposed surfaces or its nature *i.e.*, solid or liquid) on the observed uptake rates γ_{obs}.

During the course of the EUROTRAC subproject HALIPP, we studied, using the droplet train technique, the uptake rates of several gases that may influence the acidity of precipitation (*i.e.* SO_2, HNO_2, HNO_3, NH_3), activate precursors of radicals (NO_2, N_2O_5, $ClNO_2$) or represent secondary products of the degradation of CFCs, HCFCs (COF_2, CF_3COF and CF_3COCl) or perchlorethene (CCl_3COCl).

Experimental

The droplet train technique (Fig. 2) developed and used in our studies has already been described in the literature [1]. It involves a highly controlled train of mono-dispersed droplets passing through different and independent chambers at low pressure (typically between 15 and 60 Torr). These chambers are the droplet generation chamber, a separating buffer chamber, the interaction chamber (where the trace gas under study can interact with the train of droplets for very short times, typically less than 20 ms), a second buffer chamber, and finally the collection chamber where the droplets are collected in order to be analysed by means of ion chromatography.

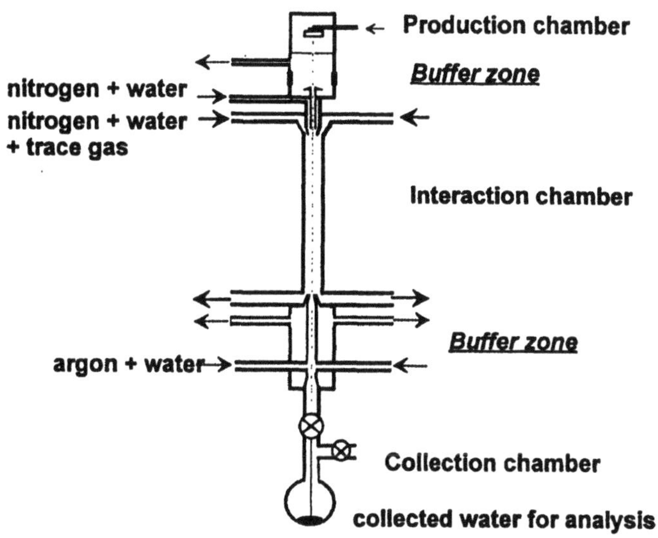

Fig. 2: Schematic diagram of the droplet train apparatus.

The rate of uptake of a gaseous component by aqueous droplets can be affected by several processes such as gas phase diffusion, interfacial transfer, chemical reaction and diffusion in the aqueous phase. It is therefore expressed as an uptake coefficient γ_{obs} which refers to the measured net flux through the interface. The trace gas density (n^{inlet}) can be monitored at the entry of the flow tube by mass spectrometry while the quantity of trace gas absorbed by the droplets (n^{abs}) is derived from the analysis of the collected droplets by means of ion chromatography. Thus the observed uptake coefficient can be calculated from the integrated loss of the trace gas in the interaction zone as:

$$\gamma_{obs} = \frac{4F_g}{\langle c \rangle N\pi d^2} \ln\left\{\frac{n^{inlet}}{n^{inlet} - n^{abs}}\right\}, \tag{1}$$

where <c> is the trace gas average thermal velocity, F_g the carrier gas volume flow rate and $N\pi d^2$ is the surface exposed by the N droplets (having a diameter d) present in the interaction zone.
The design of this technique gives very good sensitivity for high uptake rate measurements (*i.e.*, $\gamma \geq 5 \times 10^{-3}$) but also allows the determination of uptake coefficients down to 10^{-5}.

Results

As pointed out above, the uptake of a trace species by a water droplet is a multi-step process in which different limitations may occur. The major difficulty involved in these uptake rates measurements is the deconvolution of the different limiting factors. In order to perform such treatment and to extract, from the uptake coefficient, information on the accommodation process, it is necessary to experimentally investigate separately the effects of gas phase diffusion, surface saturation or aqueous phase reactions.

The effect of gas phase diffusion is obvious in the case of fast uptake rates ($\gamma_{obs} \geq 5 \times 10^{-3}$) since in this case, it represents the limiting factor for the rate at which a trace species is transported to the liquid droplet and therefore it represents also the upper limit for the net uptake rate (in given conditions). Most of the gases (*i.e.* SO_2, HNO_2, HNO_3, N_2O_5, NH_3) that we studied exhibited high uptakes rates ($\gamma_{obs} \geq 10^{-2}$) and therefore the measured net flux through the gas/liquid interface appears to be limited only by gas phase diffusion. To identify to which extent the measured kinetics is limited by mass transport, we studied the influence of the total pressure on the uptake coefficient γ_{obs}. We observed, for example, a dependence of γ_{obs} with the total pressure (P_{tot}) for HNO_2 [2]. This dependence may be attributed to the variation of the gas phase diffusion coefficient (D_g) with the total pressure (D_g is inversely proportional to P_{tot}). Such a dependence of γ with P_{tot} is clearly an evidence of a gas phase diffusion limited uptake.

Another limitations may occur with poorly water soluble gases (*e.g.* SO_2 or NH_3) for which a surface equilibrium, described by Henry's law, is very rapidly obtained (on a time scale that may be much shorter than the ms scale of our experiments). In such a case, the net uptake rate will depend on the solubility of the gases (described by the Henry's law constant H) and on the rate at which the dissolved gases are transported away from the interface by aqueous phase diffusion. It can be rapidly demonstrated that, in such a case, the uptake coefficient is time dependent since the penetration depth of the gases in the droplet will be depend upon the square root of time t. Therefore in order to verify if our measurements are affected by surface saturation, we studied the uptake rates of several gases as a function of the gas/liquid interaction time showing the two expected behaviour *i.e.* time independent for very soluble gases such as HNO_3 or time dependent such as SO_2 [1].

Finally, a careful examination of the dependence of γ_{obs} on the composition of the condensed phase (pH, presence of scavengers *etc.*) has also been made since it may alter the degree of surface saturation (by changing the degree of dissociation of the dissolved acid or alkaline gases) or may change the rate of bulk chemical transformations (*e.g.* $ClNO_2$). Both of these changes will be observed by modifying the composition of the droplets *e.g.* the pH for SO_2 [1] of the concentration of scavengers for $ClNO_2$ [3].

In order to describe the overall uptake kinetics, it is necessary to take into account all these effects. Using a simplified resistance model [4] which assumes that the processes affecting the uptake can be decoupled, it becomes possible to express each step as a resistance term which stays for its limiting effect on the overall rate (Fig. 3).

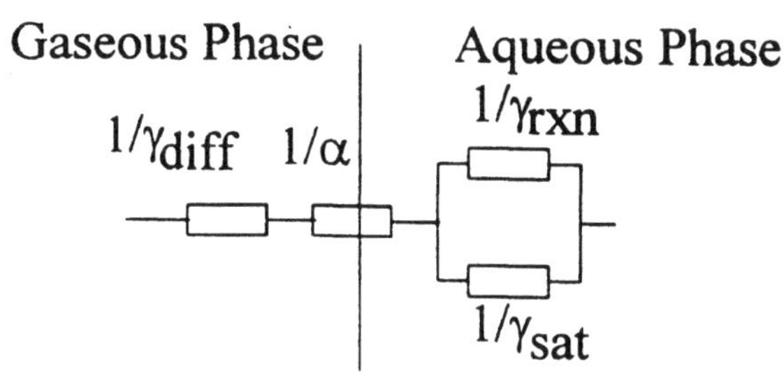

Fig 3: Resistance model for the uptake of a gas by a liquid droplet.

The values for the individual resistance are then given by Jayne *et al.* [4] and Kolb *et al.* [5].

The overall uptake coefficient is then given by:

$$\frac{1}{\gamma_{obs}}=\frac{1}{\gamma_{diff}}+\frac{1}{\alpha}+\frac{1}{\gamma_{sat}+\gamma_{rxn}}=\frac{<c>d}{8D_g}-\frac{1}{2}+\frac{1}{\alpha}+\frac{4HRT\sqrt{D_{aq}}}{<c>}\left(\frac{2}{\sqrt{\pi t}}+\sqrt{k}\right) \quad (2)$$

Using this expression, it becomes possible to extract, from measured uptake coefficient, information about the accommodation α (see Table 1). We also studied the effect of temperature on the mass accommodation coefficient for N_2O_5 and observed an anti-Arrhenius behaviour [6] as expected in regard to previously published values [7] that can be explained by a cluster model for the uptake process.

Table 1: Measured mass accommodation coefficients.

Substance	Temperature / K	α
SO_2	298	0.13 $\pm$ 0.01
HNO_3	298	0.11 $\pm$ 0.01
NH_3	290	0.097 $\pm$ 0.09
HNO_2	245	$0.04 < \alpha < 0.09$
N_2O_5	262–278	0.03–0.013
Halogenated acetyls	273–295	$\geq 10^{-3}$

For the other gases (NO_2, $ClNO_2$, COF_2, CF_3COF, CF_3COCl and CCl_3COCl), we were only able to obtain a lower limit of α ($\alpha \geq 10^{-3}$) because the last term of the right hand side of Eq.(2) is too large compared to the others. However in this case, we tried to deconvolute the saturation process and bulk reactions using the time dependent uptake of COF_2, CF_3COF, CF_3COCl and CCl_3COCl in order to estimate values of both H and k (the hydrolysis rate constant) (see Table 2) [8, 9]. It should be noted that, because of the simplified treatment of the data and because of the large scatter of the data, the values in Table 2 are subject to very large uncertainties (up to a factor 4).

Table 2: Estimated values for H and K (errors are up to a factor 4)

Substance	Temperature / K	H / M.atm^{-1}	k / s^{-1}
COF_2	273	35	< 1
CF_3COF	284	3	150
CF_3COCl	284	2	220
CCl_3COCl	274	2	500

The measured uptake of NO_2 was not described by equation (2) *i.e.* the observed kinetics was to rapid to be explained by known bulk properties of NO_2. Recently, the group in Jülich (A. Wahner) [10] investigated the uptake of NO_2 using a jet method and showed that the uptake of NO_2 is effectively faster than expected from its bulk chemistry. The enhanced uptake was attributed to a "facile and fast" surface reaction. Such a process, was also recently identified for Cl_2 and Br_2 by Davidovits and co-workers [11]. At the same time, we also obtained evidences for such a surface reaction in the case of $ClNO_2$ when reacting with iodide [3] that is not taken into account by equation (2). The state of the art explains this surface process according to an Eley-Rideal mechanism which may represent a further resistance in Fig. 2, which should be added to equation (2).

Finally, the subproject HALIPP gave an ideal framework in which collaboration have been developed. Two joint studies were performed, with the group of Schurath (uptake of HNO_2) and Zetzsch (uptake of N_2O_5 and $ClNO_2$). In both cases, we were able to settle our droplet train technique in Bonn and Hannover.

Conclusions

During the course of the subproject HALIPP, we developed a innovative and sophisticated method devoted to the study of gas/liquid interactions *i.e.* the droplet train technique. This technique represents an extremely useful tool for the measurements of uptake coefficients and was applied to different gases (*i.e.* SO_2, HNO_2, HNO_3, NH_3, NO_2, N_2O_5, $ClNO_2$, COF_2, CF_3COF, CF_3COCl and CCl_3COCl). These measurements led to information about the interfacial mass transport (*i.e.* the mass accommodation coefficient α) showing that it does limit the uptake of most gases under typical atmospheric conditions. We were also able to estimated some physical constant such as Henry's law constant or hydrolysis rate constants. Finally, we identified a new pathway for the uptake of trace species *i.e.* a surface reaction.

The subproject HALIPP also gave opportunities for collaboration with other groups involved and especially at Bonn and Hannover which led to very strong links between these different groups.

Acknowledgements

This work was supported by the Ministère de la Recherche et de la Technologie under the framework of Procope, by the CNRS and the Ministère de l'Environnement (ATP Phase Atmosphérique des grands Cycles Biogéochimiques), by the European Chlorinated Solvent Association (ECSA) and the Alternative Fluorocarbons Environmental Acceptability Study (AFEAS).

References

1. J. L. Ponche, Ch. George, Ph. Mirabel, *J. Atmos. Chem.* **16** (1993) 1.
2. A. Bongartz, Ch. George, J. Kames, Ph. Mirabel, J. L. Ponche, U. Schurath, *J. Atmos. Chem.* **18** (1994) 149.
3. Ch. George, W. Behnke, V. Scheer, C. Zetzsch, L. Magi, J. L. Ponche, Ph. Mirabel, *Geophys. Res. Lett.* in press.
4. J. T. Jayne, P. Davidovits, D. R. Worsnop, M. S. Zahniser, C. E. Kolb, *J. Phys. Chem.* **94** (1990) 6041.
5. C. E. Kolb, D. R. Worsnop, M. S. Zahniser, P. Davidovits, D. R. Hanson, A. R. Ravishankara, L. F. Keyser, M. T. Leu, L. R. Williams, M. J. Molina, M. A. Tolbert: Laboratory studies of atmospheric heterogeneous chemistry, in: J.R. Baker (ed), *Current Problems in Atmospheric Chemistry*, in press.
6. Ch. George, J. L. Ponche, Ph. Mirabel, W. Behnke, V. Scheer, C. Zetzsch, *J. Phys. Chem.* **98** (1994) 8780.
7. J. T. Jayne, S. X. Duan, P. Davidovits, D. R. Worsnop, M. S. Zahniser, C. E. Kolb, *J. Phys.Chem.* **95** (1991) 6329.
8. Ch. George, J. Lagrange, Ph. Lagrange, Ph. Mirabel, C. Pallares, J. L. Ponche, *J. Geophys. Res.* **99** (1994) 1255.
9. Ch. George, J. Y. Saison, J. L. Ponche, Ph. Mirabel, *J. Phys. Chem.* **98** (1994) 10857.
10. D. Becker, A. Wahner, Heterogeneous processes of nitrogen oxide and nitrous acid on aqueous surfaces, presented at the XXth General Assembly of the European Geophysical Society, Hamburg 1995.
11. J. H. Hu, Q. Shi, P. Davidovits, D. R. Worsnop, M. S. Zahniser, C. E. Kolb, *J. Phys. Chem.* (1995) in press.

6.4 Laboratory Determination of Physico-Chemical Rate Parameters Pertinent to Mass Transfer into Cloud and Fog Droplets

U. Schurath, A. Bongartz, J. Kames, C. Wunderlich and T. Carstens

Institut für Physikalische und Theoretische Chemie der Universität Bonn, Wegelerstr. 12, D-53115 Bonn, Germany

Summary

A jet technique has been fully developed and applied to measure the mass accommodation coefficients mass accommodation coefficient (MAKs α) of 11 trace gases on water surfaces. With the exception of $\alpha(CO_2)$, which may be an order of magnitude smaller, all MAKs determined in this and other work are of the order 10^{-3}, and often significantly larger. We conclude that deviations from Henry's equilibrium in cloud droplet systems cannot be accounted for by low MAKs. Uptake rate measurements in the presence of a surfactant which could be applied from the gas phase (1-octanol) have shown furthermore that the effect of ≈ 0.2 monolayer is insignificant.

Henry's law constants of 20 organonitrate compounds of atmospheric interest have been measured at 293 K, and temperature dependencies determined for ten compounds. While all alkyl mononitrates exhibit low solubilities, and do not hydrolyse under ambient conditions, peroxyacyl nitrates are nearly two orders of magnitude more soluble and hydrolyse slowly with typical activation energies of 54.5 kJ mol^{-1}. Although the dinitrates and hydroxy nitrates are up to three orders of magnitude more soluble, most of them are stable or exhibit very low hydrolysis rates in neutral aqueous solutions. Hydrolysis is mildly catalysed by the presence of certain anions, and increases significantly in sea water.

Aims of the research

Most natural and anthropogenic pollutants are emitted near the surface. Intense chemical processing, at least of the more reactive pollutants, occurs predominantly in the lower parts of the troposphere. Since gas phase reactions of NMHCs and NO_x give rise to ozone formation, oxidant levels are strongly influenced by these pollutants in an altitude range where the environment is immediately affected.

About 5 % by volume of the lower troposphere are permanently filled with clouds and fog [1]. Chemical processing in droplets and rainout are important loss mechanisms for pollutants which readily dissolve, or are very reactive, or both, in aqueous solution. It is clearly necessary to include aqueous phase reactions in atmospheric chemical models. While this has become routine for reactions in the homogeneous gas phase, the presence of small cloud or fog droplets not only adds numerous aqueous phase reactions whose rate parameters must be determined; it also requires that the partitioning and the transport rates of reactants and products between the phases be accurately described. This can only be achieved when a number of physico-chemical properties of the relevant trace gases, such as their Henry's law constants, diffusion coefficients in both phases, hydrolysis rates and mass accommodation coefficients (MAKs) on liquid water surfaces, are known.

It was the purpose of this subproject to identify and measure unknown physico-chemical properties of trace gases which are deemed important in this context. The subproject focused on measurements of MAKs which are among the most elusive properties of atmospheric trace gases (and thus were essentially unknown when the proposal was written). In addition, measurements of Henry's law coefficients and hydrolysis rates of organonitrate compounds (alkyl nitrates, substituted alkyl nitrates, peroxyacyl nitrates) were planned.

Inspired by the experience gained in our HALIPP subproject, companion projects were started in our laboratory which deal with diffusion coefficients and studies of adsorption and reactions of trace gases on ice. These projects, which exist outside the HALIPP/EUROTRAC framework, are supported by Deutsche Forschungsgemeinschaft.

Principal scientific results

The liquid jet technique

The *effective* uptake coefficient $\gamma = 4\Phi / ({}^1N \langle c \rangle)$ of a trace gas by a freshly exposed pure liquid water surface approaches the *true* MAK*true* mass accommodation coefficient α in the following limit (Φ = trace gas flux through the surface; ${}^1N$ = its number density at the water surface; $\langle c \rangle$ = its mean molecular speed; t = surface exposure time):

$$\lim_{t \to 0} \gamma = \alpha \tag{1}$$

This maximizes the flux Φ across the phase boundary, because diffusion control and/or surface saturation effects and/or reaction rate limitations vanish with $t \to 0$. In this limit the number density ${}^1N$ of the trace gas is homogeneous throughout the gas phase.

As was discussed elsewhere [2], the limit (1) cannot be established experimentally. It is, however possible, if fluxes Φ are measured under conditions *approaching* the ideal limit (1), to apply suitable corrections in order to obtain α *via* equation (1). Such conditions have been established by two methods: (a) the droplet train technique which was developed by the Aerodyne group [3] and by Mirabel and coworkers in Strasbourg [4], and (b) the liquid jet technique, which was developed in our laboratory. Only a brief description of method (b) is given here: a fast laminar water jet of *ca.* 100 μm, which emerges from a thin-walled quartz capillary in the centre of a miniature flow tube, is exposed under isokinetic conditions to typically 100 ppm of a trace gas in air, N_2, or He. The exposed length *z*, typically between 0.3 and 5 mm, can be varied with μm resolution by trapping the jet in a counter capillary. The solute concentration in the trapped liquid is measured as function of jet length *z* by a suitable analytical technique (ion chromatography; on-line conductometry; spectrophotometry if the solute or a reaction product is a strong optical absorber [5]).

We have shown that the time evolution of the concentration profiles $c(r)$ perpendicular to the exposed jet surface can be computed numerically if the diffusion coefficients D_g and D_{aq}, Henry's law constant *H*, and all relevant rate constants in the liquid phase are known, while a range of values of the MAK α is assumed. By dividing the properly integrated concentration profiles $c(r,t)_{aq}$ through the amount of jet water which has flown through the capillary during the surface exposure time *t*, volume-averaged concentrations $c(t)$ of the solute are computed which depend *only* on the assumed value of α, and can be directly compared with measurements. The *true* MAK is obtained by matching computed solute concentrations $c(t)$ with measured concentrations $c(z)$, treating α, as the only free fitting parameter.

While the transformation from the experimental length scale *z* to the exposure time scale *t* of the computations is straight forward in the plug flow approximation, a more realistic transformation must take account of the true velocity of the surface, which is accelerated from zero to a final velocity U_∞ when the Hagen-Poiseuille profile has relaxed to a plug profile. The transformation, which requires that the variable surface velocity U_z is known, has been described elsewhere [6]. We have written a computer program [7] which solves the pertinent Navier-Stokes equations, using a problem-adapted coordinate system [8]. The program yields velocity profiles $U_{z,r}$ and surface velocities U_z for any set of assumed boundary conditions. While the true surface velocity of the jet cannot be determined directly, the computations also show that the jet contracts more or less rapidly to about 86 % of its initial diameter, depending on the mean velocity inside the capillary, the viscosity of the liquid, and the surface tension. Careful measurements with sub-micron resolution of the jet diameter $\varnothing_z$ have shown that the numerical calculations are in reasonable agreement with observed jet profiles [9], giving credibility to the computed surface velocities. We find, however, that irregularly-shaped capillaries give rise to standing surface waves of the jet, which cause rapid disintegration of the jet into small droplets. Therefore only perfectly shaped capillaries ought to be utilized in determinations of MAKs.

Measurements of mass accommodation coefficients

Table 1 lists MAKs which have been measured in this subproject, as well as data from the literature for comparison. Our data with ≥ signs were determined in the early stages of our project when the jet method was not yet fully developed (including, *e.g.*, measurements with improperly shaped jet capillaries). While most of our MAKs, lower limits included, are in reasonable agreement with the literature, we suspect that the MAK of NO_2 published by Mirabel *et al.* may be in error, because we found in our experiments that the uptake rate of NO_2 (which seemed to occur in the form of N_2O_4 [10]) was solubility limited *for even the shortest surface exposure times*. A possible source of interference could have been formation of HONO by surface reactions of NO_2, which was observed in a similar system. We intend to reassess this system in the near future with chemical scavengers, to reduce surface saturation effects.

Table 1: Mass accommodation coefficients of gases and radicals on water at room temperature, extrapolated to 298 K where necessary.

trace gas	Worsnop *et al.* [19–21]	Mirabel *et al.* [4, 11, 22]	Tang and Lee [23, 24]	This work	others
SO_2	0.09	0.13	> 0.02	0.1	–
H_2O_2	0.11	–	–	–	–
HNO_3	0.08	0.11	–	≥ 0.01	–
HCl	0.08	–	–	≥ 0.01	–
N_2O_5	0.02	< 0.01	–	≥ 0.005	0.05 a)
NO_2	–	0.0015	0.0006	see text	–
PAN	–	–	–	≥ 0.001	–
NH_3	–	0.097	–	0.040	–
HONO	–	0.05	–	0.05	–
HCOOH	0.015		–	0.013	–
HAc	0.021		–	0.0175	–
CH_3OH	0.017		–	–	–
EtOH	0.009		–	–	–
CO_2			–	$1–2 \times 10^{-4}$	–
O_3			0.0005	0.0045	≥ 0.002 b)
HO_2			–	–	> 0.2 c)
OH					≥ 0.0035 d)

a) Mozurkevich [25]; b) Utter [26]; c) Mozurkevich [27]; d) Hanson [28]

It should be noted that the MAK of HONO was measured in co-operation with Mirabel's coworkers [11]. This was possible because large amounts of HONO in N_2 carrier gas, sufficient to run both the droplet train experiment and our own jet apparatus in parallel, could be stored in our large evacuable glass chamber. HONO concentrations in the carrier gas were measured by long path optical absorption spectrometry, using new spectral data which had been previously determined in our laboratory [12]. The good agreement between the droplet train and jet techniques, which differ considerably with respect to exposure time, total pressures, and surface geometries, lends credibility to the obtained MAK of HONO.

Most of the MAK measurements by the droplet train technique have been carried out with supercooled water droplets at very low temperatures. It has been shown in a series of papers by the Aerodyne group that most MAKs exhibit negative temperature dependencies, which explains early discrepancies between our MAKs and the often much larger values obtained by the droplet train technique. The MAKs in Table 1 for formic and acetic acid, which have been measured at room temperature in our laboratory [13], and in a much lower temperature range by the Aerodyne group (their data in Table 1 were extrapolated to room temperature), are in excellent agreement when the temperature dependencies are taken into account. Fig. 1 shows a typical set of acetic acid concentrations which were measured in our laboratory as function of jet surface exposure time. The solid lines are model calculations for a range of assumed MAKs: the best fit is obtained for an assumed MAK of 0.0175. Note that two out of four data sets were measured in the presence of 1-octanol as a surfactant, which was present in the carrier gas at a mixing ratio of 300 ppm (limited by the vapour pressure of the pure compound). We calculate that a maximum of 25 % of the jet surface was covered with 1-octanol in these experiment. The results show that the surfactant had no effect on the uptake rate of acetic acid.

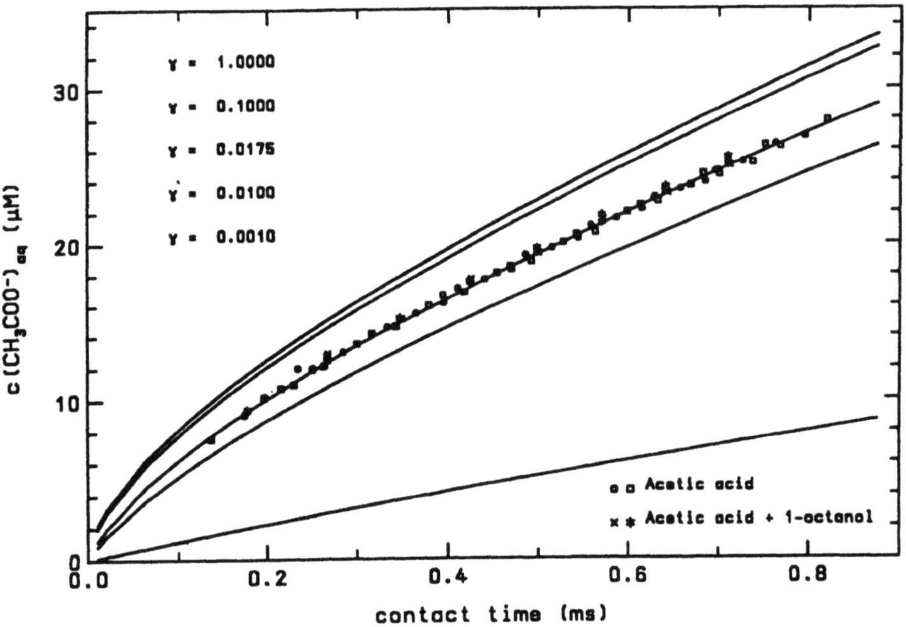

Fig. 1: Effect of co-deposited 1-octanol on the MAK of acetic acid.

MAKs of low solubility gases like CO_2 and NO_2 are difficult to measure unless an effective chemical scavenger can be added in sufficient amounts to eliminate surface saturation effects. This strategy was successful in the case of ozone [5] which reacts extremely rapidly with I^- to form IO^- as primary product. In buffered solution IOH is formed, which reacts with excess I^-, yielding I_3^- as end product. I_3^- could be measured by optical absorption spectrometry in the UV. The required diffusion coefficient of ozone in air, which has been measured for the first time, can be represented by the formula $D_{ozone/air} = 0.152 \times (T/293\ K)^{1.963}\ cm^2\ s^{-1}$.

Henry's law constants and hydrolysis rates

Table 2: Henry's law constants, enthalpies and entropies of solution, hydrolysis rate constants of organic nitrates; n.d. = not determined; "stable" means $k_{hydrol.} \ll 10^{-6}\ s^{-1}$.

compound	$\frac{H(293\ K)}{M\ atm^{-1}}$	$\frac{\Delta H^0_{soln}}{kJ\ mol^{-1}}$	$\frac{\Delta S^0_{soln}}{J\ K^{-1}\ mol^{-1}}$	$\frac{k_{hydrol}(293\ K)}{10^{-4}\ s^{-1}}$
CH_3ONO_2	2.64	−39.4 ± 0.6	−126.4 ± 3	stable
$C_2H_5ONO_2$	2.18	−44.5 ± 1.5	−145.5 ± 5	stable
1–$C_3H_7ONO_2$	1.55	−45.6 ± 0.4	−152.2 ± 2	stable
2–$C_3H_7ONO_2$	1.08	−44.5 ± 0.7	−151.3 ± 3	stable
1–$C_4H_9ONO_2$	1.42	−48.1 ± 0.9	−161.3 ± 3	stable
2–$C_4H_9ONO_2$	0.89	−45.0 ± 0.6	−154.6 ± 3	stable
2–methyl–propyl–1–nitrate	0.96	−43.6 ± 1.2	−149.2 ± 4	stable
1–$C_5H_{11}ONO_2$	1.2	n.d.	n.d.	stable
2–$C_5H_{11}ONO_2$	0.53	−52.3 ± 2.2	−187.7 ± 4	stable
1,2–ethane–dinitrate	640 ± 8	n.d.	n.d.	stable
2–nitrooxy–ethanol	39900±900	n.d.	n.d.	< 0.1
2–nitrooxy–1–propanol *)	7300 ± 300	n.d.	n.d.	stable
1–nitrooxy–2–propanol *)	6700 ± 300	n.d.	n.d.	stable
1,2–propane dinitrate	175 ± 20	n.d.	n.d.	stable
nitrooxyacetone	1010 ± 40	n.d.	n.d.	0.024
perooxyacetyl nitrate	97.9 ± 2.0	−47.4 ± 0.8	−150 ± 2.5	3.4 ± 0.1
peroxypropionyl nitrate	70.8 ± 1.5	−49.4 ± 0.9	−160 ± 3	3.1 ± 0.2
peroxy–n–butyl nitrate	54.8 ± 1.5	n.d.	n.d.	2.4 ± 0.4
peroxymethacryloyl nitr.	41.0 ± 3.8	n.d.	n.d.	5.6 ± 1.2
perroxy–*i*–butyryl nitrate	24.0 ± 0.7	n.d.	n.d.	7.4 ± 0.5

*)A mixture of the isomers was synthesized and used. A definitive assignment (which Henry's law constant belongs to which isomer?) was not possible.

The results of our solubility and hydrolysis rate studies in approximately neutral ion-free water are summarized in Table 2. Typical solvation enthalpies ΔH^0_{soln} range between −40 and −50 kJ mol^{-1}, with solvation entropies ΔS^0_{soln} in the range −130 to −198 J K^{-1} mol^{-1}. The hydrolysis rates of peroxyacetyl nitrate (PAN) and peroxypropionyl nitrate (PPN) exhibit a positive temperature dependence, $E_a \approx 54.5$ kJ mol^{-1}. The solubilities and hydrolysis rates of a few compounds were also studied as a function of ionic strength. While solubilities were found to decrease when the ionic strength increased, hydrolysis rates were significantly enhanced in the presence of phosphate and other ions. The hydrolysis rate of peroxyacetyl nitrate (PAN) was found to be strongly enhanced in filtered North Sea water; the effect was much less pronounced in synthetic sea water. These and other effects are discussed in detail elsewhere [14,15].

Conclusions

With the possible exception of CO_2 (which is notoriously difficult to measure due to its low physical solubility and slow hydrolysis rate), all MAKs listed in Table 1 are in the order of 10^{-3} and larger. According to criteria developed by Schwartz [16], mass accommodation at the air/liquid interface is unlikely to be uptake rate limiting under real atmospheric conditions. Even for very small cloud droplets, surface saturation and reaction rate limitations will be more important. Therefore, establishment of Henry's equilibrium between small cloud droplets and the interstitial air seems to be a reasonable approximation to be made in atmospheric modelling. We could also show that surfactants are unlikely to increase interface resistance to such an extent that deviations from Henry's law, which have been observed for organic acids [17] and SO_2 [18] in cloud water, could be explained. Dramatic increases of the MAK have only been observed for complete monomolecular coverage of the water surface [13]. However, one can show that the amount of surfactants needed to form a monomolecular surfactant layer on cloud droplets is unrealistically high [13].

Most of the organonitrate compounds studied in this subproject exhibit low solubilities and hydrolysis rates, and should be unaffected by rainout and/or hydrolysis in cloud and fog droplets. Notable exceptions are dinitrates, hydroxy- and ketonitrates, which are fairly soluble in water, and should thus be strongly adsorbed on surfaces. This may be one of the reasons why these compounds, although expected to be present in the atmosphere, are very elusive in field studies.

Acknowledgements

This work was supported as a subproject of HALIPP/EUROTRAC by Bundesministerium für Forschung und Technologie, Grant No. 07EU727 9, and in part by the European Community, grant No. EV4V-0070 D.

References

1. F. J. Dentener, P. J. Crutzen, J. Lelieveld, in *Physico-Chemical Behaviour of Atmospheric Pollutants*, published by the European Commission (1994) 14–17.

2. W. Kirchner, F. Welter, A. Bongartz, J. Kames, S. Schweighoefer, U. Schurath, *J. Atmos. Chem.* **10** (1990) 427–449.
3. J. H. Hu, J. A. Shorter, P. Davidovits, D. R. Worsnop, M. S. Zahniser, C. E. Kolb, *J. Phys. Chem.* **97** (1993) 11037–11042.
4. J. L. Ponche, Ch. George, Ph. Mirabel, *J. Atmos. Chem.* **16** (1993) 1–21.
5. C. Wunderlich, A. Schlemm, K. Hesse, D. Hupperich, U. Schurath, Uptake of ozone by aqueous surfaces, in: K. H. Becker (ed), *Tropospheric Oxidation Mechanisms, Air Pollution Research Report* **54**, EC, Luxembourg 1995, pp. 45–50, (ISBN 92-827-0080-1).
6. A. Bongartz, S. Schweighoefer, C. Roose, U. Schurath, *J. Atmos. Chem.* **20** (1995) 35–58.
7. A. Schlemm, Computersimulation und Messung der Kontraktionvon laminaren Flüssigkeitsstrahlen, *Diploma Thesis*, University of Bonn, 1994.
8. J. L. Duda, J. S. Vrentas, *Chem. Eng. Sci.* **22** (1967) 855–869.
9. A. Nink, Untersuchung der Temperaturabhängigkeit des Strahlprofils eines laminaren Flüssigkeitsstrahls, *Diploma Thesis*, University of Bonn, 1995.
10. W. Kirchner, Massen-Akkommodationskoeffizienten atmosphärischer Spurengase für die Grenzfläche Luft/Wasser, *Dissertation*, University of Bonn, 1989.
11. A. Bongartz, J. Kames, U. Schurath, Ch. George, Ph. Mirabel, J. L. Ponche, *J. Atmos. Chem.* **18** (1994) 149–169.
12. A. Bongartz, J. Kames, F. Welter, U. Schurath, *J. Phys. Chem.* **95** (1991) 1076–1082.
13. A. Bongartz, Massen-Akkommodationskoeffizienten schwacher atmosphärischer Säuren und Basen, *Dissertation*, University of Bonn, 1993.
14. J. Kames, U. Schurath, *J. Atmos. Chem.* **15** (1992) 79–95.
15. J. Kames, U. Schurath, *J. Atmos. Chem.* 20 (1995) 151–164.
16. S. E. Schwartz, in: *Physico-Chemical Behaviour of Atmospheric Pollutants*, EC, Luxembourg 1994, pp. 891–900.
17. W. Winiwarter, H. Fierlinger, H. Puxbaum, M. C. Facchini, B. G. Arends, S. Fuzzi, D. Schell, U. Kaminski, S. Pahl, T. Schneider, A. Berner, I. Solly, C. Kruisz, *J. Atmos. Chem.* 21 (1995) 173–188.
18. M. Preiss, R. Maser, H. Franke, W. Jaeschke, J. Graf, *Beitr. Phys. Atmos.* **67** (1994) 341–351.
19. D. R. Worsnop, M. S. Zahniser, C. E. Kolb, J. A. Gardner, L. R. Watson, J. M. Van Doren, J. T. Jayne, Paul Davidovits, *J. Phys. Chem.* **93** (1989) 1159-1172.
20. J. M. Van Doren, L. R. Watson, Paul Davidovits, D. R. Worsnop, M. S. Zahniser, C. E. Kolb, *J. Phys. Chem.* **94** (1990) 3265-2369.
21. J. T. Jayne, S. X. Duan, P. Davidovits, D. R. Worsnop, M. S. Zahniser, C. E. Kolb, *J. Phys. Chem.* **95** (1991) 6329-6336.
22. Ch. George, J. L. Ponche, Ph. Mirabel, W. Behnke, V. Scheer, C. Zetzsch, *J. Phys. Chem.* **98** (1994) 8780-8784.
23. I. N. Tang, J. H. Lee, in: R. W. Johnson, G. E. Gordon (eds), *The Chemistry of Acid Rain*, ACS Symposium Series 1987, pp 109-117.
24. J. H. Lee, I. N. Tang, *Atmos. Environ.* **22** (1988) 1147-1151.
25. M. Mozurkevich, J. G. Calvert, *J. Geophys. Res.* **93** (1988) 15889-15896.
26. R. G. Utter, J. B. Burkholder, C. J. Howard, A. R. Ravishankara, *J. Phys. Chem.* **96** (1992) 4973-4979.
27. M. Mozurkevich, P. H. McMurry, A. Gupta, J. G. Calvert, *J. Geophys. Res.* **92** (1987) 4163-4170.
28. D. R. Hanson, J. B. Burkholder, C. J. Howard, A. R. Ravishankara, *J. Phys. Chem.* **96** (1992) 4979-4985.-2

Chapter 7

Aerosol Surface Reactions and Photocatalysis

7.1 Determination of β-Pinene OH Radical Rate Constant: Possible Impact of Solid Particles

Panayotis Foster, I. Denis and V. Jacob

Groupe de Recherche sur l'Environnement et la Chimie Appliquée, I.U.T. de Chimie - U.J.F., 39-41, Bd Gambetta, 38000 Grenoble, France.

Summary

The rate constant for the gas phase reaction between and OH radicals has been determined at 293 ± 3 K and one atmosphere pressure of air, using a relative rate technique with *n*-octane and *n*-nonane as reference organics. In similar conditions, the rate constant has been also determined in the presence of solid particles (titanium dioxide and fly ash). According to the additional gas phase experiments and analytical artefact calculations which have been done on all experimental results, these particles do not seem to affect the β-pinene reactivity under our conditions; this is contrary to what we have concluded previously.

Aims of the research

The study of biogenic VOC reactivity in the atmosphere has progressed greatly over the last few years because of their importance in atmospheric biogeochemical cycles [1, 2]. Until recently, reactivity studies have been mainly based on their gas-phase chemistry.

Aerosols are important constituents of the atmosphere because of their physical properties and because of their influence on the gaseous compound atmospheric reactivity (SO_2, VOCs, PAHs) which has been demonstrated by several authors [3–11]. Previous work in our laboratory has shown the possible influence of solid aerosols on some organic compound reactivity by modification of kinetic rate constants and product distributions in heterogeneous experiments in comparison with homogeneous simulations [12–14].

New studies on the possible photocatalytic activity of the atmosphere show that the oxidation of terpenes on the surface of aerosols may provide a significant atmospheric sink [15, 16].

In our work we concluded that under our conditions titanium dioxide and fly ash seem to enhance photochemical oxidation. So we are now planning to confirm the impact of these solid particles upon the chemical phototransformation of β-pinene and to validate the homogeneous rate constants as reference.

Activities during the year

The rate constants for the reaction of OH radicals with β-pinene were determined using a relative rate method in which the relative decay rate of the studied compound and the reference organics (*n*-octane and *n*-nonane) were monitored in the presence of OH radicals. Of course, the OH radical rate constant for the reference organics was reliably known.

The atmospheric simulations were then performed in a perfectly stirred 2 m^3 altuglass-wall reactor, where irradiation, temperature, humidity and reactant concentrations were under control [12].
Our experimental conditions were:

β-pinene:	0.15 to 0.17 ppm	NO: 5 ppm to prevent of ozone formation
n-nonane:	0.13 to 0.14 ppm	H_2O: 20–25 % R.H.
n-octane:	0.15 to 0.17 ppm	temperature: 293 to 298 K
particles:	130 to 190 mg/m^3	
OH:	direct photolysis of CH_3ONO in excess	
irradiation:	12 neons 400 W	$360 < \lambda < 700$ nm
Hg lamp	400 W	

The course of the hydrocarbon gas-phase concentrations were monitored during experiments by capillary gas chromatography-flame ionisation detection (CP9000 Chrompack), using a WCOT column 50 m × 0.32 mm CP-Sil-5-CB, df = 1.2 μm (Chrompack). The reactants were sampled by preconcentration on sorbent tubes (600 ml on 100 mg Tenax TA) and then thermodesorbed on a thermodesorption cold trap system (TCT Chrompack) which was coupled with a gas chromatograph for analysis.

Additional gas phase experiments were performed in order to confirm the rate constants we measured last year.

Principal experimental results

During our experiemnts, we encountered analytical problems: the terpene studied underwent chemical degradation on the sampling adsorbent. Monoterpenes and oxygenated compounds are formed under different conditions: on β-pinene standard samplings and on β-pinene kinetic experiment samplings. Then, we decided to account for this artefact in the measurements and to calculate for each sample a corrected chromatographic area (A) in the following way:

A = (measured area) + (sum of all artefact peak areas)

We have obviously made an important approximation because we postulated that all the artefact molecules had the same chromatographic response factor as one. Several experiments, the results of which are not presented here, indeed showed a good recovery for corrected areas with theoretical areas. So we calculated the kinetic rate constant for β-pinene in two ways: with measured areas and with corrected areas.

The homogeneous gas phase reaction was first investigated. The reaction for each compound could be simplified by the following way:

β-pinene	+	OH	→	products	k_P
nonane	+	OH	→	products	k_N
octane	+	OH	→	products	k_O

providing that the reactants were only removed by reaction with OH radicals, then:

$$\mathrm{Ln}\left(\frac{(\beta\ \mathrm{pinene})o}{(\beta\ \mathrm{pinene})t}\right) = \frac{kP}{kREF}\ \mathrm{Ln}\left(\frac{(\mathrm{reference})o}{(\mathrm{reference})t}\right) \qquad \text{(A)}$$

where $(\beta\text{-pinene})_o$ and $(\text{reference})_o$ were the concentrations of the monoterpene and the reference organic (*n*-octane or *n*-nonane) respectively, at time $t = 0$, $(\beta\ \text{pinene})_t$ and $(\text{reference})_t$ were the corresponding concentrations at time t, $k_{REF} = k_O$ for n-octane and $k_{REF} = k_N$ for *n*-nonane. Hence, plots of $\mathrm{Ln}((\beta\text{-pinene})_0 / (\beta\text{-pinene})_t)$ versus $\mathrm{Ln}((\text{reference})_0 / (\text{reference})_t)$ should yield a straight line of slope k_P/k_{REF} and a zero intercept.

The first-order rate constants for β-pinene (k_P) and *n*-octane (k_O) were then calculated from the experimental curves, the reference compound was *n*-nonane (k_N = 10.2 for β-pinene 10^{-12} cm^3 $molec^{-1}$ s^{-1} and k_O = 8.7 for β-pinene 10^{-12} cm^3 $molec^{-1}$ s^{-1} [16]); the results are presented in Table 1.

Table 1

Homogeneous phase	GRECA results		GRECA results*		Values from literature
	§	§§	§	§§	
	Rate constant (10^{-12} cm^3 $molec^{-1}$ s^{-1})				
β-pinene	68.6	74.3	51.0	46.8	65.8 [17]
	100.7	105.6	100.9	106.2	77.6 [17]
	101.4	88.5	89.1	77.6	79.5 [17]
	91.9	88.6	69.0	66.6	78.9 [16]
	93.6	85.5	83.5	76.3	
	86.7	76.5	80.5	79.8	
	84.7	69.9	72.6	60.2	
	82.9	71.7	73.1	63.4	
		97.0		80.5	
n-octane	8.64				8.42 [17]
	8.50				8.89 [17]
	9.04				8.80 [17]
	9,03				8.68 [16]
	9.66				
	9.02				
	9.9				
	10.2				

§ : Reference = *n*-nonane §§ : Reference = *n*-octane
GRECA results*: results with corrected areas

Mean values and standard deviation are then calculated, see Table 2 where the first two values are excluded.

Table 2

	GRECA results		GRECA results*	
	§	§§	§	§§
β-pinene	90 ± 7	83 ± 10	78 ± 8	72 ± 9
octane	9.5 ± 0.5			

§ : Reference = *n*-nonane §§ : Reference = *n*-octane
GRECA results* : results with corrected areas

It should be noted that the corrected rate constants are in good agreement with literature values (GRECA results*), but some differences exist with an octane result which we are unable to explain at the moment. Nevertheless, these results allow us to consider the homogeneous simulations as references for the further heterogeneous kinetic studies.

The heterogeneous simulations were performed in the presence of titanium dioxide (with rutile form; average diameter: 1.06 μm; specific area: 10.1 m^2/g) and fly ash (collected from a coal-burning power plant; average diameter: 1.2 μm; specific area: 1 m^2/g). We also calculated corrected rate constants from the β-pinene corrected chromatographic areas (GRECA results*) and the results are presented in Table 3:

Table 3

Heterogeneous phase		GRECA results §	GRECA results §§	GRECA results* §	GRECA results* §§
		Rate constant ($10^{-12} \cdot cm^3 \cdot molec^{-1} \cdot s^{-1}$)			
β-pinene		92.0	94.1	81.0	82.9
	TiO_2	97.4	102.1	73.73	77.,1
		92.6	91.5	65,1	64.2
	Fly ash	88.5	88.2	67,5	67.5
		79.7	78.9	77,8	77.1
n-octane		8.59			
	TiO_2	8.56			
		8.60			
	Fly ash	8.32			
		8.79			

§: Reference = *n*-nonane §§ : Reference = *n*-octane
GRECA results* : results with corrected areas

In the same manner, mean values and standard deviations are given in Table 4:

Table 4

	GRECA results §	GRECA results §§	GRECA results* §	GRECA results* §§
β-pinene TiO_2	95 ± 4	98 ± 6	77 ± 6	80 ± 5
β-pinene fly ash	87 ± 6	86 ± 6	70 ± 7	70 ± 7
octane	8.58 ± 0,03	8.6 ± 0.,3		

In the previous report we concluded that titanium dioxide and fly ash seemed to show a photocatalytic influence upon the OH radical reaction: the heterogeneous rate constants were higher than the homogeneous ones with a 61 % rate increase in the presence of titanium dioxide and a 43 % increase with fly ash. As we encountered some problems of repeatability for additional homogeneous rate constant measurements, we could not draw any conclusions.

The calculation of corrected chromatographic areas which has been carried out for all previous experiments, allows us to set better kinetic results for homogeneous reactions. In this situation, we show that these particles do not lead to any appreciable modification to β-pinene reactivity under our experimental conditions.

Main conclusions

Contrary to our previous conclusions, these new results showed the lack of catalytic activity of particles on β-pinene photodegradation with OH radicals during our experiments. No general conclusions can be drawn yet because we need to perform experiments in different situations, with higher concentrations of particles, for example.

Aims for the coming year

Different heterogeneous simulations are also planed with other particle species in order to determine the nature and structural effects on β-pinene catalysed reactions with OH radicals. Reaction product identification and kinetic rate constant measurements will be carried out in parallel. This research will be extended to other monoterpenes like α-pinene and limonene.

References

1. V. A. Isidorov, I. G. Zenkevich, B. V. Ioffe, *Atmos. Environ.* **19** (1985) 1.
2. C. N. Hewitt, R. A. Street, *Atmos. Environ.* **17** (1992) 3069.
3. G. E. Gordon, *Environ. Sci. Technol.* **25** (1991) 1822.
4. B. L. B. Wiman, H. O. Lannefors, *Atmos. Environ.* **19** (1985) 349.
5. C. Zetzsch, W. Behnke, *Air Pollution Research, Report 33,* CEC (1990) 95.
6. R. Dlugi, H. Güsten, *Atmos. Environ.* **17** (1983) 1765.
7. K. Takeuchi, T. Yazawa, T. Ibusuki, *Atmos. Environ.* **17** (1983) 2253.
8. W. H. Griest and B. A. Tomkins, *Environ. Sci. Tech.* **20** (1986) 291.
9. D. A. Saucy, R. M. Kamens, R. W. Linton, *Atmos. Environ.* **17** (1983) 2617.
10. T. Ibusuki, K. Takeuchi, *Atmos. Environ.* **20** (1986) 1711.
11. R. W. Matthews, S. R. McEvoy, *J. Photobiol. A: Chem.* **64** (1992) 231.
12. P. Foster, V. Jacob, M. Laffond, P. Baussand, in: P. Borrell, P. M. Borrell, W. Seiler (eds), *Proc. EUROTRAC Symp. '90*, SPB Acedemic Publishing bv, The Hague 1991, p. 323.
13. M. Laffond, P. Foster, P. Baussand, I. Denis, V. Jacob, *Air Pollution Research, Report 33,* CEC (1990) 144.
14. P. Foster, V. Jacob, P. Baussand, M. Laffond, *Fresenius Environ. Bull.*,**2** (1993), 78.
15. V. A. Isidorov, *Ecological Bull.* **42** (1992) 71.
16. K. I. Zamaraev, M. I. Khramov, V. N. Parmon, *Catalysis Revs.* **36(4)** (1994) 617.
17. J. Peral, D. F. Ollis, *J. Catalysis* **136** (1992) 554.
18. R. Atkinson, *Atmos. Environ.* **24A** (1990) 1.
19. R. Atkinson, *Chem. Revs.* **85** (1985) 69.

7.2 Calorimetric Study of Lewis Acid Centres and Adsorbed Water in the Heterogeneous Reactions Involving Tropospheric Particles

Jean-Claude Petit and Jamel Thlibi

CNRS - Laboratoire de Combustion et Systèmes Réactifs, 1C Avenue de la Recherche Scientifique, F-45071 Orléans Cedex 2, France

Summary

The formation by UV illumination of Lewis acid centres upon both model compounds (SiO_2, TiO_2, Al_2O_3) and tropospheric materials (desert sand and fly ash) has been examined. The Lewis Acid centres formed have been titrated (in the 10^{-4}–10^{-6} mol/g range) by water vapour or NH_3 reactions. However, the high rate of destruction of such active sites and the high concentration of water vapour in the troposphere lead to a stationary state too low to allow the reaction with other reactive molecules to occur.

The presence of unburned carbon in fly ash has lead us to examine separately the reaction of (photoactivated or not) carbonaceous materials with water vapour. Near room temperature, in the absence of photoactivation, oxidised carbonaceous materials such as soot, activated carbon, carbon black and coal react with water vapour. They yield small but significant amounts of CO_2 and CO in the range of some µmol/g with a ratio $CO_2/CO > 10$, allowing a clear distinction from thermal decomposition ($CO_2/CO < 4$). Under UV illumination carbonaceous materials exhibit only a small increase in the CO_2 + CO formation. On the contrary, the addition of water vapour to unoxidised carbonaceous materials yields no CO_2 or CO.

Aims of the research

The aims are to examine if the UV illumination of both model compounds (SiO_2, Al_2O_3, TiO_2) and tropospheric solids (as desert sand or fly ash) are able to form high reactive sites which could then be the source of radicals in the gas phase.

Principal scientific results

It is now well established that the photocatalytic degradation of pollutants adsorbed on solid surfaces of inorganic and organic materials both of anthropogenic and natural origin plays an important role in the chemistry of the troposphere in generating oxidising radicals [1]. As a rule, laboratory studies use the direct UV illumination of either the organic adsorbed phase or of both the gaseous phase and the adsorbed phase. Under such conditions it is sometimes difficult to distinguish between the respective roles of the different phases *i.e.* adsorbent, gas phase and adsorbant.

In the present work, as an aim to simplification, we are examining the direct effect of UV illumination on the possible generation of reactive sites on the surface of both model and tropospheric materials. In other words we suppose that the activation of solids usually obtained by heating may be replaced by a photo-activation at ambient or subambient temperature and that this latter will lead to the formation of similar reactive sites. The solid contained in a silica glass vessel are subjected to an outgassing treatment then either UV illuminated for a 16 hour period with a 480 W Osram lamp filtered: $\lambda > 320$ nm or thermally activated. Then, the admission of micromole doses of gaseous reagents (NH_3 or H_2O) allows the titration of sites and the measurement of the enthalpy of reaction.

Formation of reactive sites by UV illumination of model and stratospheric compounds

Model compounds

Fig. 1 shows the titration at 100 °C with water vapour (curves A, B, B' and C) or ammonia (curve D) of sites formed upon SiO_2, TiO_2 and Al_2O_3 samples outgassed at low or high temperature (curves B and C respectively) or subjected to an UV illumination without trapping (curves A) or with an additional trapping (curve A2).

Silica

In region I corresponding to silica no important activation occurs under UV illumination and the enthalpy of titration is lower than that measured with a thermal treatment at high temperature which forms Lewis Acid centres [2, 3]. Such a result is expected since silica exhibits only very weak semiconductor properties. Note that the agreement between the number of Lewis sites dosed with water and ammonia is rather satisfactory taking into account the well-known tendency of ammonia to coordinate.

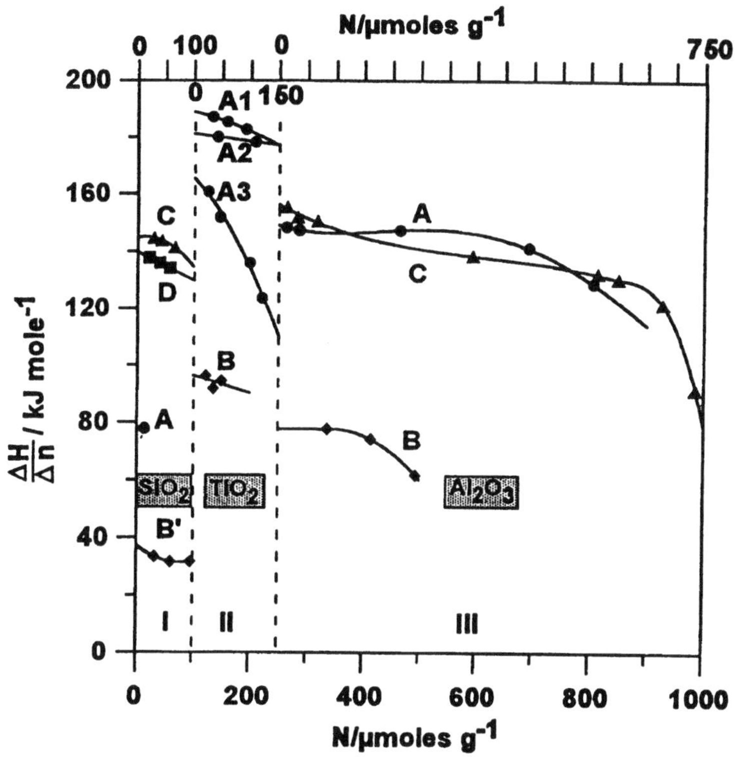

Fig 1: Titration by water vapour and ammonia of active sites formed upon model compounds SiO_2 (domain I), TiO_2 (domain II) and Al_2O_3 (domain III) by UV illumination (curves A) or thermal activation at low or high temperature (curves B and C respectively).

Titanium dioxide

The TiO_2 behaviour is reported in region II. It shows that the number of sites formed by outgassing at low temperature (210 °C) is higher than that observed with silica. However, samples subjected to an UV illumination exhibit a high enthalpy of reaction with water vapour (curves A1, A2, A3). The trapping of water vapour formed during UV illumination has only a light influence on the number of site formed (curve A2). It is interesting to note that after the destruction of sites by water vapour a subsequent UV illumination increases the number of sites generated (curve A3). Manuera *et al.* [4] have observed during the UV illumination of TiO_2 the formation of both H_2O_2 and O_2. The sites formed under UV illumination are very likely different from the real Lewis Acid Centres which are obtained with solids exhibiting rather weak semiconductor properties.

Alumina

In region II a more conspicuous site formation appears as well with UV illuminated samples (curve A) as with thermal activated samples at low or high temperature (curves B and C respectively). Furthermore, it is interesting to note that the energy of sites formed by UV illumination and by thermal treatment at high temperature are similar showing that, very likely, in both cases Lewis Acid Centres are formed. This result is rather unexpected with alumina which is known to have only weak semiconductor properties.

Tropospheric solids

Desert sand

The behaviour of desert sand samples (origin: Sahara) either outgassed at low and high temperature or UV illuminated is very similar to that of Al_2O_3. The trapping of water vapour during UV illumination increases substantially (>50 %) the number of formed sites (Fig. 2). This result indicates that the UV illumination involves an important photodesorption of water vapour which can, in the absence of trapping, overshadow the photo formation of active sites.

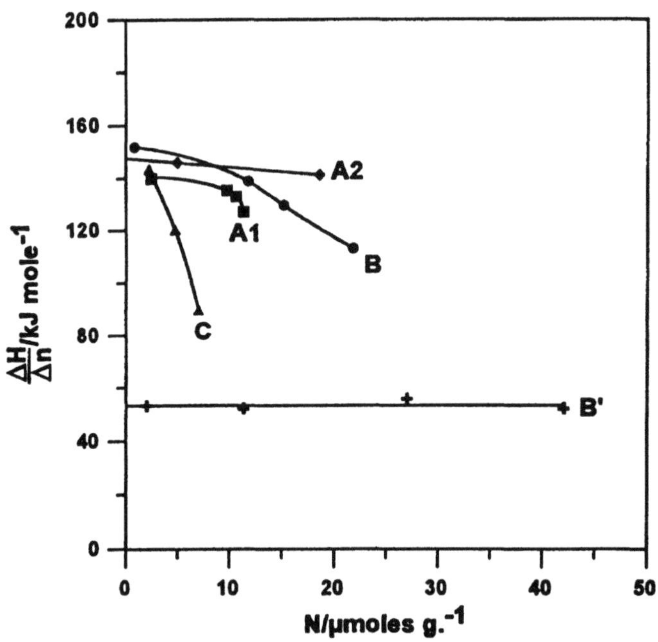

Fig 2: Water vapour titration of sites formed on desert sand by thermal treatment at high (curve C) and low temperature (B and B') or UV illumination without trapping (A1) or with trapping (A2).

Fly ash

In Fig. 3 it is clearly shown that the sites formed on fly ash (origin Gardanne plant) both under UV illumination (curves A1, A2 and A3) or by thermal activation at high or low temperature (curves C and B respectively) are able to yield high energetic sites. However, under UV illumination, in the absence of trapping of water vapour (curve A1), the formed sites have a low enthalpy of reaction. On the contrary, when the UV illumination is accompanied of water trapping, very energetic sites are formed (curve A2) and easily regenerated (curve A3) in large amount (>100 μmol site/g). Although the sites formation under UV illumination may be overshadowed by the desorbed water vapour, fly ash appear as one of the most reactive tropospheric solids which is more reactive than the model compounds studied here [5].

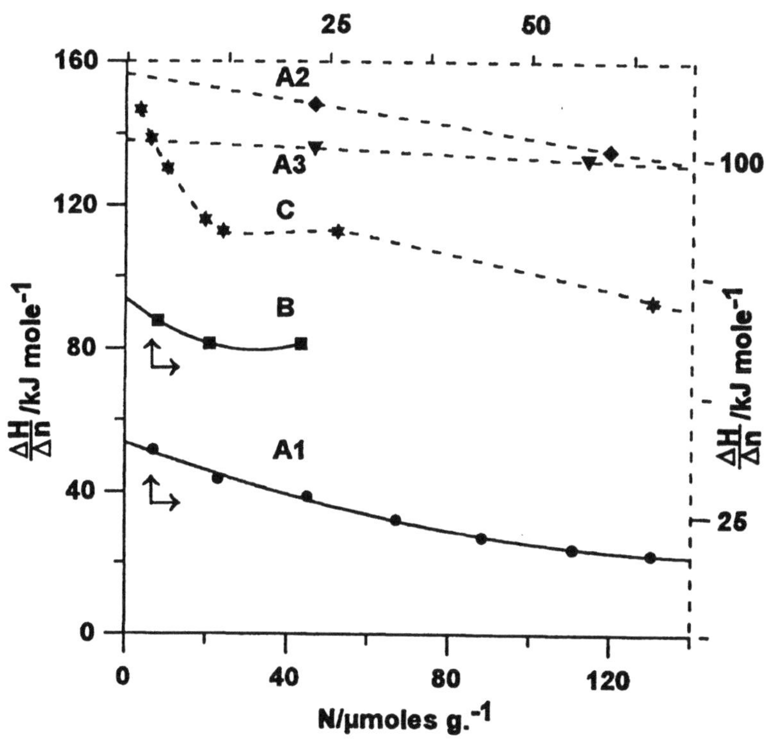

Fig. 3: Water vapour titration of sites formed upon fly ash by UV illumination (curves A) or thermal activation at low or high temperatrure (curves B and C respectively).

Because of both the high affinity of water for reactive sites and the high vapour pressure of water compared to that of the pollutants in the conditions of temperature prevailing in the troposphere, only a very low stationary state in reactive sites must be expected, leaving no chance to pollutant molecules to react with them. It is why, in the second part of this work we have researched, on the surface of fly ash, for the presence of less reactive sites towards water vapour (surface groups), especially in the carbonaceous fraction of fly ash.

Study of the interaction of water vapour with oxidised carbonaceous materials

Fig. 4 shows the cumulative amounts of gases formed at 150°C against time by reaction of water vapour with previously oxidised coal [6–7]. It is clear that the addition of water vapour (regions II and IV) leads mainly to CO_2 formation in the range of some µmol/g of carbonaceous materials and it is superimposed on a minor but significant CO formation. These CO_2 and CO formations must be distinguished from those coming from the pyrolysis treatments as exemplified in regions I and III.

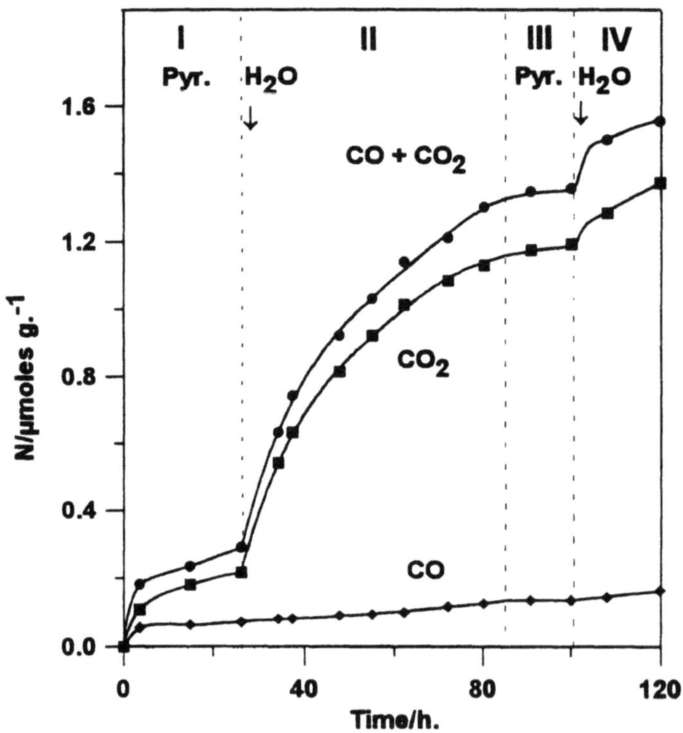

Fig. 4: CO_2+CO, CO_2, CO evolutions against time of a coal sample (La Mure anthracite) alternatively subjected to pyrolysis and water vapour treatments at 150°C. In regions I and III CO, CO_2, CO_2+CO formation corresponds to a pyrolysis reaction whereas in region II and IV the vapour reaction is superimposed on the pyrolysis (water vapour addition)

Fig. 5 shows the changes in the CO_2/CO ratio against time. It is obvious that during pyrolysis treatments (regions I and III) this ratio remains low (close to 4) and independent of time. On the contrary, in regions II and IV, corresponding to the main water vapour treatment, the ratio is high (reaching 28) and depends on the time.

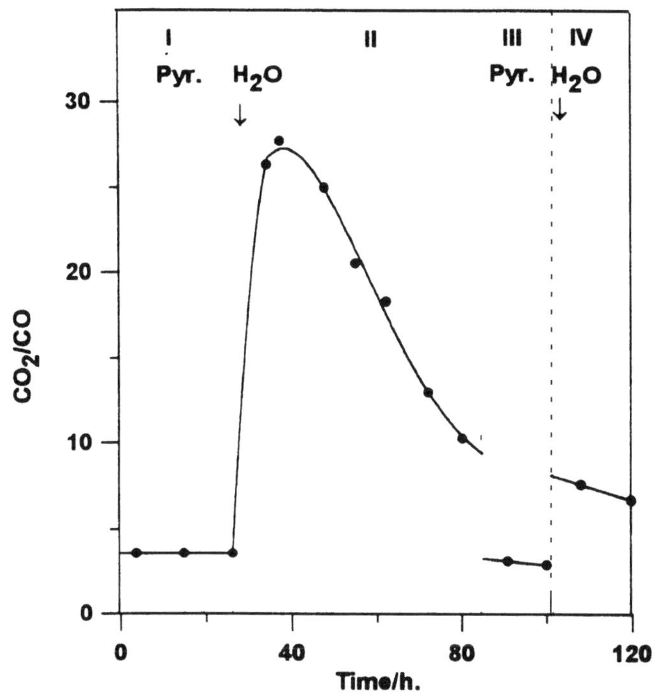

Fig. 5: Changes in the CO_2/CO ratio against time for the same run as in Figure 4.

Similar behaviour has been observed even at room temperature with other carbonaceous materials as, activated carbon, carbon black and soot (these latter two materials being considered as potential constituents of fly ash). On the contrary, when carbonaceous materials have not been previously oxidised, or more exactly when they have been subjected to an outgassing treatment at moderate temperature, the water vapour admission is not accompanied of gas formation.

When water vapour is added on previously oxidised carbonaceous materials, FTIR/DRIFT (Diffuse Reflectance Fourier Transform Infrared Spectroscopy) measurements show successively, the rapid increase of –COOH groups and their subsequent destruction. This destruction is connected with the formation of HCOOH identified by (Ion Chromatography) and its subsequent decomposition into CO_2 and CO.

Conclusions

This study shows that the very reactive sites formed by UV illumination of tropospheric solids, such as desert sand and fly ash, cannot play an important role in tropospheric chemistry. Indeed, they are quickly destroyed by background water vapour, very likely leaving only a small stationary level, which is not sufficient to allow the reaction of other molecules to proceed.

However, fly ash appears as one of the most reactive tropospheric solids. As it consists of up to 10 % of soot and unburned carbon, we have looked for the presence of sites less reactive towards water vapour. The identification of a slow hydrolysis reaction yielding mainly CO_2, attributable to the water vapour interaction with oxidised surface groups of carbonaceous materials, indicates that in the oxidative conditions encountered in the troposphere, competition for the destruction of such sites between water and pollutant molecules is conceivable.

Acknowledgements

Funding by the French EUROTRAC committee is gratefully acknowledged.

References

1. H. Güsten, in: W. Jaeschke (ed), *Chemistry of Multiphase Atmospheric Systems, NATO ASI Series*, Vol. G6, Springler Verlag, Berlin 1986, p. 567.
2. B. Fubini, V. Bolis, E. Giamello, *Inorg. Chim. Acta* **138** (1987) 193.
3. B. A. Morrow, I. A. Cody, *J. Phys. Chem.* **80** (1976) 1998.
4. G. Manuera, V. Rives-Arnau, A. Saucedo, *J. C. S. Faraday Trans. I*, **75** (1979) 736.
5. J. Fatnassi, J. C. Petit, in: P. M. Borrell, P. Borrell, T. Cvitaš, W. Seiler (eds), *Proc. EUROTRAC Symp. '92*, SPB Academic Publishing bv, The Hague 1993, p. 619.
6. J. C. Petit, *Fuel* **70** (1991) 1053.
7. J. C. Petit, Y. Bahaddi, *Carbon* **31** (1993) 821.

7.3 Laboratory Studies of the Phototransformation of Organic Compounds in the Presence of Water and/or Inorganic Particulate Solids

P. Pichat, C. Guillard, C. Hoang-Van, H. Delprat, D. Mas, F. Marme, T. Bouvier and B. Servajean

URA au CNRS "Photocatalyse, Catalyse et Environnement", Ecole Centrale de Lyon, BP 163, 69131 Ecully Cédex, France

Summary

To investigate the effects of liquid water and inorganic particulate solids in tropospheric photochemistry, reactions have been performed in the laboratory with various representative organic compounds (*o*-xylene, naphthalene, 2,4-dichlorophenoxyacetic acid (2,4-D), α-pinene and oxalic acid) and a series of minerals (SiO_2, Al_2O_3, Fe_2O_3, $FeTiO_3$, TiO_2, a clay, a mica, as well as a fly ash sample) in the presence of O_2, either in the dry state or in liquid water (with or without H_2O_2 or Fe^{3+} ions), under UV irradiation. In the dry state, the effects were important over TiO_2 which behaved as a photocatalyst; non-negligible effects were, in general, observed for the other solids, presumably because of electron transfers between the photoexcited organic and the adsorbent. In aqueous solution, large effects were found for the couples laponite-2,4-D, SiO_2-α-pinene (essentially in the presence of H_2O_2) and Fe_2O_3-oxalate ; the importance of dissolved Fe^{3+} ions and H_2O_2 was confirmed. These effects were attributed to

(i) the interactions of the 2,4-D carboxylate groups with the separated clay layers,
(ii) an increase probability of reaction between hydroxyl radicals formed from H_2O_2 and α-pinene at the surface of SiO_2, and
(iii) the formation of free and surface oxalato-iron complexes. Fe_2O_3 and $FeTiO_3$ exhibited very weak photocatalytic activities.

Even when the influence of the solid upon the disappearance rate of the organic was weak, significant changes in the nature and/or distribution of the products were generally found at equal transformation percentages. These results show that both liquid water and inorganic matter affect the tropospheric photochemistry.

Aims of the research

The general objective was to show the role of the phenomena occurring in heterogeneous phase (with respect to the gas phase, *i.e.* in the presence of liquid water and/or inorganic particulate solids) in the phototransformation of organic compounds found in the troposphere.

Principal scientific results

Introduction

With the objectives indicated above, we have intended, in this first approach, to show that heterogeneous photochemical processes can concern several kinds of atmospheric aromatic compounds, inorganic particulate solids and conditions. After a brief preliminary study concerning *o*-xylene, naphthalene (NPH), 2,4-dichlorophenoxyacetic acid (2,4-D), α-pinene, ethanal and oxalic acid were chosen to represent the products of anthropogenic and natural combustion, the pesticides, the biogenic emissions and the oxidized products formed in the final transformation stages of organic compounds in the atmosphere, respectively. The choice of the solids was based on their occurrence in the areas where the above mentioned organic compounds are generally emitted. For NPH, found principally in urban areas, a fly ash and a ferric oxide sample were selected. For 2,4-D, a herbicide, a silica and a clay were used as soil substitutes introduced in the atmosphere by erosion. The same silica sample was employed to study the phototransformation of α-pinene, since this compound is mainly emitted in rural areas. As oxalic acid forms complexes with ferric ions either free or at the surface of iron oxyhydroxides, these latter solids were chosen in the studies concerning this acid and, for comparison, ethanal was included. Titanium dioxide, a minor constituent of atmospheric inorganic particulate matter, was used as a reference, in particular to comparatively assess the photocatalytic properties of ferric oxide. As oxidants, dioxygen (or air) was always present; hydrogen peroxide was added in some cases both in the presence and absence of silica. Similarly, ferric ions, a source of hydroxyl radicals in tropospheric water droplets, was sometimes added or resulted from the dissolution of iron oxyhydroxides. Experiments were performed both in the dry state and in aqueous phase.

This variety of investigated systems was expected to illustrate the significance of the effects of various particulate solids and constituents of water droplets.

Rates and products of the phototransformation of adsorbed naphthalene

The photodegradation of naphthalene (NPH), chosen as a model of polynuclear aromatic pollutants, has been studied in the presence of a layer of four water-insoluble inorganic solids which can be found in the troposphere (TiO_2, Fe_2O_3, muscovite, and a fly ash sample). Direct photolysis of NPH was negligible at $\lambda > 340$ nm. Dark adsorption of NPH on TiO_2 (mainly anatase, nonporous, 50 $m^2\ g^{-1}$) at 293 K corresponded to a surface coverage of *ca.* 50 % at equilibrium. Under these conditions (saturated surface), the stationary-state photocatalytic degradation reached 0.4 molecule $nm^{-2}\ h^{-1}$ ($\lambda > 340$ nm, radiant flux *ca.* 22 mW cm^{-2}). Dioxygen was required. In air the NPH degradation was zero order in O_2. Water vapour markedly increased the rate. The other particulate solids had also an effect, less important than that of TiO_2, however quite noticeable with respect to surface area unit for the fly ash sample. Whether their activity came from the presence of TiO_2 in muscovite, or iron oxide in the fly ash sample, remains to be determined. Apart from 1,4-naphthoquinone, which was the main intermediate product in all cases, 2-naphthol, phthalide, phthaldialdehyde, phthalic acid, acetophenone, benzaldehyde, benzoic acid were also formed on dry TiO_2. Depending on their volatility, these compounds were transferred to the gas phase or remained principally adsorbed on the solid particles where they were further transformed. For instance, phthalic acid (or anhydride) and benzoic acid were generated from 1,4-naphthoquinone. These phototransformation, together with those occurring in homogeneous phase, can give rise to compounds that are more harmful or odoriferous than the primitive pollutant whose emissions to the atmosphere could be restricted in consequence.

Effects of a silica, a clay and H_2O_2 *on the phototransformation of 2,4-D*

In the absence of water and with or without silica or laponite

Silica or laponite (a synthetic clay) was impregnated with 2,4-D. In the absence of adsorbent, 2,4-D was deposited as a thin film on the flat bottom of the photoreactor. Under UV irradiation ($\lambda > 290$ nm), the disappearance rate r of 2,4-D was nearly the same in the presence of laponite or without a solid. By contrast, r was markedly increased in the presence of silica. This effect of silica could be due to the capture of electrons arising from 2,4-D by the Lewis acid sites, which would facilitate the decarboxylation :

$$R\text{–}O\text{–}CH_2\text{–}COO^- \rightarrow R\text{–}O\text{–}CH_2^{\circ} + CO_2 + e^-$$

In all cases, 2,4-dichlorophenol was the main intermediate; however, its concentration as a function of time depended both on the presence and on the nature of the adsorbent. Differences in the distribution of the other aromatic intermediates (chlorohydroquinone, chloro-*p*-benzoquinone, 4-chlorocatechol and an unidentified compound) were also observed.

In aqueous solution and with or without silica or laponite

The initial concentration of 2,4-D was 0.362 mmol/L (80 mg/L). In some cases 1 g/L of silica or laponite was dispersed. Under UV irradiation ($\lambda > 290$ nm) the 2,4-D disappearance rate calculated on the basis of an apparent first-order kinetic law was approximately in h^{-1}: 0.04 in the absence of solid; 0.032 with silica, and 0.018 with laponite. The accelerating effect observed with silica in the dry state did not exist, presumably because the Lewis acid sites were occupied by water molecules. The substantial effect of laponite can be explained by the separation of the layers of this solid in liquid water.

The degradation aromatic intermediates were the same as without water. In addition, the formation of 2,4-dichloroanisole was found in the absence of solid or with silica, but not with laponite.

The above results show that the presence of silica or laponite caused differences, sometimes in the disappearance rate of 2,4-D, and, in all cases, in the distribution of the intermediates and their temporal variations.

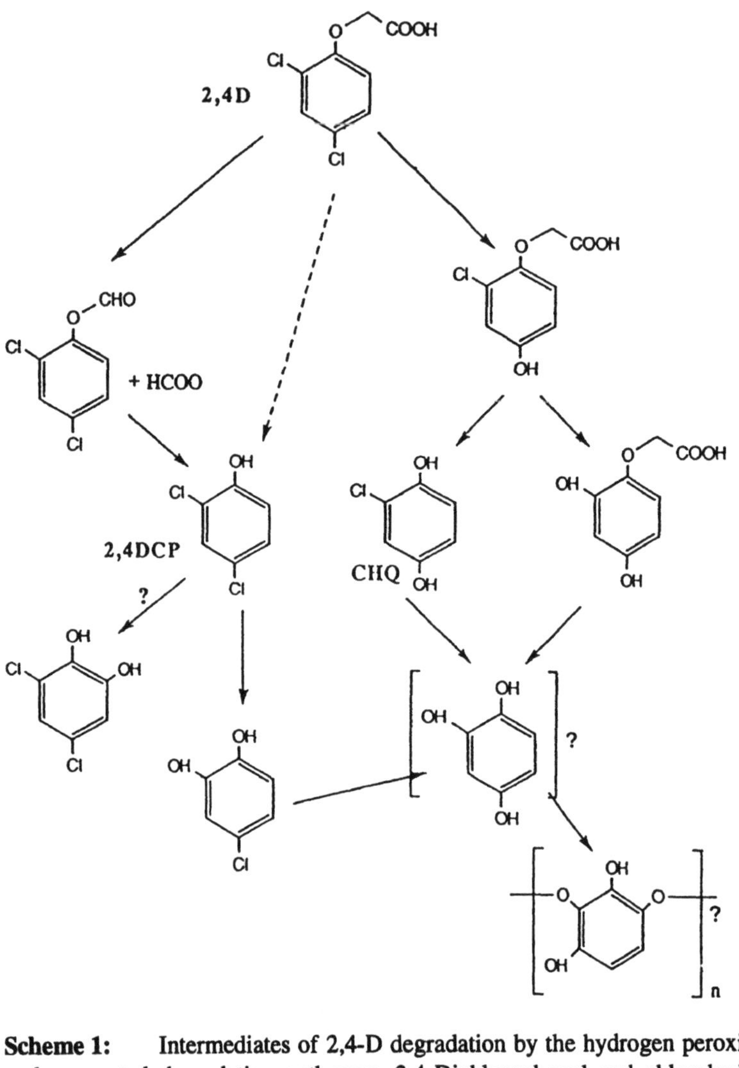

Scheme 1: Intermediates of 2,4-D degradation by the hydrogen peroxide - UV system and suggested degradation pathways. 2,4-Dichlorophenol and chlorohydroquinone were detected both by HPLC and GC and identified by their UV and mass spectrum. 4-chlorocatechol was identified by HPLC/photodiode array. The other aromatics were identified by GS/MS (comparison with authentic samples or interpretation of mass spectra (EI/CI)). Some of the compounds, formed in small amounts, can be due to direct photolysis.

In aqueous solution containing H_2O_2

The atmospheric water droplets contain hydrogen peroxide whose photodissociation is one of the three main sources of hydroxyl radicals. Our study was not undertaken to obtain quantitative data under conditions simulating those in the atmosphere, but to delineate the 2,4-D degradation pathways due to the OH° radicals and those resulting from the direct photolysis of this herbicide. Consequently, we used hydrogen peroxide concentrations higher than those found in atmospheric water, and an irradiation containing a greater fraction of short wavelengths than the solar spectrum.

In the absence of UV irradiation, H_2O_2 had no detectable effect. Under irradiation at wavelengths > 220 nm, 2,4-D disappeared according to an apparent first-order kinetics whose constant k depended linearly on the ratio $[H_2O_2]$ / [2,4-D]. Accordingly, it was possible to distinguish the degradation intermediates whose formation was principally due to the hydroxyl radicals. Scheme I shows these intermediates and suggested degradation pathways of 2-4 D in the presence of hydrogen peroxide.

Effects of a silica and H_2O_2 on the α-pinene phototransformation in water

The phototransformation of α-pinene were studied in aqueous solution at $\lambda > 220$ nm, in the presence of SiO_2 previously impregnated with α-pinene, with or without H_2O_2 and also in the presence of H_2O_2 without SiO_2. In the absence of H_2O_2, the disappearance of α-pinene impregnated on SiO_2 was very slow. However a large number of intermediates was detected (see below). The addition of H_2O_2 caused a very large increase in the degradation rate, since *ca.* 85 % of the α-pinene initially impregnated on SiO_2 had disappeared within 2 h under the conditions used. Similarly, SiO_2 enhanced the capabilities of H_2O_2 in photodegrading α-pinene because, without SiO_2, only 35 % of α-pinene was degraded within an equal irradiation period.

Table 1 lists the products of α-pinene phototransformation which we identified. The presence of silica markedly increased the number of these products. The simultaneous presence of SiO_2 and H_2O_2 caused changes in the distribution of the products and, generally, a decrease in their concentration for a given irradiation duration as was expected because of the increased transformation rate of α-pinene and its degradation intermediates.

Table 1: List of the intermediate products of the photodegradation of α-pinene.

Type of recation	Intermediate products
Isomerization	*limonene, *3 non-formally identified products (presumably cyclofenchene or occimene isomers)
Monohydroxylation	**trans*-verbenol, borneol, isopinocampheol, *trans*-carveol, perillylic alcohol, *p*-menth-3-ene-1-ol.
Epoxidation	*α-pinene oxide, limonene oxide
Dihydroxylation	*trans-p*-menth-6-ene-2,8-diol, 6-*endo*-hydroxyborneol, 2,3-bornanediol, 2,10-bornanediol.
Formation of a C=O group	*pinonaldehyde, S-(+)-S-(1)-methylethyl-2-methyl-2-cyclohexen-1-one, α-campholene aldehyde, pinocamphone
Oxidative ring opening	non-formally identified products (presumably *cis*-pinonic acid and norpinonaldehyde)

* The products indicated by an asterisk were the only ones, in this list, that were found in the presence of H_2O_2 and in the absence of SiO_2.

Respective roles of Fe_2O_3 *and* Fe^{3+} *ions in the photodegradation of aqueous oxalic acid*

As a result of soil erosion and combustion of iron-containing materials, the amounts of ferric oxide and oxyhydroxides in the atmosphere are important; for example, they often represent between 5 and 15 wt % of fly ashes. The amounts of iron ions in atmospheric water range from nmol L^{-1} to mmol L^{-1}.

Since the photocatalytic activity of ferric oxide in the degradation of aqueous oxalic acid cannot be delineated from the photodegradation of iron-oxalato complexes either at the oxide surface or in the water bulk because of iron ions dissolution, we have tried to assess the importance of heterogeneous photocatalysis both from photoconductance measurements in the dry state, or from comparisons with titanium dioxide in aqueous solutions for the oxidation of oxalic acid or ethanal. From these experiments it was concluded that heterogeneous photocatalysis played a minor role as compared with the other photoprocesses. This conclusion was also supported by the spectral sensitivity of the oxalic acid degradation. The role of the complexes formed at the solid surface was confirmed and illustrated, in particular, by the detrimental effect of sulfate ions used in the preparation of ferric oxide or added to the suspension of this oxide in water. Addition of two different concentrations of ferric ions to a ferric oxide sample (Fig. 1) allowed us to show the importance of the ratio $[Fe^{3+}]_{diss}$ / $[Fe_2O_3]$. Apart form this ratio, several other parameters were shown to be at the origin of the distinct effects of various ferric oxide samples, among which the cristallinity was noticeable.

Although the concentrations used in these laboratory studies were generally higher than those encountered in atmospheric waters, experimental curves showed that acceptable extrapolations to lower concentrations could be made. Consequently, the above conclusions may reasonably be expected to remain valid for environmental conditions.

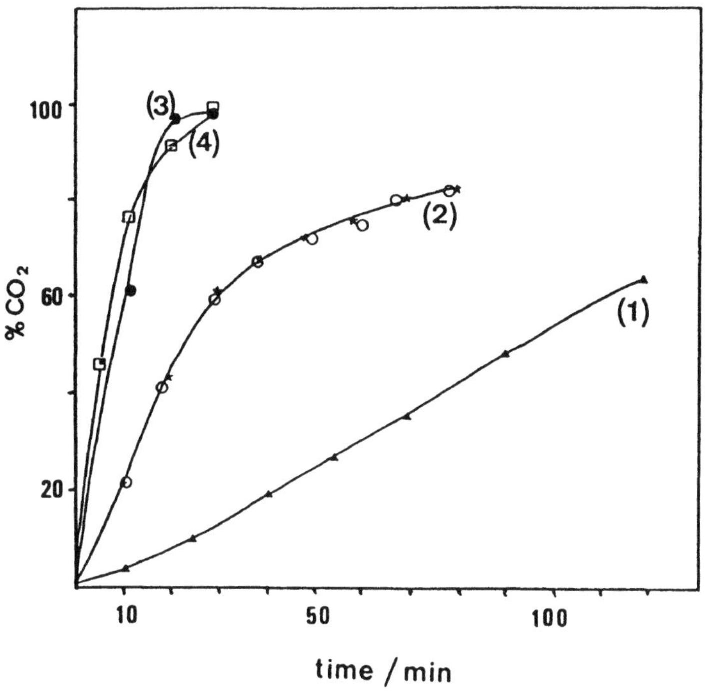

Fig. 1: Percentages of CO_2 formed from the photodegradation of oxalic acid (5 mmol/L; 10 mL) at $\lambda > 290$ nm, radiant flux ca.65 mW/cm^2 (100 % corresponds to the complete transformation) as a function of the irradiation time in the presence of (1) 10 µmol/L ferric nitrate, (2) 100 mg hematite with (circles) or without (stars) 0.2 mmol/L ferric nitrate, (3) 100 mg hematite and 20 mmol/L ferric nitrate, and (4) 20 mmol/L ferric nitrate. The hematite sample had a surface area of 19 m^2/g and, at a concentration of 10 g/L, released 53 µmol/L iron ions in an aerated solution of oxalic acid (5 mmol/L).

Conclusions

At the beginning of our HALIPP work we intended principally to study the effects of inorganic particulate solids upon the phototransformation of organic pollutants in the dry state. During the project, our investigations were extended to the aqueous phase with or without dispersed solid particles and H_2O_2. Our studies about ferric oxide samples also led us to carry out experiments with ferric ions. One of the main achievements was to show that, even in the absence of substantial kinetic effects, the heterogeneous processes can markedly modify the nature and distribution of products from a given organic compound.

In the future, a difficult task will be to select heterogeneous systems and conditions leading to quantitative data that could ultimately be included in models of atmospheric chemistry.

Our studies correspond to the general HALIPP objectives concerning "the contribution of aerosols to the chemistry" and also the "processes that generate oxidants and products in atmospheric droplets". More precisely, they conform to the specific objectives "oxidation of organic compounds in the presence of solid aerosols" and "clarifying the role of transition metals such as iron".

Acknowledgements

This work was financially supported by the "Ministère de la Recherche" (contracts 88 T 0159 (01), 88 W 1148, 89 W 0081), the "Ministère de l'Environnement" (contracts 90 304, 92 256) and the "Programme Environnement du CNRS" (contracts N 82/0055 and 0074).

7.4 Heterogeneous Reactions of Nitrogen Oxides (NO_2, N_2O_5, HNO_3, $ClONO_2$) with Surfaces Representative of Atmospheric Aerosol

Michel J. Rossi, F.F. Fenter, K. Tabor, F.Caloz and L.Gutzwiller

Laboratoire de Pollution Atmosphérique et Sols, Swiss Federal Institute of Technology (E.P.F.L.), CH-1015 Lausanne, Switzerland

Summary

Using a Knudsen cell reactor, we have studied the heterogeneous reactions of nitrogen oxides and related gas-phase species with materials representative of atmospheric aerosol. The heterogeneous kinetics of the title species with salt (chlorides, bromides, nitrates, carbonates, etc.), as well as the reactions of NO_2 and HNO_3 with amorphous carbon (soot) have been studied in detail. At the low-pressures employed in the Knudsen cell reactor, the uptake probability for a heterogeneous reaction can be determined without the need to make corrections for diffusion, and secondary gas-phase reactions are too slow to interfere with the heterogeneous process of interest. The loss of reactants and the formation of products can be followed in real time with millisecond time resolution by either mass spectrometry or by optical techniques (laser induced fluorescence, REMPI).

Aims of the Research

The research is directed by the need to understand better the reactions that take place between atmospheric trace species and the aerosols that exist in both the troposphere and the stratosphere. With our experimental technique we can determine the probability that a collision between a gas-phase molecule and a surface will lead to the irreversible loss of the molecule from the gas phase; the versatility of the detection schemes at our disposition allows us to identify products and quantify their yields. With detailed kinetic modelling of our experimental observations, we gain additional insight into the overall heterogeneous chemical reaction. The derived mechanism can be subsequently used by atmospheric modellers to assess the role of heterogeneous interactions in atmospheric chemistry.

More specifically, we have set out to determine the reaction probabilities and product yields for a series of reactions between nitrogen oxides and related species with various ionic substrates [1-5]. The goal of this research is to determine the relative contributions of the following reactions to the displacement of chloride from marine aerosols:

$$HNO_3(g) + NaCl(s) \rightarrow HCl(g) + NaNO_3(s) \quad (1)$$
$$N_2O_5(g) + NaCl(s) \rightarrow ClNO_2(g) + NaNO_3(s) \quad (2)$$
$$ClONO_2(g) + NaCl(s) \rightarrow Cl_2(g) + NaNO_3(s) \quad (3)$$
$$2\,NO_2(g) + NaCl(s) \rightarrow ClNO(g) + NaNO_3(s) \quad (4)$$

We have also studied the analogous reactions between nitrogen oxides and bromides:

$$HNO_3(g) + NaBr(s) \rightarrow HBr(g) + NaNO_3(s) \quad (5)$$
$$N_2O_5(g) + NaBr(s) \rightarrow \text{products} \quad (6)$$
$$ClONO_2(g) + NaBr(s) \rightarrow BrCl(g) + NaNO_3(s) \quad (7)$$
$$2\,NO_2(g) + NaBr(s) \rightarrow BrNO(g) + NaNO_3(s) \quad (8)$$

These reactions all have in common the production of a gaseous halogen-containing species with the formation of nitrate in the condensed phase. In reactions (2)–(4) and (6)–(8), the gaseous product can be photolysed in sunlight to yield a free halogen atom in the atmosphere. In particular, reactions (2) and (3) have been proposed as important free chlorine-atom sources in the marine troposphere or in the stratosphere after volcanic eruption [6]. Of course, the extent to which these reactions provide an atmospheric source of Cl atoms cannot be assessed without kinetic information on reactions (1) and (4) because NO_2 and HNO_3 are so much more abundant in almost all regions of the atmosphere [7]. To understand the atmospheric implications of these reactions, it is necessary to know the relative reaction probabilities as well as the product yields.

In a related study, we looked at the nitrification of calcium carbonate by nitric acid [5]. Carbonates are present in sand and tropospheric dusts:

$$HNO_3(g) + CaCO_3(s) \rightarrow H_2O(g) + CO_2(g) + Ca(NO_3)_2(s) \quad (9)$$

In addition, we have studied the interactions of NO_2 and HNO_3 with amorphous carbon[8–10]. Amorphous carbon aerosol is produced by various combustion processes and is present throughout the earth's atmosphere. The present loading of the atmosphere with carbon aerosol by civil air transport and the subsequent chemical reactions that may take place on these particles is currently a subject of concern. The reaction of NO_2 with amorphous carbon can proceed by at least two pathways:

$$NO_2 + C \rightarrow NO + C[O] \quad (10)$$
$$NO_2 + H_2O(ads) \rightarrow HONO + HNO_3 \quad (11)$$

Nitric acid is itself reactive with respect to amorphous carbon:

$$HNO_3 + C \rightarrow NO, NO_2, + \text{other products} \quad (12)$$

Our principle aim in this work is to identify the role that NO_2 and HNO_3 may play in the oxidation of amorphous carbon aerosol in the atmosphere; as above, this includes the determination of reaction probabilities and product yields. The kinetic information thus gained will be useful in modelling studies to assess the impact of carbon aerosol loading on atmospheric chemistry.

Principal scientific results

The Knudsen cell reactor is shown schematically in Fig. 1. It consists of a gas handling system, a low-pressure reactor, and the detection system. With a pressure transducer we measure the flow of reactant species from a calibrated volume into the reactor; all experimental signals are calibrated by absolute flow measurements. Gas phase reactants are admitted into the reactor either as a known flow through a capillary (steady-state experiments) or via a millisecond pulse across a solenoid valve (pulsed valve experiment). Control experiments and kinetic measurements are carried out under identical conditions by use of the isolation plunger, which seals the solid sample from the gaseous reactant. In the absence of reaction, the only loss of gas-phase molecules is by passage through an escape orifice into the vacuum chamber; the molecular beam thus formed is modulated by a tuning-fork chopper and directed to the mass spectrometer. The size of the orifice determines the characteristic residence times and densities of the gas-phase species. The modulated signal generated by the mass spectrometer is passed into a lock-in amplifier and transferred directly to a personal computer for analysis and storage.

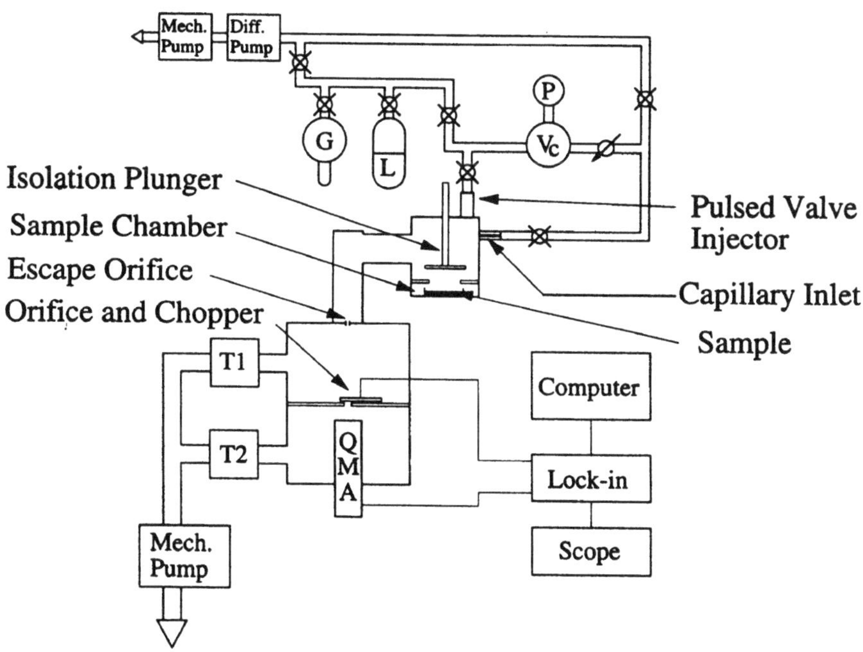

Fig. 1: Schematic drawing of the experimental apparatus. G: gas sample; L: liquid sample; P: pressure gauge; Vc: calibrated volume; T1,T2: turbomolecular pumps; QMA: mass spectrometer.

In a steady-state experiment, a known constant flow of the gas-phase reactant is introduced into the reactor with the sample plunger closed to isolate the solid sample. Because the only loss term for the gaseous species is the rate of effusive loss through the escape aperture, the reactant density can be precisely determined by solving steady-state kinetic equations. At the moment that the plunger is lifted, the density of the gas-phase reactant will decrease if there is an interaction between it and the solid sample. The rate of uptake can be determined from the ratio of the initial and final signal levels. An experimental trace of a typical steady-state experiment is shown for the reaction of $ClONO_2$ on bromide powder in Fig. 2.

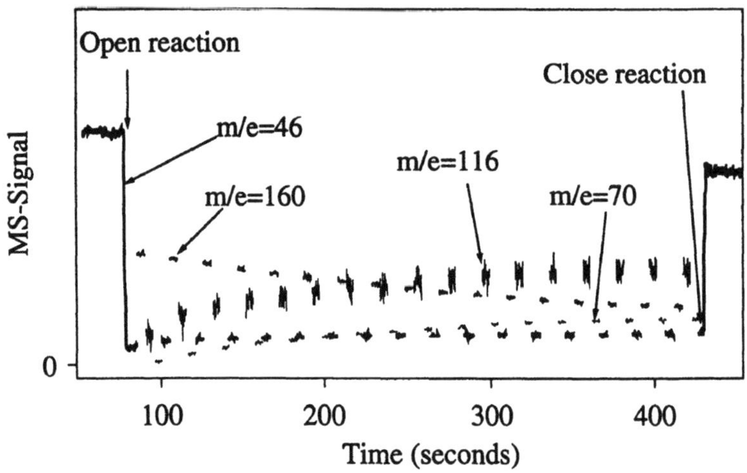

Fig. 2: An example of a steady-state experiment, KBr + $ClONO_2$ → products: At the start, the *m/e* 46 signal is followed to determine the reactant flow. At t = 60 s, the sample plunger is opened and the mass spectrometer scans between *m/e* 46 ($ClONO_2$), m/e 160 (Br_2), *m/e* 116 (BrCl), and *m/e* 70 (Cl_2). When the reaction is closed at 450 s, the mass spectrometer is reset to the *m/e* 46 setting. From the response of *m/e* 46, we calculate a reaction probability of 0.30.

In the pulsed-valve experiments, the reactant is introduced during millisecond pulses of the solenoid valve and the flux of gaseous species from the cell is monitored in real time with the mass spectrometer. A control experiment, in which the solid sample is isolated by lowering the plunger, yields a simple exponential decay of the reactant signal that corresponds to the value of k_{esc}. The plunger is then opened and a second pulse actuated. From the resulting decay trace, the first-order rate of uptake is directly determined by fitting to an exponential decay function. An example of a pulsed valve experiment for the reaction of N_2O_5 on KBr powder is shown in Fig. 3.

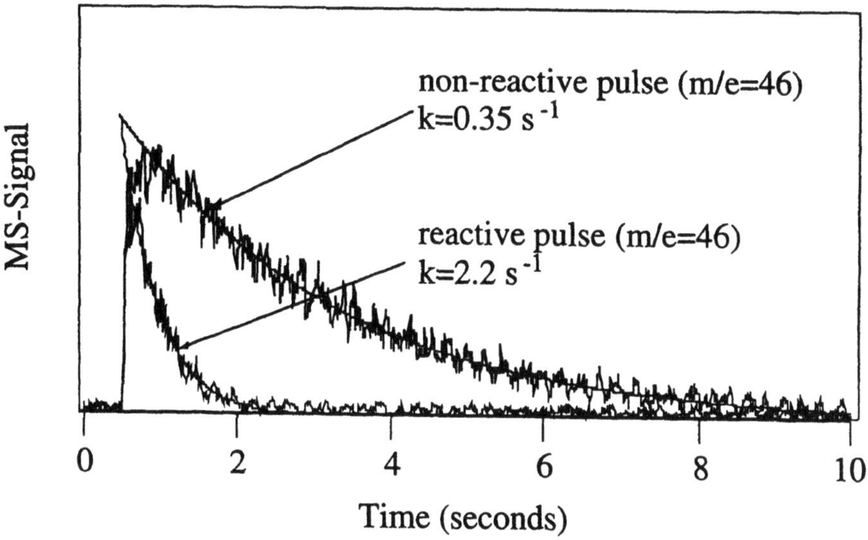

Fig. 3: Pulse valve experiment: N_2O_5 + KBr → products. 5 × 10^{14} molecule of reactant have been injected; the surface under study is a powder. The non-reactive trace shows the first-order loss of N_2O_5 from the cell due to the passage through the escape orifice. The second pulse is actuated under the same conditions with the sample plunger lifted. By fitting the resulting trace with a simple exponential decay function, we directly obtain the first-order rate constant for the reaction. The rate constant is related to the reaction probablity by the collisional frequency between an N_2O_5 molecule and the salt surface.

Fig. 2 has been selected to emphasize the often complex nature of these interactions. Then, it is seen that Br_2, BrCl, and Cl_2 are all observed as products of the $ClONO_2$ interaction with bromides. Ancillary experiments demonstrate that BrCl, the anticipated product, is itself reactive with respect to bromide, producing molecular bromine. And Cl_2 can be formed by the reaction of $ClONO_2$ with the traces of chloride that form as a product on the substrate surface. Often, an entire family of reactions, in this case those of the halogens with halide salts, must be characterised in order to understand the observations concerning the reaction of interest.

Fig. 3 underlines the advantage of the pulsed-valve technique. With this approach, first-order rate constants are determined in the limit of low exposure (far from the onset of surface saturation). The technique thus provides a second kinetic perspective which is useful for the unravelling a complex reaction mechanism (as observed for NO_2 + C) or for the assessment and identification of potential systematic effects.

The format of this report does not permit the discussion of the results in detail. A summary of the kinetic measurements is found in Table 1, in which a note on products and yields is included for each reaction.

Table 1: Summary of kinetic results.

Reaction	Uptake probability[(a)]	Products and yields
$HNO_3 + NaCl$	3.0×10^{-2}	HCl produced with 100 % yield
$N_2O_5 + NaCl$	5.0×10^{-4}	$ClNO_2$ main product, 60 % yield
$ClONO_2 + NaCl$	2.0×10^{-1}	Cl_2 produced with near 100 % yield
$2\ NO_2 + NaCl$	$< 10^{-7}$	HCl and HONO observed as products
$HNO_3 + NaBr$	3.0×10^{-2}	HBr produced with 100 % yield
$N_2O_5 + NaBr$	4.0×10^{-3}	Br_2 and HONO products with 35 % yield
$ClONO_2 + NaBr$	3.0×10^{-1}	Br_2, BrCl, and Cl_2 observed
$2\ NO_2 + NaBr$	2×10^{-7}	no products observed
$HNO_3 + CaCO_3$	1.0×10^{-1}	H_2O and CO_2 (100 % yield) produced
$NO_2 + C$	4.0×10^{-2}	NO yield varies with soot source
$HNO_3 + C$	$> 1.0 \times 10^{-2}$	NO_2 yield on the order of 1 %

[(a)]As described in the text, the uptake coefficient is derived from the first-order rate constant that is determined directly by the experiment.

A general aspect of this research is the characterisation of the surfaces under study. Keyser and coworkers have recently stressed the need to take into account the role of the internal void of the solid samples on the rate of uptake [11]. For the study of the HNO_3, N_2O_5 and $ClONO_2$ reactions with chlorides and bromides, we employed a variety of surfaces with varying degree of internal void, including powders (very large but poorly characterised internal surface), monodisperse grain (large and well-characterised internal surface), micron-thick spray coatings (little or no internal surface), and salt IR windows. We find that nitric acid is insensitive to the presence of internal surface, *i.e.* that to within experimental uncertainty, the rate of uptake is the same for all of the surfaces studied. For N_2O_5, the observed uptake depended strongly on the internal surface of the samples, and this variation is well described by the theory adopted by Keyser *et al.* Results for the reactions of $ClONO_2$ with salt indicate that the reactivity is insensitive to the presence of internal surface. In Fig. 4, we show the dependence of the observed uptake coefficient for reactions (5) and (6) on the mass of a monodisperse grain salt sample. This is a convenient means to assess the role of internal surface because, with the submillimeter grain, the rate of uptake is predicted to change significantly over the range of masses selected. It is seen that the rate of uptake for the nitric acid reaction does not fall off at low sample mass, as would be the case if diffusion into the internal void influenced significantly the observed rate of uptake. Grains and powders are frequently used in the study of heterogeneous chemistry - the lesson from this study is that the characterisation of the internal surface of the substrate is not sufficient to account for the role of diffusion into the internal void of a sample. In addition, the diffusion properties of the reactant with respect to the solid material must be characterised.

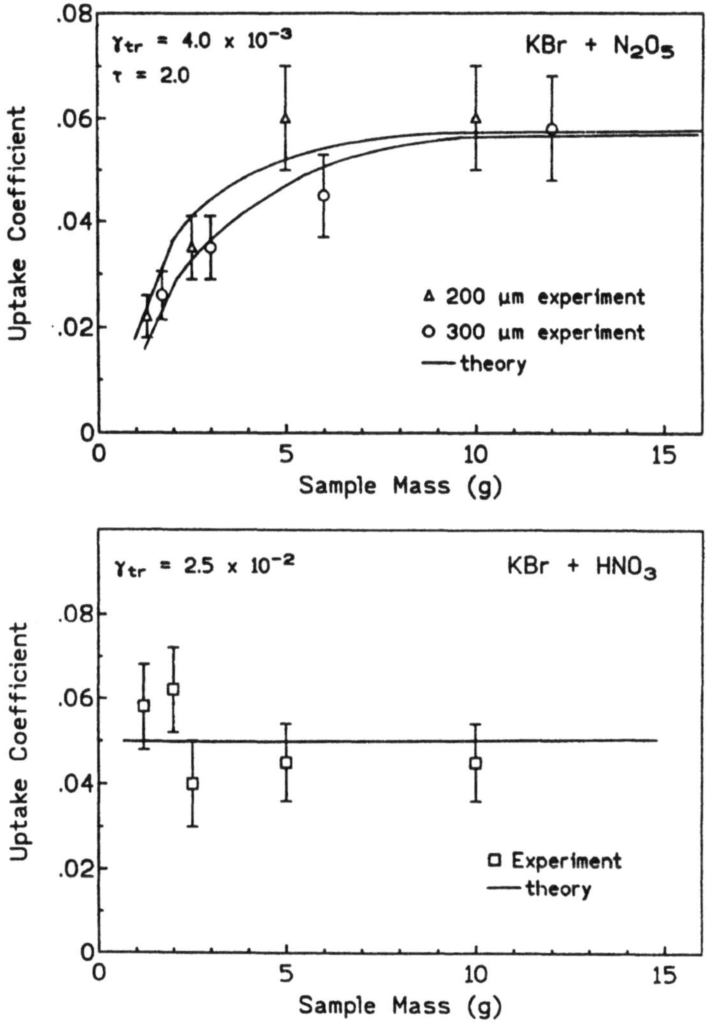

Fig. 4: Observed uptake coefficient versus sample mass for the reactions of N_2O_5 and HNO_3 with KBr monodisperse grain. For N_2O_5, a falloff in reactivity is observed for low sample mass, indicating that the internal surface area of the sample is contributing to the observed reactivity. The nitric acid reaction shows no such dependence on sample mass. The lines that pass through the data are theoretical fits based on the treatment of ref. 11. The HNO_3 curve is calculated with a diffusion coefficient that is a factor 100 smaller than its Knudsen diffusion coefficient.

Conclusions

The Knudsen cell study of the heterogeneous reactions of nitrogen oxides with ionic materials reveals that all species studied are capable, at least to some extent, of displacing chloride and bromide into the gas phase. In literature discussions of the roles of displacement reactions in atmospheric chemistry, the nitric acid reactions are frequently overlooked; we have shown that the uptake probability for HNO_3 is actually much greater than for N_2O_5 on dry salt. The reaction of $ClONO_2$ with dry salt is found to be much more efficient than previously thought. In addition, we have provided detailed kinetic information for the NO_2 reaction with amorphous carbon and have preliminary results on the interaction between nitric acid and soot. These results should provide the kinetic parameters necessary to better assess the role of aerosols on the partitioning of trace species in the earth's atmosphere.

Acknowledgements

Funding for this work was provided by the Office Fédéral de l'Education et de la Science (OFES) as part of the HALIPP subproject of the EUROTRAC programme, and by the Fonds National de la Recherche Scientifique (FNRS). In addition we thank Prof. Hubert van den Bergh for his interest in and support of this effort.

References

1. F. F. Fenter, F. Caloz, M. J. Rossi, *J. Phys. Chem.* **98** (1994) 9801.
2. F. F. Fenter, F .Caloz, M. J. Rossi: The heterogeneous kinetics of N_2O_5 uptake on salt: the influence of the internal surface on the uptake of N_2O_5 and HNO_3, *J. Phys. Chem.*, in press.
3. F. F. Fenter, F. Caloz, M. J. Rossi, in: P. M. Borrell, P. Borrell, T. Cvitaš, W. Seiler (eds), *Proc. EUROTRAC Symp. '94*, SPB Academic Publishing bv, The Hague 1994, p. 1069.
4. F. Caloz, F. F. Fenter, M. J. Rossi: The heterogeneous kinetics of the $ClONO_2$ reaction with chlorides and bromides, *J. Phys. Chem.* in press.
5. F. F. Fenter, F.Caloz, M. J. Rossi, Experimental evidence for the efficient "dry deposition" of nitric acid on calcite, *Atmos. Environ.* in press.
6. F. E. Livingston, B. J. Finlayson-Pitts, *Geophys. Res. Lett* . **18** (1991) 17.
7. B. J. Finlayson-Pitts, J. N. Pitts, Jr., *Atmospheric Chemistry*, John Wiley and Sons, New York 1986.
8. K. Tabor, L.Gutzwiller, M. J. Rossi, *Geophys. Res. Lett.* **20** (1993) 1431.
9. K. Tabor, L. Gutzwiller, M. J. Rossi, *J. Phys. Chem.* **98** (1994) 6172.
10. F. F. Fenter, M. Wolf, M. J. Rossi, unpublished data on the reaction of nitric acid with amorphous carbon.
11. L. F. Keyser, S. B. Moore, M. T. Leu, *J. Phys. Chem.* **95** (1991) 5496.

7.5 Heterogeneous Reactions of Air Pollutants on Various Solid Surfaces

Chris Vinckier[1], Naddy Van Hoof[1], Saleh Ashty[2] and Frans Compernolle[2]

[1]Laboratory of Analytical and Physical Chemistry,
[2]Laboratory of Organic Synthesis, Department of Chemistry, KU Leuven, Celestijnenlaan 200F, 3001 Heverlee, Belgium.

Summary

The capabilities and limitations of the fast-flow reactor technique will be defined for the determinations of gas-solid sticking coefficients γ. In the pressure range of a few Torr and a reactor radius of 1.5 cm, γ must be smaller than 10^{-2}. For α-pinene an upper limit of γ on quartz and sodium chloride is found to be $\gamma \leq 1.4 \times 10^{-4}$.

The rate constant k_1 for the reaction α-pinene + OH → products was determined to be $(5.7 \pm 0.5) \times 10^{-11}$ cm^3 $molec^{-1}$ s^{-1}.

A GC-MS analysis of the reaction products, formed during the oxidation of α-pinene with hydroxyl radicals in the presence of nitric oxide and oxygen, revealed the formation of a number of condensable products such as pinonaldehyde, campholene aldehyde, pinocamphone and a nitrate adduct. These products may play an important role in the heterogeneous chemistry leading to the formation of aerosols under atmospheric conditions.

Aims of the research

The aims of this project were to determine the nature and yields of the various condensable reaction products formed in the oxidation of the α-pinene with hydroxyl radicals in a fast-flow reactor. In addition the possible role of heterogeneous reactions on pre-existing particles had to be examined under atmospheric conditions.

Principal scientific results

Experimental technique

A fast-flow reactor with an initial diameter of 2.8 cm and a length of about 70 cm was built for this α-pinene/hydroxyl radical research: Fig. 1. It is made of quartz and has a number of side arms. One side arm carries a microwave cavity (MC) to generate hydrogen atoms from a microwave discharge in hydrogen/helium mixtures. By reacting these atoms with nitrogen dioxide, hydroxyl radicals can be produced. A central probe movable along the entire reactor axis serves as the α-pinene inlet. At the downstream end the reactor is equipped with windows (W) for detection of the hydroxyl radicals by resonance fluorescence. A mass spectrometer (MS) allows on-line analysis of the volatile reaction products while a liquid nitrogen (LN_2) trap is installed to collect the condensable reaction products.

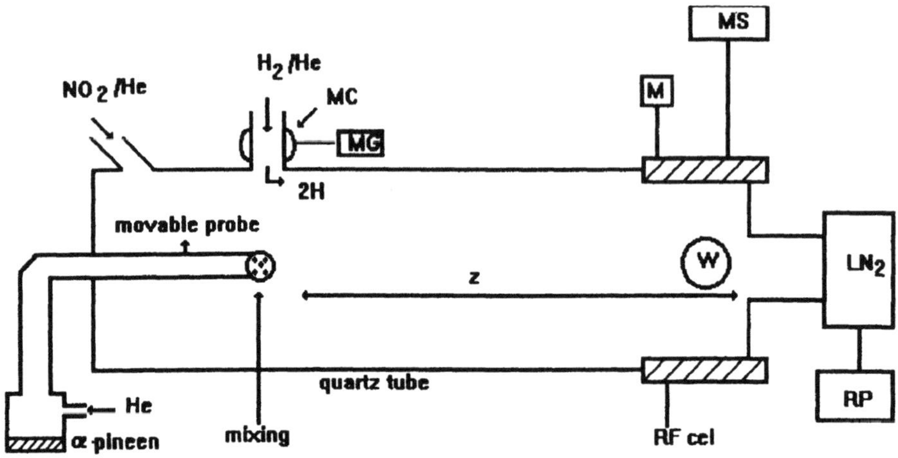

Fig. 1: Schematic view of the experimental set-up: Microwave Cavity (MC), Microwave Generator (MG), Electronic Manometer (M), Mass Spectrometer (MS), Window (W), Liquid Nitrogen Trap (LN_2), Vacuum pump (RP).

Experimental results

1. Sticking of α-pinene on surfaces

By following the concentration of α-pinene as a function of the axial distance one can determine an apparent loss constant k_w of α-pinene on the reactor surface. There are however two major limitations which must be taken into account when this technique for γ-determination is used:

a) plug-flow conditions must prevail so that radial concentration gradients are negligible. This is the case when:

$$\gamma << D\, r^{-1}(2\pi\, M / RT)^{0.5}$$

D is the binary diffusion coefficient of α-pinene in the carrier gas, *r* is the reactor radius and *M* is the molar mass of α-pinene [1]. Under these conditions γ can be expressed as:

$$\gamma = \frac{2\, k_w}{c} \qquad (1)$$

where c stands for the average thermal speed of α-pinene.

b) another important limitation is that the characteristic time to carry out an experiment τ_{exp} must be smaller than the time τ_e to reach the equilibrium for the process [α-pinene]$_g$ <=> [α-pinene]$_{ads}$:

$$\tau_e < (k_{ads} + k_{des})^{-1} \qquad (2)$$

When the desorption energy amounts to about 20 kcal mol^{-1} and at 2 Torr helium one can calculate that the limiting value for γ is $\gamma < 10^{-2}$ with this experimental method.

Running a mass spectrum of α-pinene showed that with the above instrument the best detection limit could be achieved when the ion fragment at M/e = 93 was selected instead of the parent ion M/e = 136. Following the intensity of this fragment as a function of the axial distance a value of $k_w \leq 1.01$ s^{-1} was obtained both on quartz and on sodium chloride coated reactor surface. Using equation (1) one arrives at an upper limit for γ on both surfaces of $\gamma \leq 1.4 \times 10^{-4}$.

In view of this very low value of γ it is clear that the removal of α-pinene on pre-existing quartz or sodium chloride particles will be kinetically controlled and not by its diffusion rate towards the particle. Under this condition the heterogeneous removal rate constant k_{het} of α-pinene is given by [2]:

$$k_{het} = (RT/2\pi M)^{0.5} S\gamma \qquad (3)$$

where S is the total particle surface per unit volume of air. For an average case with $S = 1.4 \times 10^{-3}$ m^2/m^3 one arrives at values of $k_{het} \leq 1.1 \times 10^{-5}$ s^{-1}. Taking into account that α-pinene also reacts in the homogeneous gas phase with hydroxyl radicals, ozone and the nitrate radical NO_3, one can also calculate a homogeneous removal rate constant k_{hom} for α-pinene equal to $k_{hom} = k_{OH}[OH] + k_{O3}[O_3] + k_{NO3}[NO_3]$. With a hydroxyl and ozone concentration of respectively 5×10^5 and 7.5×10^{11} molecules cm^{-3} and even neglecting the reaction of α-pinene with the nitrate radical, one arrives already at a value of $k_{hom} = 8.9 \times 10^{-5}$ s^{-1}.

The above allows us to conclude that heterogeneous removal of α-pinene on pre-existing quartz or sodium chloride particles is unimportant under atmospheric conditions compared to its homogeneous destruction.

2. Determination of the rate constant of the α-pinene + OH reaction

The decay of α-pinene with an excess of hydroxyl radicals was followed mass spectrometrically as a function of the axial distance. Knowing the flow velocity of the reagents, this distance scale can readily be converted to a time scale needed for the derivation of the rate constant k_1:

$$C_{10}H_{16} + OH \longrightarrow \text{adduct} \qquad \text{r1}$$

The value of k_1 was derived from kinetic modelling of the α-pinene decay curves. The model contained in essence four additional reactions:

$C_{10}H_{16}$	$\rightarrow$	wall	r2
OH	$\rightarrow$	wall	r3
adduct + OH	$\rightarrow$	adduct-OH	r4
adduct + NO	$\rightarrow$	adduct-NO	r5

The values of both k_2 and k_3 were experimentally determined to be 1 and 10 s^{-1} respectively. The values of k_4 and k_5 were not critical within the following range: $3 < k_4 < 300$ and $0.1 < k_5 < 10$ in units of 10^{-12} cm^3 $molec^{-1}$ s^{-1}.

The experimental conditions were as follows: $[C_{10}H_{16}]_i = 2.1 \times 10^{11}$ molec cm^{-3}, $[OH]_i$ ranging from 7.8 to 31.5×10^{11} molec cm^{-3}, $[NO]_i = 7.8 \times 10^{13}$ molec cm^{-3}, a reactor pressure of 2 Torr and a temperature of 298 K. The results are summarized in Table 1.

Table 1: Calculated values of k_1 as a function of $[OH]_i$.

[OH] / 10^{12} molec cm^{-3}	k_1 / 10^{-11} cm^3 $molec^{-1}$ s^{-1}
0.78	5.8 ±0 .4
1.8	5.8 ± 0.4
2.6	5.9 ± 0.6
3.2	5.3 ± 0.6

From Table 1 one can calculate a weighted mean of $k_1 = (5.7 \pm 0.5)\ 10^{-11}$ cm^3 $molec^{-1}$ s^{-1}. This is in excellent agreement with other determinations in the literature: 6.1×10^{-11} [3] and 5.5×10^{-11} [4] each in units of cm^3 $molec^{-1}$ s^{-1}.

3. Formation of condensable products in the α-pinene/OH/O$_2$ reactions

A qualitative analysis of the volatile products showed the presence of pinon-aldehyde, acetone, carbon monoxide, carbon dioxide and nitrogen dioxide. The latter compound is especially important since under real atmospheric conditions it is a direct precursor of tropospheric ozone.

Condensable products were collected on a liquid nitrogen trap and then dissolved in either dichloromethane (fraction I) or methanol (fraction II). In some experiments the reaction products were collected directly on active carbon from which they were further dissolved in dichloromethane (fraction III).

A summary of the GC-MS analysis is given in Table 2. The experimental conditions were as follows: for fraction I and II, $[C_{10}H_{16}]_i = 7 \times 10^{12}$, $[OH]_i = 2.7 \times 10^{13}$, $[NO]_i = 2.7 \times 10^{13}$, $[O_2]_i = 3 \times 10^{16}$ in units of molec cm^{-3}, the reactor pressure was 2 Torr and the reaction time t = 25 ms; for fraction III, $[C_{10}H_{16}]_i = 1.8 \times 10^{12}$, $[OH]_i = 2.7 \times 10^{13}$, $[NO]_i = 2.7 \times 10^{13}$, $[O_2]_i = 3 \times 10^{16}$ in units of molec cm^{-3}, the reactor pressure was 9.3 Torr and the reaction time t = 60 ms.

Table 2: Condensable products formed in the α-pinene/OH/O_2 reaction. (a) The NO adduct is formed as a result of the addition of an NO molecule to the primary $C_{10}H_{16}$-OH adduct. It will not be formed under real atmospheric conditions in view of the very high [O_2]/[NO] ratio. (b) This polar compound could not be identified but is certainly not pinonic acid. (c) The identification of these nitrite/nitrate-adducts is tentative.

Fraction I	Fraction II	Fraction III
pinonaldehyde	pinonaldehyde	pinonaldehyde
pinocamphone	pinocamphone	pinocamphone
carvotanacetone	carvotanacetone	carvotanacetone
campholene	campholene	pinene oxide
aldehyde	aldehyde	*trans* carveol
trans-carveol	polar compound (b)	$C_{10}H_{16}$-OH.ONO (c)
NO-adduct (a)		$C_{10}H_{16}$-OH.O_2NO (c)

The major importance of these condensable products may be their role as precursors of aerosols which are found to be formed in the photo-oxidation of α–pinene [5].

A quantitative analysis of the volatile products indicated that they could account for about 70 % of the α-pinene reacted. This would mean that the reaction of α-pinene with hydroxyl radicals is not a main source of condensable products.
It was found that the reaction products pinonaldehyde and campholene aldehyde are the major condensable products. In addition it is reasonable to assume that these aldehydes will not be stable end-products under atmospheric conditions. Indeed aldehydes are known to be reactive towards hydroxyl radicals and are also photochemically active.

In view of the above one is tempted to conclude that the α-pinene/OH/NO/O_2 reaction will not lead to an important aerosol formation in the atmosphere. This may be misleading since this conclusion is based on experiments in a fast-flow reactor with a very low water concentration. In smog chamber work the effect of the relative humidity on the aerosol formation yield of α-pinene was clearly established [6].

The formation of NO-adducts might quantitatively be not too important but is still relevant in atmospheric chemistry in view of their possible toxicity.

Conclusions

From this laboratory study, where the oxidation of α-pinene with hydroxyl radicals in the presence of nitric oxide and oxygen was investigated, one is allowed to conclude that heterogeneous reactions will only play a minor role under atmospheric conditions at least at relatively low relative humidities. To what extent the α-pinene reaction with ozone or with the nitrate radical leads to considerable amounts of condensable reaction products is something which should be looked at in a further project.

In view of the number of strict limitations the fast-flow reactor technique at low pressure is not well suited to determine sticking coefficients of gases on solid surfaces.

Acknowledgements

The authors are grateful to the Belgian Ministry of Scientific Policy which has financed the project under the National Impulse Program "Global Change".

References

1. C. Vinckier, N. Van Hoof, in: P. M. Borrell, P. Borrell, T. Cvitaš, W. Seiler (eds), *Proc. EUROTRAC Symp. '92*, SPB Academic Publishing, The Hague 1993, pp. 652–654.
2. T. E. Graedel, C. J. Weschler, Rev. *Geophys. Space Phys.* **19** (1981) 505–539.
3. T. E. Kleindienst, G. W. Harris, J. N. Pitts Jr., *Environ. Sci. Technol.* **16** (1982) 844.
4. R. Atkinson, *Chem. Rev.* **86** (1986) 69–201.
5. S. H. Zang, M. Shaw, J. H. Seinfeld, R. C. Flagan, *J. Geophys. Res.* **97** (1992) 20717–20729.
6. D. Kotzias, B. Nicollin, M. Duane, C. Sparta, J. Van Eijk, L. Viras, P. Kaluzny, P. Foster, *Proc. VOC Workshop*, RIVM, Amersfoort, 1993, pp. 144–156.

Chapter 8

HALIPP Publications: 1988–1995

1988

Behnke, W., C. Zetzsch; The generation of radicals on aerosol surfaces, in: R. A. Cox (ed), *Mechanisms of Gas-phase and Liquid-phase Chemical Transformations in Tropospheric Chemistry, Proc.*COST 611 Working Party 2, CEC, Brussels 1988, pp. 119-124.

McElroy, W. J.; A laser photolysis studiy of the reaction of SO_4^- with Cl^- and the subsequent decay of Cl_2 in aqueous solution, *CEGB Report* No. TPRD/L/3274 /R88 (1988).

McElroy, W. J.; Kinetic and spectroscopic studies of the SO_4^- and Cl_2^- radicals in aqueous solution, in: R. A. Cox (ed), *Mechanisms of Gas-phase and Liquid-phase Chemical Transformations in Tropospheric Chemistry, Proc.*COST 611 Working Party 2, CEC, Brussels 1988.

Zetzsch, C., G. Pfahler, W. Behnke; Heterogeneous formation of chlorine atoms from NaCl in a photosmog system, *J. Aerosol Sci.,*19 (1988) 1203-1206.

Zetzsch, C.; Einfluß von heterogenen Reaktionen auf die Bildung und Verteilung von Photooxidantien, in: S. Schulte-Hostede, M. Kirchner, M. Reuther (eds), *Verteilung und Wirkung von Photooxidantien im Alpenraum. GSF-Bericht* 17/88, GSF München 1988, pp. 152-165.

1989

Behnke, W., C. Zetzsch; Heterogeneous formation of chlorine atoms from various aerosols in the presence of O_3 and Hcl, *J. Aerosol Sci.* 20 (1989) 1167-1170.

Behnke, W., C. Zetzsch; Photochemical formation of Cl atoms from NaCl and HCl in the presence of ozone and aerosol, in: P. J. Crutzen, J.-C. Gerard, R. Zander (eds) *Proc. 28th Liege Internat. Astrophys. Coll.,Our Changing Atmosphere,* University of Liege 1989, p. 493.

Behnke, W., C. Zetzsch; Smog chamber investigations of the influence of NaCl aerosol on the concentration of O_3 in a photosmog system, in: R. Bojkov, P. Fabian (eds) *Proc. Internat. Ozone Symp., Ozone in the atmosphere,* Deepak, Hampton, VA 1989, pp. 519-522.

Herrmann, H., R. Zellner; Absolute OH quantum yields of NO_2^- and NO_3^- photolysis at 308 nm in aqueous solutions, in: R. A. Cox (ed), *Mechanisms of Gasphase and Liquid-phase Chemical Transformations in Tropospheric Chemistry* Air Pollution Rep. 17 EUR 12035, CEC, Brussels,1989

Kraft, J., R. van Eldik; Evidence for multistep reactions in the iron(III) catalyzed autoxidation of sulfur(IV)-oxides: possible steps during acid rain formation, *J. Chem. Soc., Chem. Commun.* (1989) 790.

Kraft, J., R. van Eldik; Kinetics and mechanism of the iron(III) catalyzed autoxidation of sulfur(IV)-oxides in aqueous solution, I. Formation of transient iron(III)-sulfur(IV) complexes, *Inorg. Chem.* 28 (1989) 2297.

Kraft, J., R. van Eldik; Kinetics and mechanism of the iron(III) catalyzed autoxidation of sulfur(IV)-oxides in aqueous solution, II. Decomposition of transient iron(III)-sulfur(IV) complexes, *Inorg. Chem.* 28 (1989) 2306.

McElroy, W. J., S. J. Waygood; Kinetics of the reactions of the SO_4^- radical with SO_4^-, $S_2O_8^{2-}$, H_2O and Fe^{2+}, *CEGB Report* No. ESTD/L/0028/R89 (1989).

1990

Behnke, W., C. Zetzsch; Heterogeneous formation of Cl atoms from NaCl aerosol, NO_x and ozone, *J. Aerosol Sci.* 21, Suppl. 1 (1990) S229-S232.

Behnke, W., C. Zetzsch; Heterogeneous production of Cl atoms under simulated tropospheric conditions in a smog chamber, in: G. Restelli, G. Angeletti (eds), *Proc. 5th Europ. Symp. on Physico Chemical Behaviour of Atmospheric Pollutants,* Kluwer Acad. Publ., Dordrecht 1990, pp. 277-282.

Benkelberg, H. J., U. Deister, P. Warneck; OH quantum yields for the photodecomposition of Fe(III)-hydroxo complexes in aqueous solution and the reaction of OH with hydroxymethane sulfonate, in: G. Restelli, G. Angeletti (eds), *Proc. 5th Europ. Symp. on Physico Chemical Behaviour of Atmospheric Pollutants*, Kluwer Acad. Publ., Dordrecht 1990, pp. 263-269.

Berglund, J.; Manganese(II) catalyzed oxidation of sulfur(IV) by oxygen in aqueous solution, in: *Proc. Conf. Coordination Chem. in Solution*, Stockholm 1990, p. 15. Buxton.

G. V., G. A. Salmon, N. D. Wood; A pulse radiolysis study of the chemistry of oxysulphur radicals in aqueous solution, in: G. Restelli, G. Angeletti (eds), *Proc. 5th Europ. Symp. on Physico Chemical Behaviour of Atmospheric Pollutants*, Kluwer Acad. Publ., Dordrecht 1990, pp. 245-256.

Casado, J., J. M. Herrmann, P. Pichat; Phototransformations of o-xylene over atmospheric solids in the presence of oxygen and water, in: G. Restelli, G. Angeletti (eds), *Proc. 5th Europ. Symp. on Physico Chemical Behaviour of Atmospheric Pollutants* 1989, Kluwer, Dordrecht 1990, pp. 283-288.

Deister, U., P. Warneck, C. Wurzinger; OH radicals generated by NO_3 photolysis in aqueous solution: Competition kinetics and a study of the reaction OH + $CH_2(OH)SO_3$, *Ber. Bunsenges. Phys. Chem.* 94 (1990) 594-599.

Deister, U., P. Warneck; On the photo-oxidation of SO_2 in aqueous solution, *J. Phys. Chem.* 94 (1990) 2191-2198.

Foster, P., V. Jacob, M. Laffond, R. Perraud; Etude de enceinte de simulation de la photodegradation du naphtalene en presence de particules, in: G. Restelli,
G. Angeletti (eds), *Proc. 5th Europ. Symp. on Physico Chemical Behaviour of Atmospheric Pollutants*, Kluwer Acad. Publ., Dordrecht 1990, pp. 289-299.

Kirchner, W., F. Welter, A. Bongartz, J. Kames, S. Schweighoefer, U. Schurath; Trace gas exchange at the air/water interface: measurements of mass accommodation coefficients, *J. Atmos. Chem* 10, 427-449 (1990).

Kirchner, W., F. Welter, U. Schurath; Trace gas exchange kinetics at the air/water interface, in: G. Restelli, G. Angeletti (eds), *Proc. 5th Europ. Symp. on Physico Chemical Behaviour of Atmospheric Pollutants*, Kluwer Acad. Publ., Dordrecht 1990, pp. 270-276.

Lagrange, J., P. Lagrange; Evaluation des trois méthodes de détermination de H_2O_2 dans les pluies, *Analysis* 18 (1990) 37.

Lagrange, P., J. Lagrange; Mechanism of homogeneous liquid oxidation of sulphur (IV) to sulphur (VI) by hydrogen peroxide, in: G. Restelli, G. Angeletti (eds), *Proc. 5th Europ. Symp. on Physico Chemical Behaviour of Atmospheric Pollutants*, Kluwer Acad. Publ., Dordrecht 1990, pp. 257-262.

Lagrange, P., J. Lagrange; Three methods of hydrogen peroxide determination in rain water, in: G. Restelli, G. Angeletti (eds), *Proc. 5th Europ. Symp. on Physico Chemical Behaviour of Atmospheric Pollutants*, Kluwer Acad. Publ., Dordrecht 1990, pp. 109-113.

McElroy, W. J., S. J. Waygood; Kinetics of the reactions of the SO_4^- radical with SO_4^-, $S_2O_8^-$, H_2O and Fe^{2+}, in: G. Restelli, G. Angeletti (eds), *Proc. 5th Europ. Symp. on Physico Chemical Behaviour of Atmospheric Pollutants*, Kluwer Acad. Publ., Dordrecht 1990, pp. 251-256.

McElroy, W. J., S. J. Waygood; Kinetics of the reactions of the SO_4^- radical with SO_4^-, $S_2O_8^-$, H_2O and Fe^{2+}, *J. Chem. Soc. Faraday Trans.* 86 (1990) 2557.

McElroy, W. J., S. J. Waygood; The oxidation of formaldehyde by the hydroxyl radical in aqueous solution, *NP Report* No TEC/L/0221/R90.

Zang, V., R. van Eldik; Influence of the polyaminocarboxylate chelating ligand (L) on the kinetics and mechanism of the formation of FeII(L)NO in the system FeII(L)/NO/HONO/NO_2 in aqueous solution, *Inorg. Chem.* 29 (1990) 4462.

Zang, V., R. van Eldik; Kinetics and mechanism of the autoxidation of iron (II) induced through chelation by ethylenediaminetetraacetate and related ligands, *Inorg. Chem.* 29 (1990) 1705.

Zellner, R., M. Exner, H. Herrmann; Absolute OH quantum yields in the laser photolysis of aqueous solutions of nitrate, nitrite and H_2O_2 at 308 and 351 nm, in: G. Restelli, G. Angeletti (eds), *Proc. 5th Europ. Symp. on Physico Chemical Behaviour of Atmospheric Pollutants*, Kluwer Acad. Publ., Dordrecht 1990, pp. 300-306.

Zellner, R., M. Exner, H. Herrmann; Absolute OH quantum yields in the laser photolysis of nitrate, nitrite and dissolved H_2O_2 at 308 and 351 nm in the temperature range 278-353 K, *J. Atmos. Chem.* 10 (1990) 411-425.

1991

Bal-Reddy, K., N. Coichev, R. van Eldik; Redox cycling of iron in atmospheric water: the important role of sulfite, *J. Chem. Soc., Chem. Commun.* (1991) 481.

Behnke, W., H.-U. Krüger, V. Scheer, C. Zetzsch; Formation of atomic Cl spray via photolysis of nitryl chloride: Determination of the sticking coefficient of N_2O_5 on NaCl aerosol, *J. Aerosol Sci.* S22 (1991) 609-612.

Benkelberg, H. J., A. Schaefer, P. Warneck; Product quantum yields for the photodecomposition of H_2O_2 and Fe(III) complexes in aqueous solution, in: K.-H. Becker (ed), *Atmospheric Oxidation Processes, COST 611/EUROTRAC Joint Workshop*, CEC, Brussels 1991, p. 130.

Berglund, J., L. I. Elding; Manganese (II) and nickel(II) catalyzed oxidation of S(IV) by oxygen in aqueous solution, in: P. Borrell, P.M. Borrell, W. Seiler (eds), *Proc. EUROTRAC Symp. '90*, SPB Academic Publishing bv, The Hague 1991, pp. 295-298.

Bongartz, A., J. Kames, F. Welter, U. Schurath; Absorption cross sections of gaseous nitrous acid in the 300-400 nm range, in: P. Borrell, P. M. Borrell, W. Seiler (eds), *Proc. EUROTRAC Symp. '90*, SPB Academic Publishing bv, The Hague 1991, pp. 299-302.

Bongartz, A., J. Kames, F. Welter, U. Schurath; Near UV absorption cross sections and trans/cis equilibrium of nitrous, *J. Phys. Chem.* 95 (1991) 1076.

Buxton, G. V., G. A. Salmon, N. D. Wood; Spectra and reactivity of SO_3^-, SO_4^- and SO_5^-, in: P. Borrell, P. M. Borrell, W. Seiler (eds), *Proc. EUROTRAC Symp. '90*, SPB Academic Publishing bv, The Hague 1991, pp. 303-305.

Buxton, G. V., G. A. Salmon, N. D. Wood; A radiolysis study of the chemistry of oxysulfur radicals in aqueous solution, in: K.-H. Becker (ed), *Atmospheric Oxidation Processes, COST 611/EUROTRAC Joint Workshop*, CEC, Brussels 1991, p. 156.

Casado, J., J. M. Herrmann , P. Pichat; Effect of some atmospheric particulate oxides on the photodegradation of o-xylene, in: P. Borrell, P. M. Borrell, W. Seiler (eds), *Proc. EUROTRAC Symp. '90*, SPB Academic Publishing bv, The Hague 1991, pp. 307-311.

Coichev, M., R. van Eldik; A mechanistic study of the sulfite-induced autoxidation of Mn(II) in aqueous azide medium, *Inorg. Chim. Acta* 185 (1991) 69.

Coichev, N., K. Bal Reddy , R. van Eldik; Sulfite-catalyzed autoxidation of metal ions. An important step in metal-catalyzed oxidation cycles of sulfur (IV)-oxides, in: P. Borrell, P. M. Borrell, W. Seiler (eds), *Proc. EUROTRAC Symp. '90*, SPB Academic Publishing bv, The Hague 1991, p. 313.

Coichev, N., R. van Eldik; Kinetics and mechanism of the sulfite-induced autoxidation of Co(II) in aqueous azide medium, *Inorg. Chem.* 30 (1991) 2375.

Deister, U., W.J. McElroy, P. Warneck; Kinetics of the oxidation of sulphur (IV) by peroxymonosulfate, in: P. Borrell, P. M. Borrell, W. Seiler (eds), *Proc. EUROTRAC Symp. '90*, SPB Academic Publishing bv, The Hague 1991, pp. 315-316.

Dellert-Ritter, M., R. van Eldik; Fe(III) chelate complexes as models to investigate the catalyzed autoxidation of S(IV)-oxides, in: P. Borrell, P. M. Borrell, W. Seiler (eds), *Proc. EUROTRAC Symp. '90*,SPB Academic Publishing bv, The Hague 1991, p. 317.

Exner, M., H. Herrmann, R. Zellner; A laser photolysis study of reactions of the carbonate radical anion (CO_3^-) in aqueous solution, in: K.-H. Becker (ed), *Atmospheric Oxidation Processes, COST 611/EUROTRAC Joint Workshop*, CEC, Brussels 1991, p. 134.

Exner, M., H. Herrmann, R. Zellner; Absolute OH quantum yields in the 308 and 351 nm laser photolysis of NO_3^-, NO_2^- and H_2O_2, in: P. Borrell, P. M. Borrell, W. Seiler (eds), *Proc. EUROTRAC Symp. '90*, SPB Academic Publishing bv, The Hague 1991, pp. 319-321.

Exner, M., H. Herrmann, R. Zellner; The NO_3 radical in aqueous solution: absorption spectrum and Arrhenius parameters for the reaction with chloride,in: K.-H. Becker (ed), *Atmospheric Oxidation Processes, COST 611/EUROTRAC Joint Workshop*, CEC, Brussels 1991, p. 139.

Foster, P., V. Jacob, M. Lafond, P. Baussand; Influence of particles on napthalene degradation under irradiation, in: P. Borrell, P. M. Borrell, W. Seiler (eds), *Proc. EUROTRAC Symp. '90*, SPB Academic Publishing bv, The Hague 1991, pp. 323-326.

Herrmann, H., M. Exner, R. Zellner A Laser-Spectrometric Study of the NO_3 Radical in the Aqueous Phase, *Fresenius J. Anal. Chem.* 340 (1991) 638-640.

Herrmann, H., M. Exner, R. Zellner; The absorption spectrum of the nitrate (NO_3) radical in aqueous solution, *Ber. Bunsenges. Phys. Chem.* 95 (1991) 598.

Kames, J., R. W. H. Schmidt; NO_x- Analytik in der Atmosphärenchemie, *Labor 2000* (1991) 12.

Kames, J., U. Schurath; Henry's law constant and hydrolysis of peroxyacetyl nitrate, *J. Atmos. Chem.* 10 (1991) 427.

Laffond, M., P. Foster, P. Baussand, I. Denis, V. Jacob; Solid aerosol impact on the benzaldehyde - OH rate constant, in: K.-H. Becker (ed), *Atmospheric Oxidation Processes, COST 611/EUROTRAC Joint Workshop*, CEC, Brussels 1991, p. 144.

Lagrange, J., G. Wenger , P. Lagrange; A laboratory study of the oxidation kinetics of HSO_3^- by H_2O_2, in: P. Borrell, P.M. Borrell, W. Seiler (eds), *Proc. EUROTRAC Symp. '90*, SPB Academic Publishing bv, The Hague 1991, pp. 327-330.

Lagrange, J., G. Wenger, P. Lagrange; Electrolyte effects on aqueous oxidatin of SO_2 by hydrogen peroxide, in: K.-H. Becker (ed), *Atmospheric Oxidation Processes, COST 611/EUROTRAC Joint Workshop*, CEC, Brussels 1991, p. 151.

Lagrange, J., P. Lagrange; Voltammetric method for the determination of H_2O_2 in rainwater, *Fresenius J. Anal. Chem.* 339 (1991) 452.

Piechowski, M. von, J. Hoigné, R. Bühler; Pulse radiolysis: A method to simulate the transformation of reactants in atmospheric water by copper, in: P. Borrell, P. M. Borrell, W. Seiler (eds), *Proc. EUROTRAC Symp. '90*, SPB Academic Publishing bv, The Hague 1991, p. 333.

Piechowski, M. von, T. Nauser, J. Hoigné, R. E. Bühler; The reactivity of copper ions towards short-lived species in atmospheric waters. A pulse radiolysis study, in: K.-H. Becker (ed), *Atmospheric Oxidation Processes, COST 611/EUROTRAC Joint Workshop*, CEC, Brussels 1991, p. 171.

Ponche, J. L., P. Mirabel; Determination of sticking coefficients,in: K.-H. Becker (ed), *Atmospheric Oxidation Processes, COST 611/EUROTRAC Joint Workshop*, CEC, Brussels 1991, p. 162.

Ponche, J. L., Ph. Mirabel; Effect of some trace gases or surface active films on sticking coefficients, in: P. Borrell, P. M. Borrell, W. Seiler (eds), *Proc. EUROTRAC Symp. '90*, SPB Academic Publishing bv, The Hague 1991, pp. 331-332.

Warneck, P.; Chemical reactions in clouds, *Fresenius J. Anal. Chem.* 340 (1991) 585.

Warneck, P.; Introduction to HALIPP: Laboratory studies on atmospheric heterogeneous and liquid phase processes, in: P. Borrell, P. M. Borrell, W. Seiler (eds), *Proc. EUROTRAC Symp. '90*, SPB Academic Publishing bv, The Hague 1991, p. 293.

Welter, F., S. Schweighoefer, U. Schurath; The mass accommodation coefficient of SO_2 on water measured by the liquid jet technique, in: P. Borrell, P. M. Borrell, W. Seiler (eds), *Proc. EUROTRAC Symp. '90*, SPB Academic Publishing bv, The Hague 1991, pp. 335-339.

Zetzsch, C.; Photochemischer Abbau in Aerosolphasen, *UWSF-Z. Umweltchem. Ökotox.* 4 (1991) 59-64 .

Zetzsch, C., A. Alebic-Juretic, W. Behnke , F. Nolting; Photocatalysis of air chemistry by sea spray and acidifying compounds investigated using an aerosol smog chamber, in: P. Borrell, P. M. Borrell, W. Seiler (eds), *Proc. EUROTRAC Symp. '90*, SPB Academic Publishing bv, The Hague 1991, pp. 341-342.

Zetzsch, C., M. Elend, H. U. Krueger, P. Kuehn, V. Scheer, W. Behnke; Formation of atomic Cl from NaCl aerosol as a model of sea spray in a simulated photochemical smog, in: K.-H. Becker (ed), *Atmospheric Oxidation Processes, COST 611/EUROTRAC Joint Workshop*, CEC, Brussels 1991, pp. 167.

Zetzsch, C.; Experimental simulation of the influence of aerosols on atmospheric chemistry, *Pollution Atmosphérique* 33S (1991) 89-105.

Zuo, Y., J. Hoigné; Formation of hydrogen peroxide and depletion of oxalic acid in atmospheric water by photolysis of iron(III)-oxalato complexes, in: *Proc. Int. Conf. Interfacial Phenomena in the Environment*, Davos 1991.

1992

Bal-Reddy, K., R. van Eldik; Kinetics and mechanism of the sulfite-induced autoxidation of Fe(II) in acidic aqueous solution, *Atmos. Environ.* 26A (1992) 661.

Behnke, W., H.-U. Krüger, V. Scheer, C. Zetzsch, Formation of $ClNO_2$ and HONO in the presence of NO_2, O_3 and wet NaCl aerosol, *J. Aerosol Sci.* 23S (1992) 933-936.

Behnke, W., H.-U. Krüger, V. Scheer, C. Zetzsch; An estimation of the source strength of atomic Cl from aerosol smog chamber experiments on the reaction of N_2O_5 with NaCl at various humidities, in: R. A. Cox (ed), *Laboratory Studies on Atmospheric Chemistry, Air Pollution Research Report* 42, CEC, Brussels 1992, pp. 177-181.

Buxton, G. V., G. A. Salmon, S. McGowan The free radical chain oxidation of S(IV) in aqueous solution, in: R. A. Cox (ed), *Laboratory Studies on Atmospheric Chemistry, Air Pollution Research Report* 42, CEC, Brussels 1992, pp. 27-31.

Buxton, G. V., S. Croft, G. A. Salmon; A pulse radiolysis study of some reactions of HMSA, in: R. A. Cox (ed), *Laboratory Studies on Atmospheric Chemistry, Air Pollution Research Report* 42, CEC, Brussels 1992, pp. 19-22.

Coichev, N., K. Bal Reddy, R. van Eldik; The synergistic effect of manganese(II) in the sulfite-induced autoxidation of metal ions and complexes in aqueous solution, *Atmos. Envion.* 26A (1992) 2295-2300.

Dellert-Ritter, M., R. van Eldik; Kinetics and mechanism of the complex formation of ethylenediaminetetraacetate-iron(III) with sulfur(IV)-oxides in aqueous solution, *J. Chem. Soc., Dalton Trans.* (1992) 1037-1044.

Dellert-Ritter, M., R. van Eldik; Kinetics and mechanism of the redox behaviour of the ethylenediaminetetraacetate-iron(III)-sulfite system in aqueous solution,*J. Chem. Soc. Dalton Trans.* (1992) 1045-1052.

Drexler, C., H. Elias, B. Fecher, K. J. Wannowius; Kinetics and mechanism of sulfur(IV) oxidation by hydrogen peroxide in aqueous phase: the non-linear parts of the pH profile, *Ber. Bunsenges. Phys. Chem.* 96 (1992) 481-485.

Eldik, R. van, N. Coichev, K. Bal Reddy, A. Gerhard; Metal ion catalyzed autoxidation of sulfur(IV)-oxides: Redox cycling of metal ions induced by sulfite,*Ber. Bunsenges. Phys. Chem.* 96 (1992) 478-481.

Eldik, R. van; Metal-catalyzed autoxidation of S(IV). The importance of sulfite-induced oxidation of metal ions, in: R. A. Cox (ed), *Laboratory Studies on Atmospheric Chemistry, Air Pollution Research Report* 42, CEC, Brussels 1992, pp. 13-17.

Exner, M., H. Herrmann, J. Michel, R. Zellner Hydrolysis rate of CF_3COF and the decarboxylation of CF_3COO^- anions, in: *AFEAS Workshop Proc.* Brussels 1992, pp. 50-56.

Exner, M., H. Herrmann, R. Zellner; Laser-based studies of the nitrate radical in aqueous solution, *Ber. Bunsenges. Phys. Chem*: 96 (1992) 470-475.

Fischer, M., P. Warneck; Product quantum yields of the wavelength dependent photodecopmosition of NO_2^- and HONO in aqueous solution, in: R. A. Cox (ed), *Laboratory Studies on Atmospheric Chemistry, Air Pollution Research Report* 42, CEC, Brussels 1992, pp. 33-36.

Geissler, M., R. van Eldik; Development of a gradient ion-pair chromatographic procedure for the simultaneous detection of nitrogen-sulfur-oxides produced during the reaction of SO_x and NO_y species in aqueous solution, *Anal. Chem.* 64 (1992) 3004-3006.

Geissler, M., R. van Eldik; Development of analytical procedures for the detection of nitrogen-sulfur-oxides produced during the reaction of SO_x and NO_y species in aqueous solution, in: *Precipitation scavenging and atmosphere-surface exchange*, Vol. 1, Hemisphere Publ. Co., Washington 1992, p. 189.

George, Ch., J. L. Ponche, P. Mirabel; Experimental determination of mass accomodation coefficients, in: Fukuta and Wagner (eds), *Nucleation and atmospheric aerosols*, Deepak Publ., Hampton, Virginia 1992.

Guillard, C., H. Delprat, C. Hoang-Van, P. Pichat; Laboratory studies of the phototransformation of organic micropollutants over some atmospheric particulate solids, in: R. A. Cox (ed), *Laboratory Studies on Atmospheric Chemistry, Air Pollution Research Report* 42, CEC, Brussels 1992, pp. 157-161.

Kames, J., U. Schurath; Alkyl nitrates and bifunctional nitrates of atmospheric interest: Henry's law constants and their temperature dependencies, *J. Atmos. Chem.* 15 (1992) 79-95.

Kyuil, J., J. C. Petit; The HNO_3-H_2 Phase diagram revisited in the temperature and concentration ranges of PSC formation, in: R. A. Cox (ed), *Laboratory Studies on Atmospheric Chemistry, Air Pollution Research Report* 42, CEC, Brussels 1992, pp. 162-163.

Lagrange, J., C. Pallares, G. Wenger, P. Lagrange; Aqueous oxidation of sulfur dioxide by ozone and hydrogen peroxide, in: R. A. Cox (ed), *Laboratory Studies on Atmospheric Chemistry, Air Pollution Research Report* 42, CEC, Brussels 1992, pp. 37-40.

Pasiuk-Bronikowska, W., T. Bronikowski, M. Ulejczyk; Mechanism and kinetics of autoxidation of calcium sulfite slurries, *Environ. Sci. Technol.* 26 (1992) 1976-1981.

Piechowski, M. von, M.-A. Thelen, J. Hogné, R. E. Bühler; *tert*-Butanol as an OH scavenger in the pulse radiolysis of oxygenated systems, *Ber. Bunsenges. Phys. Chem.* 96 (1992) 1448-1454.

Ponche, J. L., C. George, P. Mirabel; Mass transfer at the air/water interface: mass accomodation coefficients of SO_2, HNO_3, NO_2, NH_3 and HONO, in: R. A. Cox (ed), *Laboratory Studies on Atmospheric Chemistry, Air Pollution Research Report* 42, CEC, Brussels 1992, pp. 147-151.

Schurath, U., A. Bongartz, J. Kames, S. Schweighoefer; Modelling support measurements of NH_3 transfer through air/water interface, in: R. A. Cox (ed), *Laboratory Studies on Atmospheric Chemistry, Air Pollution Research Report* 42, CEC, Brussels 1992, pp. 153-156.

Warneck, P.; Chemistry and photochemistry in atmospheric water drops, *Ber. Bunsenges. Phys. Chem.* 96 (1992) 454-460.

Waygood, S. J.; A laser flash photolysis study of the SO_3^-, SO_4^- SO_5^- radicals in aqueous solution, in: R. A. Cox (ed), *Laboratory Studies on Atmospheric Chemistry, Air Pollution Research Report* 42, CEC, Brussels 1992, pp. 23-26.

Zetzsch, C., W. Behnke; Heterogeneous photochemical sources of atomic Cl in the troposphere, *Ber. Bunsenges. Phys. Chem.* 96 (1992) 488-493.

Zuo, Y., J. Hoigné; Formation of hydrogen peroxide and depletion of oxalic acid in atmospheric water by photolysis of iron(III)-oxalato complexes, *Environ. Sci. Technol.* 26 (1992) 1014-1022.

1993

Barlow, S., G. V. Buxton, S. McGowan, G. A. Salmon; The reaction of SO_3^- radical with Fe(II). A pulse radiolysis study, in: J. Peeters (ed), *Chemical Mechanisms Describing Tropospheric Processes, CEC Air Pollution Research Report* 45, Guyot, Brussels 1993, pp. 279-283.

Behnke, W., H.-U. Krüger, V. Scheer, C. Zetzsch; Formation of $ClNO_2$ and NOCl in the presence of NO_2, O_3 and wet NaCl aerosol, in: P. M. Borrell, P. Borrell, T. Cvitaš, W. Seiler (eds), *Proc. EUROTRAC Symp. '92*, SPB Academic Publishing bv, The Hague 1993, pp. 565-570.

Behnke, W., H.-U, Kruger, V. Scheer, C. Zetzsch; Production of $ClNO_2$ from the reaction of N_2O_5 with NaCl- and seaspray aerosol, in: J. Peeters (ed), *Chemical Mechanisms Describing Tropospheric Processes, CEC Air Pollution Research Report* 45, Guyot, Brussels 1993, pp. 249-252.

Behnke, W., V. Scheer, C. Zetzsch; Formation of $ClNO_2$ and HNO_3 in the presence of N_2O_5 and wet pure NaCl- and wet mixed NaCl/$Na_2SO_4^-$ aerosol, *J. Aerosol Sci.* S24 (1993) 115-116.

Berglund, J., L. I. Elding; Manganese(III/II) catalyzed oxidation of S(IV) and the synergistic effect between Fe(III) and Mn(II), in: P. M. Borrell, P. Borrell, T. Cvitaš, W. Seiler (eds), *Proc. EUROTRAC Symp. '92*, SPB Academic Publishing bv, The Hague 1993, p. 576.

Berglund, J., S. Fronaeus, L. I. Elding; Kinetics and mechanism for the manganesecatalyzed oxidation of sulfur(IV) by oxygen in aqueous solution,.*Inorg. Chem.* 32 (1993) 4527-4538.

Bongartz, A., Ch. George, J. Kames, Ph. Mirabel, J. L. Ponche, U. Schurath; Experimental determination of HONO mass accomodation coefficients using two different techniques, in: P. M. Borrell, P. Borrell, T. Cvitaš, W. Seiler (eds), *Proc. EUROTRAC Symp. '92*, SPB Academic Publishing bv, The Hague 1993, pp. 634-638.

Bongartz, A., Schurath; Liquid jet studies of trace gas fluxes through air/water phase boundary, in: J. Peeters (ed), *Chemical Mechanisms Describing Tropospheric Processes, CEC Air Pollution Research Report* 45, Guyot, Brussels 1993, pp. 29-34.

Bongartz, A., U. Schurath; Recent determinations of mass accommodation coefficients on liquid water with an improved liquid jet technique, in: P. M. Borrell, P. Borrell, T. Cvitaš, W. Seiler (eds), *Proc. EUROTRAC Symp. '92*, SPB Academic Publishing bv, The Hague 1993, pp. 639-643.

Brandt, C., N. Coichev, E. Hostert, R. van Eldik; Rapid measurements of oxygen concentrations in aqueous soution. Reaction kinetics of the transition metal catalyzed oxidation of sulfur(IV) oxides, *GIT Fachzeitschrift Lab.* 37 (1993) 277-283.

Brandt, C., R. van Eldik; Iron(III)-catalyzed oxidation of sulfur(IV)-oxides: Evidence for a novel reaction step in the presence of oxygen, in: P. M. Borrell, P. Borrell, T. Cvitaš, W. Seiler (eds), *Proc. EUROTRAC Symp. '92*, SPB Academic Publishing bv, The Hague 1993, pp. 593-597.

Buxton, G. V., J. L. Eccles, G. A. Salmon; The NO_3 radical in aqueous solutions, in: P. M. Borrell, P. Borrell, T. Cvitaš, W. Seiler (eds), *Proc. EUROTRAC Symp. '92*, SPB Academic Publishing bv, The Hague 1993, pp. 610-614.

Buxton, G. V., S. Croft, G. A. Salmon; A pulse radiolysis study of some free radical reactions in systems containing HMSA, in: P. M. Borrell, P. Borrell, T. Cvitaš, W. Seiler (eds), *Proc. EUROTRAC Symp. '92*, SPB Academic Publishing bv, The Hague 1993, pp. 605-609.

Buxton, G. V., S. McGowan, G. A. Salmon; The free radical chain oxidation of S(IV) in aqueous solution, in: P. M. Borrell, P. Borrell, T. Cvitaš, W. Seiler (eds), *Proc. EUROTRAC Symp. '92* SPB Academic Publishing bv, The Hague 1993, pp. 599-604.

Coichev, N., R. van Eldik; Redox cycling and synergistic effect of manganese in the metal-catalyzed autoxidation of sulfur(IV)-oxides, in: P.M. Borrell, P. Borrell, T. Cvitaš, W. Seiler (eds), *Proc. EUROTRAC Symp. '92*, SPB Academic Publishing bv, The Hague 1993, pp. 589-592.

Exner, M., H. Herrmann, J. W. Michel, R. Zellner; Laser pulse initiated measurements of NO_3 reactions with S(IV) and organic compounds in aqueous solution, in: P. M. Borrell, P. Borrell, T. Cvitaš, W. Seiler (eds), *Proc. EUROTRAC Symp. '92*, SPB Academic Publishing bv, The Hague 1993, pp. 615-618.

Fatnassi, J., J. C. Petit; A calorimetric study of reactive sites formed by UV illumination of tropospheric and model materials, in: P. M. Borrell, P. Borrell, T. Cvitaš, W. Seiler (eds), *Proc. EUROTRAC Symp. '92*, SPB Academic Publishing bv, The Hague 1993, pp. 619-623.

Fischer, M., P. Warneck, J. Ziajka; Photolytic and metal ion-catalysed formation of sulphite radicals from HSO_3^- in aqueous solution, in: J. Peeters (ed), *Chemical Mechanisms Describing Tropospheric Processes, CEC Air Pollution Research Report* 45, Guyot, Brussels 1993, pp. 17-22.

George, C., J. L. Ponche, P. Mirabel; Experimental determination of uptake coefficients for acid halides, in: *Proc. STEP-HALOCSIDE/AFEAS Workshop*, Dublin 1993, pp. 196-198.

George, C., J. L. Ponche, P. Mirabel; Mass transport at the air/water interface: application for the atmospheric heterogeneous modelling, in: J. Peeters (ed), *Chemical Mechanisms Describing Tropospheric Processes, CEC Air Pollution Research Report* 45, Guyot, Brussels 1993, pp.23-27.

George, Ch., J. L. Ponche, Ph. Mirabel; Experimental determination of the mass accommodation coefficient for NH_3, in: P. M. Borrell, P. Borrell, T. Cvitaš, W. Seiler (eds), *Proc. EUROTRAC Symp. '92*, SPB Academic Publishing bv, The Hague 1993, pp. 644-647.

Grgic, I., V. Hudnik, M. Bizjak, J. Levec; Aqueous S(IV) oxidation III. Catalytic effect of soot particles, *Atmos. Environ.* 27A (1993) 1409-1416.

Guillard, C., H. Delprat, C. Hoang-Van, P. Pichat; Laboratory studiy of the rates and products of the phototransformation of naphtalene adsorbed on samples of titanium dioxide, ferric oxide, muscovite and fly ash, *J. Atmos. Chem.* 16 (1993) 47-59.

Herrmann, H., A. Reese, R. Zellner, Investigations of the sulfur species SO_x^- (x = 3,4,5) with a new laser photolysis/time-resolved optical multichannel absorption experiment, in: J. Peeters (ed), *Chemical Mechanisms Describing Tropospheric Processes, CEC Air Pollution Research Report* 45, Guyot, Brussels 1993, pp. 253-258.

Hoigné, J., Y. Zuo, L. Nowell; Photochemical reactions in atmospheric waters; role of dissolved iron species, in: G. R. Helz, R. G. Zepp, D. G. Crosby (eds), *Aquatic and Surface Photochemistry*, Lewis Publ., Ann Arbor 1993, pp. 75-84.

Kames, J., U. Schurath, F. Flocke, A. Volz.Thomas; Preparation of organic nitrates from alcohols and N_2O_5 for species identification in atmospheric samples, *J. Atmos. Chem.* 16 (1993) 349-359.

Kames, J., U. Schurath; Henry's law hydrolysis rate constants for various PAN's using a homogeneous gas phase source, in: J. Peeters (ed), *Chemical Mechanisms Describing Tropospheric Processes, CEC Air Pollution Research Report* 45, Guyot, Brussels 1993, pp. 259-264.

Kames, J., U. Schurath; Properties of mono- and bi-functional organic nitrates pertinent to atmospheric clouds, in: P. M. Borrell, P. Borrell, T. Cvitaš, W. Seiler (eds), *Proc. EUROTRAC Symp. '92*, SPB Academic Publishing bv, The Hague 1993, pp. 629-633.

Lagrange, J., C. Pallares, G. Wenger, P. Lagrange; Electrolyte effects on aqueous atmospheric oxidation of sulphur dioxide by hydrogen peroxide, *Atmos. Environ.* 27A (1993) 129-137.

Lagrange, J., C. Pallares, P. Lagrange; Kinetics and mechanisms of the oxidation of dissolved SO_2 by hypochlorite ion ClO^-, in: J. Peeters (ed), *Chemical Mechanisms Describing Tropospheric Processes, CEC Air Pollution Research Report* 45, Guyot, Brussels 1993,, pp. 265-268.

Lagrange, J., C. Pallares, G. Wenger, P. Lagrange; Aqueous oxidation of sulfur dioxide by ozone and hydrogen peroxide, in: P. M. Borrell, P. Borrell, T. Cvitaš, W. Seiler (eds), *Proc. EUROTRAC Symp. '92*, SPB Academic Publishing bv, The Hague 1993, p. 598.

Lagrange, J., P. Lagrange; Mécanisme de l'oxydation en phase aqueuse de S(IV) en S(VI) par des oxydants forts, in: *Rapport du contract STRETIE/MERE 90301, Pollution Atmosphèrique: transfert et transformation des polluants dans l'atmosphere, pollution de fond,* Paris 1993, No 112.

Pasiuk-Bronikowska, W., K. J. Rudzinski; Development of quasistationary kinetics of co-catalyzed auto-oxidation of sulfite, in: P. M. Borrell, P. Borrell, T. Cvitaš, W. Seiler (eds), *Proc. EUROTRAC Symp. '92*, SPB Academic Publishing bv, The Hague 1993, pp. 577-580.

Pasiuk-Bronikowska, W., K.J. Rudzinski; Rate constants in Co(II) oxidation from nostationary autoxidation of sulfite, in: J. Peeters (ed), *Chemical Mechanisms Describing Tropospheric Processes, CEC Air Pollution Research Report* 45, Guyot, Brussels 1993, pp. 273-278.

Pasiuk-Bronikowska, W., T. Bronikowski; Oxidation of cobalt (II) in the presence of sulfite under quasi-stationary conditions, in: J. Peeters (ed), *Chemical Mechanisms Describing Tropospheric Processes, CEC Air Pollution Research Report* 45, Guyot, Brussels 1993, pp. 11-16.

Pichat, P., J.-C. D'Oliveira, J.-F. Maffre, D. Mas; Destruction of 2,4dichlorophenoxyethanoic acid (2,3-D) in water by TiO_2-UV, H_2O_2-UV or direct photolysis, in: D. F. Ollis, H. Al-Ekabi (eds), *Photocatalytic Purification and Treatment of Water and Air*, Elsevier, London 1993, pp. 683-688.

Piechowski, M. von, Th. Nauser, J. Hoigné, R. E. Bühler; O_2^- decay catalyzed by Cu^{2+} and Cu^+ ions in aqueous solutions; a pulse radiolysis study for atmospheric chemistry, *Ber. Bunsenges. Phys. Chem.* 97 (1993) 1762-1771.

Ponche, J. L., C. George, P. Mirabel; Mass transfer at the airlwater interface: mass accomodation coefficients of SO_2, HNO_3, NO_2 and NH_3, *J. Atmos. Chem.* 16 (1993) 1-21.

Schurath, U., A. Bongartz, C. Roose, J. Karnes, S. Schweighoefer, R. Rubröder; Effect of sticking coefficients, Henry's law constants and hydrolysis rates on mass transfer into water droplets, in: *Proc. AFEAS Workshop, Atmospheric Wet and Dry Deposition of Carbonyl and Haloacetyl Halides*, Solvay Research and Technology Center, Brussels 1993, pp. 44-49.

Schweighoefer, S., R. Rubröder, U. Schurath; Determinations of gas phase diffusion coefficients: Arrested flow and twin-tube methods, in: P. M. Borrell, P. Borrell, T. Cvitaš, W. Seiler (eds), *Proc. EUROTRAC Symp. '92*, SPB Academic Publishing bv, The Hague 1993, pp. 625-628.

Sedlak, D., J. Hoigné; The role of copper and oxalate in the redox cycling of iron in atmospheric waters, *Atmos. Environ.* 27A (1993) 2173-2185.

Tabor, K., L. Gutzwiller, M. J. Rossi; The heterogeneous interaction of NO_2 with amorphous carbon, *Geophys. Res. Lett.* 20 (1993) 1431-1434.

Tabor, K., L. Gutzwiller, M.J. Rossi; The interaction of NO_2 with amorphous carbon: Heterogeneous chemical kinetics, in: J. Peeters (ed), *Chemical Mechanisms Describing Tropospheric Processes, CEC Air Pollution Research Report* 45, Guyot, Brussels 1993, pp. 35-40.

Vinckier, C., F. Compernolle, N. Van Hoof, S. Ashty; Role of a-pinene in the formation of greenhouse gases in the atmosphere, in: *Proc. Belgian Global Change Symposium*, Belgian Science Policy Office, Brussels 1993, pp. 173-188.

Vinckier, C., N. Van Hoof; A method for measuring sticking coefficients of gases on solids in a fast-flow reactor, in: J. Peeters (ed), *Chemical Mechanisms Describing Tropospheric Processes, CEC Air Pollution Research Report* 45, Guyot, Brussels 1993, pp. 41-44.

Vinckier, C., N. Van Hoof; Determination of the sticking coefficient of a-pinene on quartz, in: P. M. Borrell, P. Borrell, T. Cvitaš, W. Seiler (eds), *Proc. EUROTRAC Symp. '92*, SPB Academic Publishing bv, The Hague 1993, pp. 652-654.

Warneck, P.; Cloud chemical oxidation mechanisms, in: P. M. Borrell, P. Borrell, T. Cvitaš, W. Seiler (eds), *Proc. EUROTRAC Symp. '92*, SPB Academic Publishing bv, The Hague 1993, pp. 477-480.

Waygood, S. J.; Kinetic studies of reactions between oxysulphur radicals and transition metal ions in aqueous solution, in: P. M. Borrell, P. Borrell, T. Cvitaš, W. Seiler (eds), *Proc. EUROTRAC Symp. '92*, SPB Academic Publishing bv, The Hague 1993, pp. 585-588.

Zang, V., R. van Eldik; The reaction of nitric oxide with sulfur(IV) oxides in the presence of iron(II) complexes in aqueous solution, *J. Chem. Soc. Dalton Trans.* (1993) 111-118. Zetzsch, C., W. Behnke; Heterogeneous reactions of chlorine compounds,in: H. Niki, K.-H. Becker (eds), *The Tropospheric Chemistry in the Polar Regions, NATO ASI Series* 17 (1993) 291-306.

Ziajka, J., P. Warneck; Oxidation of manganese(II) by the sulphate radical ion, in: P. M. Borrell, P. Borrell, T. Cvitaš, W. Seiler (eds), *Proc. EUROTRAC Symp. '92*, SPB Academic Publishing bv, The Hague 1993, pp. 581-584.

1994

Amels, P., H. Elias, K. J. Wannowius; Kinetics and mechanism of the oxidation of DMS by peroxo compounds in the aqueous phase, in: P. M. Borrell, P. Borrell, T. Cvitaš, W. Seiler (eds), *Proc. EUROTRAC Symp. '94*, SPB Academic Publishing bv, The Hague 1994, pp. 1048.

Barlow, S., G. V. Buxton, G. A. Salmon; The nature and reactions of NO_3 in aqueous solution, in: P. M. Borrell, P. Borrell, T. Cvitaš, W. Seiler (eds), *Proc. EUROTRAC Symp. '94*, SPB Academic Publishing bv, The Hague 1994, pp. 1005-1009.

Beer, F. M., P. Warneck, J. Ziajka; Mechanism for the iron catalysed oxidation of bisulphite in aqueous solution: A free radical chain reaction, in: G. Angeletti, G. Restelli (eds) *Proc. 6th Europ. Symp. on Physico-Chemical Behaviour of Atmospheric Pollutants, CEC Air Pollution Rep.* 50 Luxembourg 1994, pp. 924-929.

Behnke, W., Ch. George, V. Scheer, C. Zetzsch; Formation and fate of $ClNO_2$ from the reaction between N_2O_5 and NaCl solution and its possible kinetic mechanism, in: P. M. Borrell, P. Borrell, T. Cvitaš, W. Seiler (eds), *Proc. EUROTRAC Symp. '94*, SPB Academic Publishing bv, The Hague 1994, pp. 1064-1068.

Behnke, W., H.-U. Krüger, V. Scheer, C. Zetzsch; Formation of different photolytic precursors of atomic Cl from pure and contaminated NaCl aerosol, in: G. Angeletti, G. Restelli (eds) *Proc. 6th Europ. Symp. on Physico-Chemical Behaviour of Atmospheric Pollutants, CEC Air Pollution Rep.* 50 Luxembourg 1994, pp. 960-965.

Behnke, W., V. Scheer, C. Zetzsch; Production of $BrNO_2$, Br_2 and $ClNO_2$ from the reaction between sea spray aerosol and N_2O_5, *J. Aerosol Sci.* S25 (1994) 277-278.

Berlund, J., L. I. Elding; Kinetcs and mechanism for manganese catalysed oxidation of sulphur(IV) by oxygen in aqueous solution, in: P. M. Borrell, P. Borrell, T. Cvitaš, W. Seiler (eds), *Proc. EUROTRAC Symp. '94*, SPB Academic Publishing bv, The Hague 1994, pp. 1021-1025.

Bongartz, A. Ch. George, J. Kames, Ph. Mirabel, J. L. Ponche, U. Schurath; Experimental determination of HONO mass accommodation coefficients using two different techniques, *J. Atmos. Chem.* 18 (1994) 149-169.

Brandt, C., I. Fabian, R. van Eldik; Kinetics and mechanism of the iron(III) catalyzed autoxidation of sulfur(IV) oxides in aqueous solution. Evidence for the redox cycling of iron in the presence of oxygen and modeling of the overall reaction mechanism, *Inorg. Chem.* 33 (1994) 687.

Brandt, C., R. van Eldik; pH dependence of the Fe(III) catalyzed autoxidation of S(IV) oxides, in: P. M. Borrell, P. Borrell, T. Cvitaš, W. Seiler (eds), *Proc. EUROTRAC Symp. '94*, SPB Academic Publishing bv, The Hague 1994, p. 1035.

Brandt, C., V. Lepentsiotis, A. Paul, H. Hohmann, R. van Eldik; Investigation of rain and snow water samples by AAS and IC, *GIT Fachz. Lab.* 38 (1994) 770.

Breytenbach, L., W. van Pareen, J. J. Pienaar, R. van Eldik; The influence of organic acids and metal ions on the kinetics of the oxidation of sulfur(IV) by hydrogen peroxide, *Atmos. Environ.* 28 (1994) 2451.

Buxton, G. V., G. A. Salmon, J. E. Williams; The reaction of SO_5^- with iron(II), in: P. M. Borrell, P. Borrell, T. Cvitaš, W. Seiler (eds), *Proc. EUROTRAC Symp. '94*, SPB Academic Publishing bv, The Hague 1994, pp. 1044-1047.

Buxton, G.V.; Mechanisms for chemical reactions in cloud droplets., in: P. M. Borrell, P. Borrell, T. Cvitaš, W. Seiler (eds), *Proc. EUROTRAC Symp. '94*, SPB Academic Publishing bv, The Hague 1994, pp. 978-983.

Coichev, N., R. van Eldik; A fascinating demonstration of sulfite-induced redox cycling of metal ions initiated by shaking, *J. Chem. Educ.* 71 (1994) 767.

Coichev, N., R. van Eldik; Metal catalyzed atmospheric oxidation reactions. A challenge to coordination chemists, *New J. Chem.* 18 (1994) 123.

Elias, H., U. Götz, J. Wannowius; Kinetics and Mechanism of the Oxidation of Sulfur(IV)by Peroxomonosulfuric Acid Anion *Atmos. Environ.* 28(3) (1994) 439-448.

Exner, M., H. Herrmann, G. Raabe, R. Zellner; A kinetic study of nitrate (NO_3) radical reactions of interest for atmospheric aqueous phase chemistry within continental airmasses, in: P. M. Borrell, P. Borrell, T. Cvitaš, W. Seiler (eds), *Proc. EUROTRAC Symp. '94*, SPB Academic Publishing bv, The Hague 1994, pp. 149-152.

Exner, M., H. Herrmann, R. Zellner Rate constants for the reactions of the NO_3 radical with $HCOOH/HCOO^-$ and CH_3COOH/CH_3COO^- in aqueous solution between 278 and 328 K, *J. Atmos. Chem.* 18 (1994) 359-378.

Fenter, F. F., F. Caloz, M. J. Rossi; The uptake kinetics of HNO_3 and N_2O_5 on a series of salt samples, in: P. M. Borrell, P. Borrell, T. Cvitaš, W. Seiler (eds), *Proc. EUROTRAC Symp. '94*, SPB Academic Publishing bv, The Hague 1994, pp. 1069-1073.

Foster, P., I. Denis, V. Jacob; Solid aerosol impact on the b-pinene OH radical reaction ate constant, in: P. M. Borrell, P. Borrell, T. Cvitaš, W. Seiler (eds), *Proc. EUROTRAC Symp. '94*, SPB Academic Publishing bv, The Hague 1994, p. 1079.

Geißler, M., R. van Eldik; Product identification and kinetic data for the reaction of S(IV) with N(III) oxoacids in aqueous solution in the absence and presence of metal ions, *Polyhedron* 13 (1994) 2983.

George, Ch., J. Lagrange, P. Lagrange, Ph. Mirabel, C. Pallares, J. L. Ponchie; Kinetics of mass transfer of trichloroacetyl chloride at the air/water interface,in: G. Angeletti, G. Restelli (eds) *Physico-chemical behaviour of atmospheric pollutants,* CEC, Luxembourg 1994, pp. 948-953.

George, Ch., J. Lagrange, P. Lagrange, Ph. Mirabel, C. Pallares, J. L.Ponchie; Heterogeneous chemistry of trichloroacetyl chloride in the atmosphere, *J. Geophys. Res.* 99 (1994) 1255-1262.

George, Ch., J. L. Ponche, Ph. Mirabel, W. Behnke, V. Scheer, C. Zetzsch; Study of the uptake of N_2O_5 by water and NaCl solutions, *J. Phys. Chem.* 98 (1994) 8780-8784.

George, Ch., J. L. Ponche, Ph. Mirabel, W. Behnke, V. Scheer, C. Zetzsch; Mass accommodation coefficients of N_2O_5 on pure water and on NaCl containing droplets, in: P. M. Borrell, P. Borrell, T. Cvitaš, W. Seiler (eds), *Proc. EUROTRAC Symp. '94*, SPB Academic Publishing bv, The Hague 1994, pp. 1090-1095.

George, Ch., Y. J. Saison, J. L. Ponche, Ph. Mirabel; Kinetics of mass transfer of carbonyl fluoride, trifluoroacetyl fluoride and trifluoroacetyl chloride at the air/water interface, *J. Phys.Chem.* 98 (1994) 10857-10862.

Granby, K., O. Hertel, O. J. Nielsen, T. Nielsen, A. H. Egeløv; Organic sulphur compounds: Atmospheric chemistry and occurence, in: P. M. Borrell, P. Borrell, T. Cvitaš, W. Seiler (eds), *Proc. EUROTRAC Symp. '94*, SPB Academic Publishing bv, The Hague 1994, pp. 163-166.

Grgic, I., M. Kocutar, M. Novic, M. Bizjak, V. Hudnik; Iron(II/III) catalysed oxidation of S(IV) by oxygen in aqueous solution, in: G. Angeletti, G. Restelli (eds) *Proc. 6th Europ. Symp. on Physico-Chemical Behaviour of Atmospheric Pollutants, CEC Air Pollution Rep.* 50 Luxembourg 1994, pp. 907-911

Grgic, I., M. Poje, M. Novic, M. Bizjak, V. Hudnik; Catalytic effects of atmospheric aerosol particles in aqueous solution: Iron(II,III) catalyzed oxidation of S(IV), in: P. M. Borrell, P. Borrell, T. Cvitaš, W. Seiler (eds), *Proc. EUROTRAC Symp. '94*, SPB Academic Publishing bv, The Hague 1994, pp. 1037-1040.

Herrmann, H., A. Reese, R. Zellner; Spectroscopy of SO_x (x = 3, 4, 5) radical anions and their kinetics in the conversion of S(IV) in aqueous solution,in: P. M. Borrell, P. Borrell, T. Cvitaš, W. Seiler (eds), *Proc. EUROTRAC Symp. '94*, SPB Academic Publishing bv, The Hague 1994, pp. 1017-1020.

Herrmann, H., M. Exner, G. Raabe, A. Reese, R. Zellner; Formation and reactions of oxidants in the aqueous phase, in: G. Angeletti, G. Restelli (eds) *Proc. 6th Europ. Symp. on Physico-Chemical Behaviour of Atmospheric Pollutants, CEC Air Pollution Rep.* 50 Luxembourg 1994, pp. 936-941.

Herrmann, H., M. Exner, R. Zellner; Reactivity trends in reactions of the nitrate radical (NO_3) with inorganic cloud water constituents, *Geochim. et Cosmochim. Acta* 58(15) (1994) 3239-3244.

Herrmann, H., M. Exner, H.-W. Jacobi, R. Zellner; Influence of ionic strengths on reactions in the tropospheric aqueous phase, in: P. M. Borrell, P. Borrell, T. Cvitaš, W. Seiler (eds), *Proc. EUROTRAC Symp. '94*, SPB Academic Publishing bv, The Hague 1994, pp. 1056-1059.

Hesse, K., A.Schlemm, U. Schurath; Flow characteristics of extremely thin jets: Mass accommodation coefficient of ozone on water, in: P. M. Borrell, P. Borrell, T. Cvitaš, W. Seiler (eds), *Proc. EUROTRAC Symp. '94*, SPB Academic Publishing bv, The Hague 1994, pp. 1085-1089.

Hjorth, J., F. Cappellani, H. Fabe-Hansen, S. Sørensen, G. Restelli; Peroxyradicals and peroxynitrates in the tropospheric oxidation of DMS, in: P. M. Borrell, P. Borrell, T. Cvitaš, W. Seiler (eds), *Proc. EUROTRAC Symp. '94*, SPB Academic Publishing bv, The Hague 1994, pp. 1049-1051.

Hoffmann, P., T. Sinner, A. N. Dedik, S. Weber, J. Ensling, H. M. Ortner; Iron in atmospheric aqueous and particulate samples, in: P. M. Borrell, P. Borrell, T. Cvitaš, W. Seiler (eds), *Proc. EUROTRAC Symp. '94*, SPB Academic Publishing bv, The Hague 1994, pp. 1041-1043.

Lagrange, P., C. Pallares, G. Wenger, J. Lagrange; Electrolyte effects on aqueous atmospheric oxidation of sulphur dioxide by strong oxidants, in: G. Angeletti, G. Restelli (eds) *Proc. 6th Europ. Symp. on Physico-Chemical Behaviour of Atmospheric Pollutants, CEC Air Pollution Rep.* 50 Luxembourg 1994, pp. 930-935.

Lagrange, P., C. Pallares, J. Lagrange; Electrolyte effects on aqueous atmospheric oxidation of sulphur dioxide by ozone, *J. Geophys. Res.* 99 (1994) 14595.

Lagrange, P., G. Wenger, J. Lagrange; Kinetic study of the aqueous SO_2 oxidation by H_2O_2 at low concentration and without buffer, in: P. M. Borrell, P. Borrell, T. Cvitaš, W. Seiler (eds), *Proc. EUROTRAC Symp. '94*, SPB Academic Publishing bv, The Hague 1994, pp. 1011-1016.

Langenberg, S., U. Schurath; A new method to study trace gas interactions with ice at very low temperature, in: G. Angeletti, G. Restelli (eds) *Proc. 6th Europ. Symp. on Physico-Chemical Behaviour of Atmospheric Pollutants, CEC Air Pollution Rep.* 50 Luxembourg 1994, pp. 942-947.

Lepentsiotis, V., F. F. Prinsloo, R. van Eldik; Kinetic and mechanistic study of the influence of mixed sulfur-nitrogen oxides on the redox behaviour of manganese and iron ions in aqueous solution, in: P. M. Borrell, P. Borrell, T. Cvitaš, W. Seiler (eds), *Proc. EUROTRAC Symp. '94*, SPB Academic Publishing bv, The Hague 1994, p. 1036.

Michel, J. W., H. Herrmann, R. Zellner; Rates of liquid phase hydrolysis reactions of partially oxidized halogenated compounds of atmospheric interest, in: P. M. Borrell, P. Borrell, T. Cvitaš, W. Seiler (eds), *Proc. EUROTRAC Symp. '94*, SPB Academic Publishing bv, The Hague 1994, pp. 1052-1055.

Pasiuk-Bronikowska, W., K. J. Rudzinski; Nonstationary autoxidation of metal ions in oxidation of sulphite, in: P. M. Borrell, P. Borrell, T. Cvitaš, W. Seiler (eds), *Proc. EUROTRAC Symp. '94*, SPB Academic Publishing bv, The Hague 1994, pp. 1026-1029.

Pasiuk-Bronikowska, W., T. Bronikowski, M. Ulejczyk; Manganese catalysed autoxidation of sulphite under steady state conditions, in: P. M. Borrell, P. Borrell, T. Cvitaš, W. Seiler (eds), *Proc. EUROTRAC Symp. '94*, SPB Academic Publishing bv, The Hague 1994, pp. 1030-1034.

Prinsloo, F. F., J. J. Pienaar, R. van Eldik, H. Gutberlet; Influence of mixed sulfur-nitrogen-oxides on the redox kinetics of manganese ions in aqueous solution, *J. Chem. Soc., Dalton Trans.* (1994) 2373.

Schweighoefer, S., B. Walter, R. Rubröder, U. Schurath; Diffusion coefficients of atmospherically rrelevant trace gases in air, in: G. Angeletti, G. Restelli (eds) *Proc. 6th Europ. Symp. on Physico-Chemical Behaviour of Atmospheric Pollutants, CEC Air Pollution Rep.* 50 Luxembourg 1994, pp. 954-959.

Schweighoefer, S., B. Walter, T. Carstens, D. Hupperich, U. Schurath; Determination of gas phase diffusion coefficients for application in multiphase atmospheric chemistry, in: P. M. Borrell, P. Borrell, T. Cvitaš, W. Seiler (eds), *Proc. EUROTRAC Symp. '94*, SPB Academic Publishing bv, The Hague 1994, pp. 1080-1084.

Sedlak, D. L., J. Hiogné; Oxidation of S(IV) in atmospheric water by photooxidants and iron in the presence of copper *Environ. Sci: and Tech.* 18 (1994) 1898-1906.

Sehested, K., T. Løgager, J. Holcman, O. J. Nielsen; Formation and reactions of the NO_3 radical in aqueous solution, in: P. M. Borrell, P. Borrell, T. Cvitaš, W. Seiler (eds), *Proc. EUROTRAC Symp. '94*, SPB Academic Publishing bv, The Hague 1994, pp. 999-1004.

Tabor, K., L. Gutzwiller, M. J. Rossi; Heterogeneous chemical kinetics of NO_2 on amorphous carbon at ambient temperature, *J. Phys. Chem.* 98 (1994) 6172-6186.

Vinckier, C., N. Van Hoof, F. Compernolle, S. Ashty; Mechanistic study of a-pinene reactions with hydroxyl radicals, in: *Proc. 6th European Symp. Physico-Chemical Behaviour of Atmospheric Pollutants*, Kluwer Academic. Publ., Dordrecht 1994, pp. 47-53.

Vinckier, C., N. Van Hoof, F. Compernolle, S. Ashty; Product distribution of the reaction of a-pinene with hydroxyl radicals, in: P. M. Borrell, P. Borrell, T. Cvitaš, W. Seiler (eds), *Proc. EUROTRAC Symp. '94*, SPB Academic Publishing bv, The Hague 1994, pp. 120-123.

Vinckier, C., N. Van Hoof; Rate constant of the a-pinene + atomic hydrogen reaction at 295 K, *Int. J. Chem. Kinet.* 26 (1994) 528-534.

Warneck, P., J. Ziajka, W. Pasiuk-Bronikowska; Scavengers of SO_4^- in S(IV) autoxidation catalysed by Fe, in: G. Angeletti, G. Restelli (eds) *Proc. 6th Europ. Symp. on Physico-Chemical Behaviour of Atmospheric Pollutants, CEC Air Pollution Rep.* 50 Luxembourg 1994, pp. 901-906.

Warneck, P.; Aqueous solution chemistry, in: G. K. Mootgat, A. J. Barnes, G. LeBras, J. R. Sodeau (eds) *Low temperature chemistry of the atmosphere,* NATO ASI Series I21, Springer, Berlin 1994, pp.175-196.

Warneck, P.; Tropospheric oxidation chemistry, in: P. M. Borrell, P. Borrell, T. Cvitaš, W. Seiler (eds), *Proc. EUROTRAC Symp. '94*, SPB Academic Publishing bv, The Hague 1994, pp. 37-46.

Warneck,P.; Clouds, rain and aerosols, in: G. K. Mootgat, A. J. Barnes, G. LeBras, J. R. Sodeau (eds) *Low temperature chemistry of the atmosphere,* NATO.ASI Series I21, Springer, Berlin 1994, pp. 175-196.

Ziajka,J., F. M. Beer, P. Warneck; Iron catalysed oxidation of bisulphite aqueous solution: Evidence for a free radical chain mechanism, *Atmos.Environ:* 28 (1994) 2549-2552.

1995 + in press

Behnke, W., Ch. George, V. Scheer, C. Zetzsch; Kinetic mechanism for the formation of $ClNO_2$ from the reaction of gaseous N_2O_5 with NaCl solution, in: K.H. Becker (ed.), *Proc. Joint LACTOZ/HALIPP, EC/EUROTRAC/GDCh on tropospheric oxidation mechanisms,* Leipzig, in press.

Benkelberg, H.-J., P. Warneck; Photodecomposition of Fe(III) hydroxo- and sulfato-complexes in aqueous solution: Wavelength dependence of OH and SO_4^- quantum yields, *J. Phys. Chem.* 99 (1995) in press.

Benkelberg, H.-J., S. Hamm, P. Warneck; Henry's law coefficients for aqueous solutions of acetone, acetaldehyde and acetonitrile, and equilibrium constants for the addition compounds of acetone and acetaldehyde with bisulphite, *J. Atmos. Chem.* 14 (1995) in press.

Bongartz, A., S. Schweighoefer, Ch. Rose, U. Schurath; The mass accomodation coefficient of ammonia on water, *J. Atmos. Chem.* (1995) in press.

Botha, C. F., J. Hahn, J. J. Pienaar, R. van Eldik; Kinetics and mechanism of the oxidation of sulfur(IV) by ozone in aqueous solution, *Atmos. Environ.* (1995) in press.

Brandt, C., R. van Eldik; Transition metal catalyzed oxidation of sulfur(IV) oxides. Atmospheric-relevant processes and mechanisms, *Chem. Rev.* (1995) in press.

Fenter, F. F., F. Caloz, K. Tabor, L. Gutzwiller, M. J. Rossi; Kinetics of heterogeneous reactions relevant to tropospheric chemistry, in: K. H. Becker (ed.), *Proc. Joint LACTOZ/HALIPP, EC/EUROTRAC/GDCh on tropospheric oxidation mechanisms,* Leipzig, in press.

Fenter, F. F., F. Caloz, M. J. Rossi; Experimental evidence for the efficient dry deposition of nitric acid on calcite, *Atmos. Environ.* (1995) in press.

Fenter, F. F., F. Caloz, M. J. Rossi; The heterogeneous kinetics of N_2O_5 uptake on salt: The influence of the internal surface on the uptake of N_2O_5 (and HNO_3), *J. Phys. Chem.* (1995) in press.

George, Ch. W. Behnke, V. Scheer, C. Zetzsch, L. Magi, J. Y. Saison, J. L. Ponche, Ph. Mirabel; Heterogeneous reactions of N_2O_5 and $ClNO_2$ over aqueous solutions,in: K. H. Becker (ed.), *Proc. Joint LACTOZ/HALIPP, EC/ EUROTRAC/GDCh on tropospheric oxidation mechanisms,* Leipzig, in press.

Grgic, I., M. Poje, M. Novic, V. Hudnik Catalytic effects of atmospheric aerosol particles in aqueous phase oxidation of S(IV) oxides, in: K. H. Becker (ed.), *Proc. Joint LACTOZ/HALIPP, EC/ EUROTRAC/GDCh on tropospheric oxidation mechanisms,* Leipzig, in press.

Hiogné, J., D. L. Sedlak, R. E. Bühler; The sources, sinks and role of hydroperoxy radicals (HO_2/O_2^-) in atmospheric waters, in: K. H. Becker (ed.), *Proc. Joint LACTOZ/HALIPP, EC/EUROTRAC/GDCh on tropospheric oxidation mechanisms,* Leipzig, in press.

Kames, J., U. Schurath; Henry's law and hydrolysis rate constants for peroxyacyl nitrates (PAN's) using a homogeneous gas phase source, *J. Atmos. Chem.* (1995) in press.

Prinsloo, F. F., J. J. Pienaar, R. van Eldik; Kinetic evidence for the redox cycling of Mn(II/III) in the presence of *N*-nitrosohydroxyl-amine-*N*-sulfonate in aqueous medium, *J. Chem. Soc., Dalton Trans.* (1995) 293.

R. Zellner, H. Herrmann Free Radical Chemistry of the Aqueous Phase,in: R. J. H. Clark, R. E. Hester (eds) *Adv. In Spectroscopy, Vol. 24, Spectroscopy in Environmental Science*, Wiley London 1995, pp. 381-451.

Warneck, P., J. Ziajka; Reaction mechanism of the iron(III)-catalysed autoxidation of bisulphite in aqueous solution: Steady state description for benzene as radical scavenger, *Ber. Bunsenges. Phys.Chem.* 99 (1995) 59-65.

Theses

M. Sc. / Diploma

Barth, S.; Kinetik der S(IV)-Oxidation durch Peroxoacetylnitrat (=PAN) und Peroxocarbonsäuren in wäßriger Lösung, Technische Hochschule Darmstadt, 1991.

Beer, F.; Zur radikalischen Oxidation von Hydrogensulfit in wäßrigen Lösungen, Johannes-Gutenberg University of Mainz, 1993.

Carstens, Th.; Aufbau und Optimierung eines NOy-Detektors für Labormessungen kinetischer Parameter in atmosphärischen Multiphasen-Systemen, University of Bonn, 1994.

de Sombre, R.; Kinetik und Mechanismus der Oxidation von Salpetriger Säure mit dem Anion der Peroxomonoschwefelsäure, Technische Hochschule Darmstadt, 1995.

Götz, U.; pH-Profil der Kinetik der Schwefel(IV)-Oxidation mit Peroxomonoschwefelsäure in wäßriger Lösung, Technische Hochschule Darmstadt, 1991.

Hesse, K.; Präzisionsvermessung eines dünnen Wasserstrahls, University of Bonn, 1994.

Hupperich, D.; Temperaturabhängigkeit von Diffusionskoeffizienten in der Gasphase - ein Methodenvergleich, University of Bonn, 1994.

Langenberg, S.; Kapillar-Gaschromatographie mit Eis als stationäre Phase: Modelluntersuchungen zur Adsorption von SO_2 in Kondensstreifen, University of Bonn, 1993.

Michel, J.; Stopped-Flow-untersuchung der Oxidation von S(IV) durch O_2, H_2O_2 und O_3 in wässriger Lösung, University of Göttingen, 1991.

Paydar, M.; Nachweis von Zwischenprodukten bei der metall-katalysierten Zersetzung von Peroxisulfaten, Johannes-Gutenberg University of Mainz, 1992.

Poje, M.; Heterogeneous reactions in the atmosphere: The role of iron(II) and iron(III) in catalytic oxidation of S(IV) in aqueous solution, University of Ljubljana, 1995.

Rubröder, R.; Bestimmung der Temperaturabhängigkeit von Diffusionskoeffizienten in der Gasphase mit der "arrested flow"-Methode, University of Bonn, 1993.

Schlemm, A.; Computersimulation und Messung der Kontraktion von laminaren Flüssigkeitsstrahlen, University of Bonn, 1994.

Steingens, U.; Kinetische Untersuchung zur Oxidation von Schwefel(IV) mit den organischen Peroxiden tert-Butylhydroperoxid, Ethylhydroperoxid und 1-Hydroxyethylhydroperoxid in wäßriger Lösung, Technische Hochschule Darmstadt, 1993.

Walter, B.; Untersuchungen der Temperaturabhängigkeit binärer Diffusionskoeffizienten reaktiver und adsorbierender Spurengase mit der "arrested flow"-Methode, University of Bonn, 1994.

Wunderlich, C.; Analytik in der Gas- und Flüssigphase zur Bestimmung der Massen-Akkommodationskoeffizienten von Ozon auf Wasser, University of Bonn, 1994.

Ph. D.

Benkelberg, H. J.; Absorptionskoeffizienten, Gleichgewichtskonstanten und OH und SO_4^- Quantenausbeuten für Eisen(III)-Komplexe in wässrigen Lösungen, Johannes-Gutenberg University of Mainz, 1992.

Bongartz, A.; Massen-Akkommodationskoeffizienten schwacher atmosphärischer Säuren und Basen, University of Bonn, 1993.

Brandt, C.; Kinetic and mechanistic studies of the iron(III) catalyzed autoxidation of sulfur(IV) oxides, University of Witten/Herdecke, 1994.

Dellert-Ritter, M.; A mechanistic study on model iron(III) complexes to resolve the metal-catalyzed autoxidation of sulfur(IV) in aqueous solution, University of Witten/Herdecke, 1991.

Drexler, Ch.; Kinetik und Mcehanismus der S(IV)-Oxidation mit Peroxosäuren des Stickstoffs und Salpetriger Säure, Technische Hochschule Darmstadt, 1993.

Fecher, B.; kinetik und Mechanismus der Sulfit-Oxidatiion mit Wasserstoffperoxid und organischen Peroxiden in wäßrigen Lösung, Technische Hochschule Darmstadt, 1995.

Fischer, M.; Photolyse neutraler und acider wäßriger Lösungen von Nitrit und Sulfit im ultravcioletten Spektralbereich, Johannes-Gutenberg University of Mainz, 1992.

Fournier Bidoz, V.; Contribution a l'étude du lavage de l'iode moléculaire gazeoux par les pluies, University of Grenoble, 1991.

George, Ch.; Physico-chimie hétérogène de l'atmosphère: Etude expérimentale et modélisation des processus de transfert gaz/liquide. University of
Strasbourg, 1993.

Geissler, M.; Mechanistic study of the interaction of sulfur and nitrogen oxides in aqueous solution. Application of a newly developed ion pair chromatographic procedure and kinetic techniques, University of Witten/Herdecke, 1992.

Kames, J.; Henry-Koeffizienten und Hydrolysekonstanten organischer Nitrate in der Atmosphäre - physikochemische Parameter zur Bestimmung heterogener Verluste, University of Bonn, 1993.

Pallares, C.; Etude cinétique de l'oxidation du dioxide de soufre par des oxydants forts dans l'eau troposphérique. University of Strasbourg, 1993.

Piechowski, M. von Der EinfluB von Kupferionen auf die Redoxchemie des atmosphärischen Wassers: Kinetische Untersuchungen, University of Zurich, 1991.

Ponche, J. L.; Détermination des coefficients de tranfert gaz-lquide; application à la physicochimie de l'atmosphère, University of Strasbourg, 1991.

Schäfer, A.; Photolyse von Nitrat, Bromat und H_2O_2 in wässrigen Lösungen: Bestimmung von Quantenausbeuten mit Hilfe von Radikalfängern, Johannes-Gutenberg University of Mainz, 1991.

Schweighoefer, S.; Messung binärer Diffusionskoeffizienten in der Gasphase - ein Methodenvergleich, University of Bonn, 1994.

Welter, F.; Massenfluß des Spurengases SO_2 in einen sehr dünnen Wasserstrahl - ein Beitrag zur Kinetik heterogener Prozesse in der Atmosphäre, University of Bonn, 1991.

Welter, F.; Massenfluß des Spurengases SO_2 in einen sehr dünnen Wasserstrahl - Ein Beitrag zur Kinetik heterogener Prozesse in der Atmosphare, University of Bonn, 1991.

Zuo, Y.; Photochemistry of iron(III)/iron(II) complexes in atmospheric liquid phases and its environmental significance, University of Zurich, 1992.

Subject Index

G

H

I

K

L

M

N

O

U

X

Springer-Verlag and the Environment

We at Springer-Verlag firmly believe that an international science publisher has a special obligation to the environment, and our corporate policies consistently reflect this conviction.

We also expect our business partners – paper mills, printers, packaging manufacturers, etc. – to commit themselves to using environmentally friendly materials and production processes.

The paper in this book is made from low- or no-chlorine pulp and is acid free, in conformance with international standards for paper permanency.

MIX
Papier aus verantwortungsvollen Quellen
Paper from responsible sources
FSC® C105338

If you have any concerns about our products,
you can contact us on
ProductSafety@springernature.com

In case Publisher is established outside the EU,
the EU authorized representative is:
Springer Nature Customer Service Center GmbH
Europaplatz 3, 69115 Heidelberg, Germany

Printed by Libri Plureos GmbH
in Hamburg, Germany